Thomas W. Wieting
1997

THE INFLATIONARY UNIVERSE

ALAN H. GUTH

THE INFLATIONARY UNIVERSE

THE QUEST FOR A NEW THEORY OF COSMIC ORIGINS

WITH A FOREWORD BY
ALAN LIGHTMAN

HELIX BOOKS

ADDISON-WESLEY PUBLISHING COMPANY, INC.

Reading, Massachusetts   Menlo Park, California   New York
Don Mills, Ontario   Harlow, England   Amsterdam   Bonn
Sydney   Singapore   Tokyo   Madrid   San Juan
Paris   Seoul   Milan   Mexico City   Taipei

Library of Congress Cataloging-in-Publication Data
Guth, Alan H.
The inflationary universe : the quest for a new theory of cosmic origins / Alan H. Guth.
p. cm.
Includes bibliographical references and index.
ISBN 0-201-14942-7
1. Inflationary universe. I. Title.
QB991.I54G88 1997 96-46117
523.1'8—dc21 CIP

Jacket design by Suzanne Heiser
Text design by Jean Hammond
Set in 10-point Sabon by Argosy

1 2 3 4 5 6 7 8 9-MA-0100999897
First printing, March 1997

To my family

CONTENTS

FOREWORD

Every culture has had its cosmology, its story of how the universe came into being, what it is made of, and where it is going. Astoundingly, cosmology in this century has become a science, a well-defined area of research within astronomy and physics.

In the 1970s, the study of cosmology went through a major conceptual change. Prior to this time, modern cosmologists asked such questions as: What is the composition of galaxies and where are they located in space? How rapidly is the universe expanding? What is the average density of matter in the cosmos? After this time, in the "new cosmology," cosmologists began seriously asking questions like: Why does matter exist at all, and where did it come from? Why is the universe as homogeneous as it is over such vast distances? Why is the cosmic density of matter such that the energy of expansion of the universe is almost exactly balanced by its energy of gravitational attraction? In other words, the nature of the questions changed. The questions became more fundamental. "Why?" was added to "What?" and "How?" and "Where?". Alan Guth was one of the young pioneers of the new cosmology, asking the Whys, and his Inflationary Universe theory provided many answers.

Some of the why questions had been asked earlier but not pursued with great stamina, mainly because there seemed no hope of answers, no explanatory theories had been proposed. The standard Big Bang theory, in hand since the work of Einstein and Friedmann and Lemaître in the 1920s, offered no explanations. As a result, working cosmologists simply assumed as givens some of the most interesting observed features of the universe, such as its homogeneity and its very precise balance of energies. Indeed, before the 1970s and early 1980s, many cosmologists took the attitude that some of these features of the universe had to be accepted as "initial conditions," present at the beginning, perhaps mere accidents, perhaps incalculable even in principle. In any case, these features, observed but unexplained, were not taken as serious shortcomings of the standard big bang theory.

The benign neglect of certain observed facts, or anomalies, that cannot be explained within the reigning theory is not uncommon in science. For example, prior to the work of Alfred Wegner in 1912 and his theory of "continental drift," the conceptual framework in which geologists worked held that landmasses could move only vertically. In this picture there was no explanation of the observed remarkable similarity of shapes of the opposite coasts on the two sides of the Atlantic. South America and Africa, in particular, are shaped as if they were two fitting pieces of a jigsaw puzzle. Geologists commented on this peculiar feature of geography but mainly accepted it as a curiosity, scarcely a puzzle waiting for a solution. After Wegner proposed his new theory, which offered a natural mechanism for how the continents could have drifted horizontally apart after once being joined in a single landmass, the fit of the continents was recognized as a major problem in the old theory.

Another example comes from biology, namely the lack of adaptation of certain organisms to their environments. There are ducks with feet designed for swimming that do not swim; there are animals, like the dark-cave rat *Neotoma,* that have eyes but never use them; there are birds, like the penguin, that have wings but never fly. Before Darwin's theory of natural selection, these anomalous facts were observed but swept under the rug in the prevailing creationist framework, which held that each species is exquisitely designed for its habitat and never changes. However, nonadaptation as well as adaptation is explained in Darwin's theory because natural selection requires evolution and change. An organism that has adapted to a particular environment may change environments but still pass down the previously adapted trait, such as a long line of rats that move from the daylight to a dark cave. After Darwin's work, the nonadaptation of some animals was seen as a major flaw in the creation theory. Such is the conservative nature and the psychological forces of the scientific enterprise.

Similarly, the observed homogeneity of the universe and the precise balance of its energies were considered by most of the scientific community to represent serious deficiencies in the standard Big Bang model only after the Inflationary Universe theory was proposed. The conception of that new theory was made possible by a new stock of ideas from theoretical physicists who, in the past, had concerned themselves with the fundamental particles and forces of nature. In the early 1970s, new theories of particle physics called Grand Unified Theories proposed that three of the four fundamental forces of nature—the strong and weak nuclear forces and the electromagnetic force—were actually manifestations of a single, underlying force. (Unification and simplicity have been the eternal Holy Grail of physicists and artists. Aristotle wanted to reduce all substance to five elements; Picasso said that a painter should work with as few elements as possible.) According to the new theories, the temperature at which the single force would become evident was an incredible 10^{29} degrees centigrade, far higher than any temperature that existed on earth or even at the centers of stars. Such a fantastically high temperature could have existed at only one place, or more precisely at one time: in the very young universe, shortly after its big bang beginning. Thus the particle physicists became interested in cosmology. One of the most well known and accomplished particle physicists to combine Grand Unified Theories with cosmology at this time was Steven Weinberg. One of the younger scientists was Alan Guth.

Guth's great achievement was the proposal that, as a natural consequence of the properties of the Grand Unified Theories, the very young universe would have gone through a brief period of extremely rapid expansion, after which it would return to the more leisurely rate of growth dictated by the standard Big Bang theory. Guth's proposed modification of the older theory has been called the Inflationary Universe theory. That theory, and its various improved versions, provide explanations for some of the previously unexplained "anomalies" observed by cosmologists. In my view, the Inflationary Universe theory is the most significant new development in cosmological thinking since the foundational work in the 1920s.

Because the key phenomenon of the Inflationary Universe theory, the extremely rapid cosmic expansion, purportedly happened so long ago, we will probably never know with certainty whether that event in fact occurred. (The theory also makes some precise predictions about the universe today but cannot be proven true on the basis of these predictions.) However, like the notion that the continents were once joined in a single landmass, the "inflation" idea has such explanatory power that it has become an underlying assumption of many working cosmologists today. Thus, an intellectual revolution has occurred in cosmological thinking. In the language of the late MIT historian of science Thomas Kuhn, we have a new paradigm in cosmology.

This book that Alan Guth has written is extraordinary for several reasons. It explains complex scientific ideas without the use of equations. It is written with the same clarity and elegance as Guth's original, pathbreaking scientific paper in the *Physical Review* of 1981. It is thorough and thoughtful. It gives an honest portrait of a personal career in science, as well as the human and scientific history of a major discovery. But what is most extraordinary is that this book, like James Watson's *Double Helix*, is unique in its field. It has been written by a key scientist involved with the discovery.

Although other physicists played various roles in the discovery and development of the Inflationary Universe theory, Alan Guth is internationally recognized as the leading actor. Other popular books may be written (and have been written) about the subject, but only one book can be written by Alan Guth. This is the book.

Alan Lightman
Cambridge, Massachusetts May 1997

PREFACE

The known is finite, the unknown infinite; intellectually we stand on an islet in the midst of an illimitable ocean of inexplicability. Our business in every generation is to reclaim a little more land.

—T. H. Huxley, 1887

The traditional big bang theory has become widely accepted because, as far as we can tell, it gives an accurate picture of how our universe has evolved. The description has been tested, either directly or indirectly, for times as early as one second after the beginning, all the way up to the present. However, although the standard big bang theory is very successful, there is good reason to believe it is incomplete. Despite its name, the big bang theory is not really a theory of a bang at all. It is really only a theory of the *aftermath* of a bang. The equations of the theory describe how the primeval fireball expanded and cooled and congealed to form galaxies, stars, and planets, all of which is a tremendous accomplishment. But the standard big bang theory says nothing about what banged, why it banged, or what happened before it banged. The inflationary universe is a theory of the "bang" of the big bang. The theory of inflation modifies our understanding of just the first tiny fraction of a second of the history of the universe, and then the description merges with that of the standard big bang theory, preserving all the successes of the older theory. But, because inflation explains the bang itself, it is a richer description, providing answers to a number of questions that the traditional big bang theory cannot address. It even allows us to consider such fascinating questions as whether other big bangs are continuing to happen far away, and whether it is possible in principle for a super-advanced civilization to recreate the big bang.

The inflationary theory is a product of modern science, constructed from many ideas, in both particle physics and cosmology, that were contributed by many people. I was in the lucky position of being able to assemble the different pieces of this puzzle, to see for the first time how they could be put together to possibly answer some of the most fundamental riddles of existence. These were exhilarating times. In writing this book, my goal was to give the reader the chance to absorb this new view of cosmic origins, and to join me in reliving this exciting period of discovery.

While I would not complain if some of my fellow scientists found this book enjoyable, the book is aimed primarily at the nonscientist with an interest in science. I avoided equations, but have attempted instead to explain the underlying science in words and diagrams alone. I tried hard, however, not to lose the logical connections that make science so fascinating, but which often slip away when equations are omitted. Without actually showing any mathematics, I have attempted to explain how particle theorists use the mathematics of fields (such as electric or magnetic fields) to describe quantum particles, and how the properties of these fields lead ultimately to a mechanism—inflation—that can explain the origin of essentially all the matter and energy in the observed universe.

No special scientific knowledge is expected on the part of the reader, although presumably the reader knows about atoms, protons, neutrons, and electrons. Any more specialized scientific terms and concepts are explained as they are introduced, and a rather thorough glossary is included to help keep the definitions in mind. Nonetheless, the book is demanding in other ways. I found that I could not write this book without assuming that the reader shares with me a vague craving to understand the universe—a craving that has been part of human consciousness from the writing of Genesis to the scientific era of relativity and quantum theory. Some of the ideas described in the book are complex, so the reader will need some patience. In particular, I suspect that many will find the discussion of Higgs fields in Chapters 8 and 9 to be a bit challenging to follow. I urge these readers to get as much as they can from these chapters and then read on. The most important ideas will come across, even if some of the logical connections are left behind for another day.

In this book the development of inflation is recounted from a personal vantage point, since that is what I know best. It is definitely not intended as a thorough history of the research relevant to this work, a task which would require several volumes and a more technical level of discussion. I have attempted, nonetheless, to honestly describe the roles of the people with whom I interacted, since these were the players in the drama that I wanted to recreate. Even in these cases, however, I have probably committed some important oversights, so I apologize in advance. I have completely omitted a number of related lines of research [1], such as those of Gliner, Starobinsky, Kazanas, Sato, Brout, Englert, and Gunzig; I have also omitted the early work of Mukhanov and Chibisov on the density nonuniformity calculations discussed in Chapter 13. The omission of these contributions is by no means intended as a judgment of their importance, but is merely a consequence of the personal viewpoint that unifies the book.

As with the theory of the inflationary universe itself, there were many people who contributed significantly to the book. My agent, John Brockman, convinced me to write the book in the first place, and he and Katinka Matson have worked hard to maintain publication arrangements despite my delays in completing the manuscript. Randy Fenstermacher has acted as my assistant, chasing down reference materials and providing very helpful criticisms of my drafts. Andrea Schulz went through the entire manuscript, making numerous suggestions to improve its readability. Finally, Jeffrey Robbins of Addison-Wesley improved the manuscript with a final round of polishing, and then he, Lynne Reed, and the rest of the production and design staff transformed my computer files into an attractive book.

In the course of my writing there were numerous scientists and nonscientists who read part or all of the draft and provided crucial advice. In alphabetical order, this group included Bill Bardeen, Gregory Benford, Martin Bucher, Catalina Buttz, Richard Carrigan, Sean Carroll, Rosanne DiStefano, Eddie Farhi, George Field, Fred Goldhaber, Larry Guth, Susan Guth, Jemal Guven, Bhuvnesh Jain, Neal Jimenez, David Kaiser, Evelyn Fox Keller, Pawan Kumar, Andrew Lange, Janna Levin, Alan Lightman, Andrei Linde, John Mather, Ken Olum, Willy Osborn, Dennis Overbye, Jim Peebles, Arno Penzias, So-Young Pi, Lily Recanati, Paul Richards, Paul Schechter, Arlene Soifer, Paul Steinhardt, W. Peter Trower, Michael Turner, Henry Tye, Alexander Vilenkin, Erick Weinberg, Rainer Weiss, Victor Weisskopf, David Wilkinson, Bob Wilson, and Joshua Winn. Apologies to any others who may have been inadvertently omitted from this list. I would also like to thank Paul Langacker and Craig Copi for supplying data that was needed for some of the graphs.

Finally, I want to thank my wife, Susan, and children, Jenny and Larry, who provided much encouragement and tolerated several years of chaotic family life so that this book could be completed. Without their support, the book could never have been written.

THE INFLATIONARY UNIVERSE

CHAPTER 1

THE ULTIMATE FREE LUNCH

The universe is big. We often say that we live in a small world, but the 25,000-mile trek around the planet Earth is still a longer trip than most of us have ever attempted. The farthest location ever reached by humans is the moon, about 240,000 miles from the earth. While the lunar landing was a spectacular achievement, we would have to travel 400 times farther if we wished to extend our exploration to the sun. This glowing sphere is so large that if a map of the earth were drawn to cover the sun's surface, the entire area of our planet would fit comfortably within the outline of the Dominican Republic. The sun is not unique, but is one of many stars. A journey to our nearest stellar neighbor, a three-star system called Alpha Centauri, would carry our astronauts a hundred million times farther than a trip to the moon, a distance so great that even light requires four years to traverse it. If the astronauts looked homeward from Alpha Centauri, the separation between the earth and moon would look no bigger than a thumbtack viewed from 400 miles away.

The sun and Alpha Centauri are lazily circling in a spiral galaxy of stars, making four revolutions every billion years. The galaxy, known as the Milky Way, has a diameter about 20,000 times larger than the distance from earth to Alpha Centauri. The sun would be hard to find on a complete map of the galaxy, since the sun is only one of the hundred billion stars that make up the swirling disk. To see how large this number is, we can imagine trying to count the stars. At one star every second, day and night, we and our descendants would be rattling off integers for over three thousand years before the last star of the Milky Way would be tallied. If we stacked a hundred billion pieces of paper, the height of the pile would reach 6000 miles into space.

Even with its hundred billion stars, however, our galaxy is only a puny dot in the realm of the cosmos. Within the range of our telescopes there are a hundred billion galaxies, each one comparable in size to the Milky Way. When our descendants finished counting the stars of our own galaxy, they could begin the three-thousand-year task of counting the multitude of galaxies that are scattered across the vast expanse of the visible universe.

While there are many questions about the universe that cosmologists would like to answer, probably the most fascinating is the most fundamental question of all: Where did all this come from? Almost every human civilization in history has offered an answer to this question in the context of mythology or religion, but until recently the question had been thought to be outside the scope of science. Although the generally accepted big bang theory holds that the observable universe emerged from an explosion some ten to twenty billion years ago, the theory nonetheless assumes that all the matter in the universe was present from the start. The form of the matter may have been different, but it was all there. The classic big bang theory describes the aftermath of the bang, but makes no attempt to describe what "banged," how it "banged," or what caused it to "bang." Nothing can be created from nothing, we were always taught, so there was no hope for a scientific explanation for the actual origin of the matter in the universe.

The difficulty in constructing a scientific theory for the origin of matter stems from a set of rules, called conservation principles, that trace their origin to the very roots of science itself.

"Being is ungenerable and imperishable," wrote Parmenides in about 500 B.C., in a passage that helped to create the philosophical approach that we call science today. This basic idea, that things which exist continue to exist, became a cornerstone of the concept of natural order. Objects would not appear and disappear unpredictably, but instead would evolve continuously according to principles of nature. This notion of continuity in existence became more concrete a century later in the work of Leucippus and Democritus, who advanced the theory that all matter is composed of eternal, indivisible atoms which move through an otherwise empty space. These ideas are reflected strongly in Lucretius' *De Rerum Natura (On the Nature of Things)*, written during the first century B.C., which includes the statement that "Nothing can be created from nothing." Lucretius went on to explain that "Material objects are of two kinds, atoms and compounds of atoms. The atoms themselves cannot be swamped by any force, for they are preserved indefinitely by their absolute solidity."

While the fundamental idea of continuity in existence can be traced to the ancient Greeks, it was not until much later that this line of inquiry evolved into the conservation laws of modern science. An important step occurred in the eighteenth century, with the work of Antoine-Laurent Lavoisier and his contemporaries. In 1772 Lavoisier discovered that when sulphur or phosphorus are burned, the products weigh more than the original material. While this might have been taken as proof that matter is not conserved, Lavoisier instead traced the added matter to a component of the air, which he named oxygen. When the oxygen was taken into account, the initial and final masses were equal. In Lavoisier's words, "nothing is created in the operations either of art or of nature, and it can be taken as an axiom that in every operation an equal quantity of matter exists both before and after the operation." These ideas were soon solidified in the work of the English scientist John Dalton, who in 1803 reintroduced the atomic theory of Leucippus and Democritus, and shortly thereafter produced the first table of atomic weights. By this time the idea of conservation of mass was firmly embedded in the developing scientific tradition.

The other great conservation principle developed during this period was the conservation of energy. This principle is more subtle, and it understandably took longer to develop. Unlike mass, the definition of energy—the conserved quantity—is far from obvious. The word itself is derived from the Greek *energeia*, constructed from the roots *en* (in) and *ergon* (work). Roughly speaking, energy is the capacity of any system to do work.

The concept of energy is intimately linked to the study of motion, the main subject of Isaac Newton's legendary *Philosophiae Naturalis Principia Mathematica (Mathematical Principles of Natural Philosophy),* published in 1687. The laws of motion described by Newton lead directly to the conservation of energy, provided that forces such as friction are excluded from consideration. Newton himself was clearly aware of what we now call the conservation of energy, but he apparently did not appreciate the deep importance that this idea would later acquire. It was the German philosopher Gottfried Wilhelm Leibniz (1646–1716), a contemporary of Newton's, who first realized that energy was sufficiently important to merit a name. He defined the term *Vis Viva* (Living Force) to measure what is now called kinetic energy, the energy of motion. The kinetic energy of a moving object is equal to half its mass times the square of its velocity. (If the mass is measured in grams and the velocity is measured in centimeters per second, the resulting energy is given in a unit called an erg. An erg is a small amount of energy—a 150-pound person walking at 2 miles per hour has a kinetic energy of 272 million ergs. The conversion to other units of energy is described in Appendix D.)

In most circumstances, however, kinetic energy itself is not conserved, as one can see by imagining a baseball that is held stationary at arm's length and then released. The kinetic energy starts from zero and increases as the ball gains speed on its downward plunge. Where does this energy come from? The answer can be found by a careful examination of the motion. To eliminate the complication of air friction, consider a weight that is dropped in a vacuum. If the mass is one gram, the kinetic energy is 980 ergs after one centimeter of fall, twice this amount after two centimeters of fall, three times this amount after three centimeters, etc. Since the kinetic energy is determined by the height of the ball, the height must be related to a new form of energy. It is called *gravitational potential energy*, or just *gravitational energy* for short. For each centimeter of fall the gravitational energy decreases by 980 ergs, just as the kinetic energy increases by 980 ergs, so the total energy is constant. (If the mass is more or less than one gram, the gravitational energy varies in proportion to the mass.) As long as the gravitational energy is properly included, then the total energy is conserved.

The skeptical reader might be wondering whether the conservation of energy is really a principle of nature, or just a tautological proclamation of obstinate physicists. Suppose, for comparison, that somebody claimed that a child's height was really conserved. Like the conservation of energy, this conservation principle could be made to work by inventing a new form of "height." One could define a "potential height" for the child, which decreases by one inch for each inch the child grows. In this way the "potential height" plus the ruler height will always have the same value. There is an important difference, however, between the conservation of energy and the conservation of a child's height. Both principles are technically correct, but the former can be used to make predictions while the latter cannot. If a ball is thrown into the air, then at any time during its flight one could determine the total energy by measuring both the speed and the height. At any later time one might measure only the speed, and the conservation of energy would allow the height to be calculated. Alternatively, one could measure only the height, and the speed could be calculated. By contrast, the potential height of the child is not directly measurable, and so the child-height conservation principle leads to no predictions whatever.

If one deals with the motion of planets instead of the motion of baseballs, then the force of gravity cannot be treated as constant, as it was in the previous example. The gravitational force between two planets becomes weaker as they are separated, decreasing in inverse proportion to the square of the distance between them. That is, if the distance is doubled, then the gravitational force that each exerts on the other is decreased by a factor of four. In this case the formula for the gravitational energy is less apparent, but it was discovered in 1847 by the German physiologist and physicist, Hermann von Helmholtz [1]. Helmholtz showed that, in the absence of friction, the sum of the kinetic and gravitational potential energies does not change with time.

In the real world, however, there is always at least a small amount of friction, so the sum of the kinetic and potential energies is never quite conserved. A pendulum with friction will gradually come to a halt, so it appears that energy is lost. This is no doubt one of the reasons why the conservation of energy was not given much importance in the early development of mechanics by Newton and his followers. However, once again it is possible to make the conservation principle work, by realizing that there is another form of energy that has to be defined and included in the bookkeeping. The new form of energy is heat, and it was Benjamin Thompson, one of the more colorful scientists of the eighteenth century,* who first showed that mechanical energy can be converted to heat energy with a fixed conversion rate.

While supervising a military arsenal in Munich, Thompson became impressed by the surprising amount of heat that was produced when a brass cannon was bored by a horse-turned machine. He performed an experiment, which he reported to the Royal Society of London in 1798, in which a metal cylinder enclosed in a box containing about two gallons of water was bored until the water became hot. He measured the temperature of the water as it rose with time, until the water started to boil after about two and a half hours. According to Thompson's report to the Royal Society,

> It would be difficult to describe the surprise and astonishment expressed in the countenances of the bystanders, on seeing so large a quantity of cold water heated, and actually made to boil, without any fire.

These observations constituted the first measurement of the conversion rate of mechanical energy to heat. Thompson's measurements were improved in 1843 by the English physicist James Prescott Joule, who also measured the conversion of electrical energy into heat.† By the end of the 1840s, the law

* Benjamin Thompson was born in Woburn, Massachusetts, in 1753. An unmitigated opportunist, at the age of 19 he married an extremely wealthy 33-year-old widow. He remained loyal to the British crown at the outbreak of the American Revolution, and for a short period served as a British spy. His espionage career had all the flair of an Ian Fleming novel, including the use of secret messages written in invisible ink, and a brief affair with the wife of a leading revolutionary newspaper-publisher. In March 1776 he abandoned his own wife and daughter and fled to London, where his credentials as an important loyalist from Massachusetts earned him the position of undersecretary of state for the colonies. He was later knighted by King George III, and then with the Crown's permission served as an advisor to the government of Bavaria. There he performed his famous experiments on heat, and he also brought James Watt's steam engine into common use. He produced inventions including an improved fireplace, a double boiler, a kitchen range, and a drip coffeemaker. In 1791 he became Count von Rumford of the Holy Roman Empire, taking the name from Rumford, Massachusetts (now Concord, New Hampshire), where he had lived for a short time [2].

† Today Joule's name is used for one of the standard units of energy, equal to 10 million ergs. The well-known unit of electrical power, the watt, is equal to one joule per second.

Figure 1.1 **A model of Rumford's cannon-boring experiment.** Although shown one above the other to fit the page, the two rooms were actually side by side. The wheel turned by the horses was attached by gear wheels to the shaft at the upper right, which extended into the cannon-boring room through a hole in the wall.

of conservation of energy had been enunciated clearly by Joule, Helmholtz, and also the German physicist Julius Robert von Mayer. If all forms of energy are included, then the total amount of energy cannot be changed by any physical process.

Thus, by the middle of the nineteenth century the two most basic conservation laws—the conservation of mass and the conservation of energy—were firmly established. In terms of Newton's laws of motion, it was now understood that fundamental forces, such as gravity, always conserve energy. Those forces which fail to conserve energy, such as friction, do so only because the description is incomplete. As the motion of a pendulum dissipates due to friction, the energy is not really lost. Instead it shows up as the kinetic energy of the random motion of atoms—in the pendulum and in the air surrounding the pendulum—energy which we recognize as heat. If Newton's laws were applied not to the pendulum as a whole, but instead to the individual atoms in the pendulum and in the air, then energy would always be conserved.

No one in the middle of the nineteenth century, however, could have anticipated the dramatic leap that was to be made by Albert Einstein in 1905. In that year Einstein published his two papers on the subject that we now call special relativity. The first laid out the foundations of the subject, while the second was a short follow-up, only three pages long. In this second paper Einstein derived the now-famous formula, $E = mc^2$. Here E denotes the total energy of any body, m denotes its mass, and c^2 denotes the square of the speed of light. According to this formula, mass and energy are *equivalent*, in the sense that they measure exactly the same thing. The only difference is the factor of c^2, which really just means that mass and energy refer to different systems of units. We get a bigger number if we measure a distance in inches than if we measured it in miles, but we understand that they are merely two ways to express the same thing. Similarly, the mass/energy of an object is a single property that we can equally well choose to measure in grams or in ergs.

Thus, there are no longer separate conservation laws for mass and energy. With the advent of special relativity, the two conservation laws meld into one—the conservation of mass/energy. For brevity, the combined conservation principle is often called simply the "conservation of energy." It follows that any particle—an electron, proton, or neutron, for example—has a significant energy even when it is at rest. This energy is called the *rest energy* of the particle.

Since c is such a large number, relativity implies that a small amount of mass is equivalent to a very large amount of energy. If m is measured in grams and c is measured in centimeters per second ($c = 2.998 \times 10^{10}$ centimeters per second*), then the energy $E = mc^2$ is obtained in ergs.† If one pound of matter could be converted entirely to energy, the result would be eleven billion kilowatt-hours—enough energy to supply the annual electric power consumption of a million average Americans, two million Europeans or Japanese, or eight million Latin Americans. In an article for *Science Illustrated* in 1946, Einstein himself gave a nontechnical explanation of how such a large amount of energy could have gone undetected for so long:

> But if every gram of material contains this tremendous energy, why did it go so long unnoticed? The answer is simple enough: so long as none of the energy is given off externally, it cannot be observed. It is as though a man who is fabulously rich should never spend or give away a cent; no one could tell how rich he was.

In a nuclear reactor only about one tenth of one percent of the mass of fissionable uranium is converted to energy, but the energy output is nonetheless enormous.

Probably the most dramatic challenge in this century to the conservation of energy was the discovery in 1914 by the British physicist, James Chadwick, that energy appears to be lost in a process called *beta decay.* In this process a radioactive nucleus decays by emitting an electron, which was found to have a range of possible energies extending up to the value predicted by energy conservation. Physicists searched for other ways in which the energy might escape, but by the late 1920s all reasonable possibilities were ruled out [3].

Physicists were puzzled, but in 1931 the Austrian-born physicist Wolfgang Pauli suggested what he called a "desperate solution." He proposed that beta decay produces a new type of particle that interacts so weakly that it escaped all efforts at detection. Two years later this suggestion was elaborated by the Italian physicist Enrico Fermi, who developed a detailed theory of the decay process. Fermi called the unseen particle the "neutrino," a word constructed from Italian roots which mean "little neutral one." (Curiously, Fermi's now-classic paper on beta decay was rejected by the prominent British journal *Nature.*) Since neutrinos interact with normal matter only very rarely, their detection remained a formidable challenge that required incredible ingenuity and patience. It was not until June 14, 1956, a year and half after Fermi's death, that Pauli received a telegram informing him that his prediction had at last been experimentally confirmed. Twenty-five years after its prediction, the neutrino had been detected by Clyde L.

* Note on scientific notation (for the reader who is not familiar with this way of writing numbers): 10^{10} means a 1 followed by 10 zeros, or 10,000,000,000. Here 10 is called the exponent. For an example with a negative exponent, 10^{-4} means a decimal point, 3 zeros, and then a 1, or .0001.

† The conversion to other units of energy is described in Appendix D.

Cowan, Jr. and Frederick Reines, working at the Savannah River reactor in South Carolina. (In 1995, after Cowan's death, Reines was awarded the Nobel Prize in physics for this work.) Nonetheless, in spite of this long gap between the prediction and the observation, physicists had never seriously doubted Pauli's proposal. Their confidence in the conservation of energy was so strong that they were willing to accept for a quarter of a century, the existence of a particle that was totally unseen.

Finally then, we come to the key question: Given the present understanding of conservation laws, is there any hope for a scientific description of the creation of the universe? If the conservation laws imply that "nothing can be created from nothing," as Lucretius put it, then how could the universe have come into being? We would be forced to believe either that the universe is eternal, or that it was created by some force that worked outside the restrictions of physical laws. If, however, we seek to understand how the universe might have been created within the context of physical laws, then some loophole has to be found in the age-old dictum of Lucretius.

If the creation of the universe is to be described by physical laws that embody the conservation of energy, then the universe must have the same energy as whatever it was created from. If the universe was created from nothing, then the total energy must be zero. But the universe is clearly filled with energy: the Earth, the Sun, the Milky Way, and the hundred billion galaxies that make up the observable universe clearly contain an unfathomable amount of mass/energy. How, then, is there any hope that the creation of the universe might be described scientifically?

An answer to this question arises from an extraordinary feature of one particular form of energy—gravitational potential energy, which was discussed earlier in the chapter. To understand this feature, however, we must delve just a bit deeper into the description of gravity.

According to Newton, any two objects exert an attractive gravitational force on each other. To find the total gravitational force that pulls on a given object, one computes the sum of the forces exerted on it by every other object.* In principle one has to include the forces exerted by every other object in the universe, but in practice only nearby objects are significant. Newton's description of gravity is often called an *action-at-a-distance* formulation, since gravity is interpreted as a force that one object exerts on a distant object.

* Forces are added by a procedure called vector addition, which takes into account the directions as well as the strengths of the forces. If two forces point in opposite directions, for example, their effects tend to cancel rather than to reinforce each other.

Most modern physicists, however, think about gravity using an alternative formulation in which forces at a distance are avoided. While maintaining the underlying content of Newton's law, this newer formulation replaces the action-at-a-distance with the notion of a *gravitational field*. In some contexts this phrase is familiar—an interplanetary space probe, for example, can be said to escape the gravitational field of the earth. More precisely, the physicist uses the concept of a gravitational field to characterize the effect of gravity at each point in space. Whether or not any object is actually located at a given point, one can always ask what force would be experienced by a one-gram mass, if the mass were placed at that point. The answer to this question, usually expressed in a unit of force called a dyne,* defines the gravitational field. The force on a one-gram mass near the surface of the earth is about 980 dynes, so the gravitational field is 980 dynes per gram. At an altitude of 50,000 feet the same one-gram mass would experience a force of only 975 dynes, so the gravitational field at this altitude is 975 dynes per gram.†

The gravitational field is often depicted by drawing *gravitational field lines* in space, as in Figure 1.2. The field lines point in the direction of the gravitational field, and the strength of the field can be indicated by how closely the lines are spaced. Figure 1.2 shows the gravitational field lines for a sphere: a collection of evenly spaced lines, each pointing directly toward the center of the sphere. The spacing between the lines is closest near the surface of the sphere, indicating that the field is strongest in this region.

In this formulation each particle produces a gravitational field in the space around it, similar to the illustration in Figure 1.2. The field in turn exerts a force on distant particles.

In the discussion of gravitational potential energy earlier in this chapter, we saw that energy can be stored by lifting a weight in the gravitational field of the earth. We know that the energy is not lost, even though it has disappeared from view, because it can be recovered by allowing the weight to fall. This argument, however, tells us nothing about *where* the energy is stored. Using the field formulation, however, the question has a simple answer: The energy is stored in the gravitational field. In any region of space with a gravitational field, there is a density of energy proportional to the square of the field. By using a beautiful technique first published in 1828 by George

* A dyne is the force necessary to accelerate a mass of one gram at an acceleration of one centimeter per second per second.

† The reader might wonder whether the one-gram mass remains a one-gram mass at 50,000 feet, since its weight—the gravitational pull toward the earth—decreases by about half of a percent. By the standard definition used in physics, mass is a measure of the inertia of an object. The mass of an object is in principle determined by observing the acceleration that is produced by applying a known force. By this definition a one-gram mass remains a one-gram mass, so far as we know, no matter where in the universe it is transported.

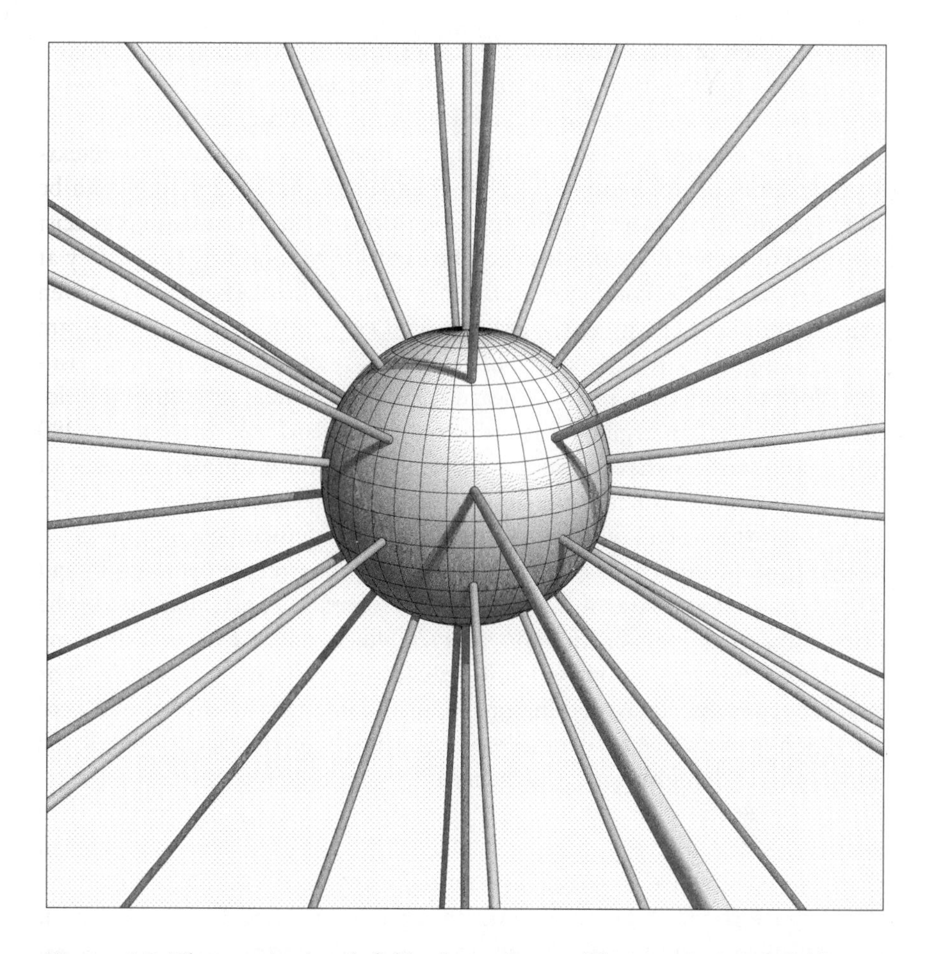

Figure 1.2 **The gravitational field of a sphere.** The gravitational field at a specified point measures the force that would be experienced by a mass if it were located at that point. A gravitational field can be depicted by gravitational field lines, as illustrated in the diagram. The lines are shown here as thick rods to make it easier to see their direction.

Green, a British baker and self-educated mathematician, one can show that all of the energy that is stored in lifting a weight can be attributed to the increased energy in the gravitational field. Thus, gravitational potential energy is really just the energy of the gravitational field. The energy of the gravitational field is therefore part of the energy of the universe, and must be included in the bookkeeping of the energy conservation law.

Now, we can return to the key question: How is there any hope that the creation of the universe might be described by physical laws consistent with energy conservation? Answer: *the energy stored in the gravitational field is represented by a negative number!* That is, the energy stored in a gravitational field is actually less than zero. If negative numbers are allowed, then zero can be obtained by adding a positive number to a negative number of equal magnitude; for example, 7 plus –7 equals zero. It is therefore conceivable that the total energy of the universe is zero. The immense energy that we observe in the form of matter can be canceled by a negative contribution of equal magnitude, coming from the gravitational field. There is no limit to the magnitude of energy in the gravitational field, and hence no limit to the amount of matter/energy that it can cancel.

For the reader interested in learning why the energy of a gravitational field is negative, an argument is presented in Appendix A. A specific example is used to show that no energy is needed to produce a gravitational field, but instead energy is *released* when a gravitational field is created.

Given this peculiar property of gravity, a scientific description of the creation of the universe is not precluded by the conservation of energy. Other conservation laws also need to be considered, in particular the conservation of a quantity called *baryon number*, which will be discussed in Chapter 6. But the conclusion will not be changed: The universe could have evolved from absolutely nothing in a manner consistent with all known conservation laws. While no detailed scientific theory of creation is known, the possibility of developing such a theory now appears open.

In the late 1960s, a young assistant professor at Columbia University named Edward P. Tryon attended a seminar given by Dennis Sciama, a noted British cosmologist. During a pause in the lecture, Tryon threw out the suggestion that "maybe the universe is a vacuum fluctuation." Tryon intended the suggestion seriously, and was disappointed when his senior colleagues took it as a clever joke and broke into laughter. It was, after all, presumably the first *scientific* idea about where the universe came from.

By a *vacuum fluctuation,* Tryon was referring to the very complicated picture of the vacuum, or empty space, that emerges from relativistic quantum theory. The hallmark of quantum theory, developed to describe the behavior of atoms, is the probabilistic nature of its predictions. It is impossible, even in principle, to predict the behavior of any one atom, although it is possible to predict the average properties of a large collection of atoms. The vacuum, like any physical system, is subject to these quantum uncertainties. Roughly speaking, *anything* can happen in the vacuum, although the probability for a digital watch to materialize is absurdly small. Tryon was advancing the outlandish proposal that the entire universe materialized in this fashion!

Edward Tryon, who likes to fish when he is not thinking about the universe.

Tryon put the idea of a vacuum fluctuation universe out of his mind for a while, but returned to it several years later while he was preparing a popular review of cosmology. In 1973 Tryon published an article in the journal *Nature*, with the title "Is the Universe a Vacuum Fluctuation?" [4]. He had understood the crucial point: the vast cosmos that we see around us could have originated as a vacuum fluctuation—essentially from nothing at all—because the large positive energy of the masses in the universe can be counterbalanced by a corresponding amount of negative energy in the form of the gravitational field. "In my model," Tryon wrote, "I assume that our Universe did indeed appear from nowhere about 10^{10} years ago. Contrary to popular belief, such an event need not have violated any of the conventional laws of physics."

A weak point of Tryon's paper was its failure to explain why the universe had become so large. While the scale of vacuum fluctuations is typically subatomic, Tryon was asking us to believe that all the matter in the universe appeared in a single vacuum fluctuation. He pointed out that the laws of physics place no strict limit on the magnitude of vacuum fluctuations, but he did not estimate the probability of such an unusually large fluctuation. "In answer to the question of why it happened," he wrote, "I offer the modest proposal that our Universe is simply one of those things which happen from time to time." Although the creation of a universe might be very unlikely, Tryon emphasized that no one had counted the failed attempts. Nonetheless, the immensity of the observed universe remained a striking feature for which Tryon's proposal had no explanation. For a number of years Tryon's work was largely ignored, as most other physicists apparently believed that any universe produced from a quantum fluctuation would, with overwhelming probability, be much smaller than the one that we observe.

It is the dream of every young scientist to be caught up in an important discovery, a scientific revolution which changes the way that people think about some fundamental problem. The problem of the origin of matter is about as fundamental as any that one can imagine. Prompted by an unlikely series of events beginning in 1978, I became embroiled in the writing of a new chapter about this subject, a chapter which involved many other physicists and which culminated in the development of the inflationary universe theory. This theory is a new twist on the big bang theory, proposing a novel picture of how the universe behaved for the first minuscule fraction of a second of its existence.

The central feature of the theory is a brief period of extraordinarily rapid expansion, or *inflation,* which lasted for a time interval perhaps as short as 10^{-30} seconds. During this period the universe expanded by at least a factor of 10^{25}, and perhaps a great deal more. After the stupendous growth spurt of inflation the description merges smoothly with the standard big bang theory, which for several decades has been the generally accepted picture of cosmic evolution.

Working within the general framework of accepted laws of physics, the inflationary theory can explain how the universe might have evolved from an initial seed as small as Tryon's vacuum fluctuations. Inflation provides a natural mechanism for tapping the unlimited reservoir of energy that can be extracted from the gravitational field—energy that can evolve to become the galaxies, stars, planets, and human beings that populate the universe today. While the standard big bang theory assumes that all the matter in the universe was present in some form since the beginning, the inflationary theory shows how all the mass could evolve from an initial seed weighing only about an ounce, with a diameter more than a billion times smaller than a proton.

The status of the theory of cosmic inflation today can be described by the same words used by Tryon in 1973 to describe his theory:

> "My model is admittedly speculative, and is still in an early stage of development. . . . I am encouraged to believe that the origin and properties of our Universe may be explicable within the framework of conventional science, along the lines indicated here."

While the final verdict on inflationary cosmology is not yet in, the basic outline of the theory seems very persuasive. The theory not only accounts for the vast amount of matter in the universe, but it also offers plausible explanations for a number of features of the universe that otherwise remain unexplained. These features will be described later in the book. There is no doubt that inflation has caught the attention of the scientific community, since there are now over 3000 publications on it in the scientific literature.

If inflation is correct, then the inflationary mechanism is responsible for the creation of essentially all the matter and energy in the universe. The theory also implies that the observed universe is only a minute fraction of the entire universe, and it strongly suggests that there are perhaps an infinite number of other universes that are completely disconnected from our own.

Most important of all, the question of the origin of the matter in the universe is no longer thought to be beyond the range of science. After two thousand years of scientific research, it now seems likely that Lucretius was wrong. Conceivably, *everything* can be created from nothing. And "everything" might include a lot more than what we can see. In the context of inflationary cosmology, it is fair to say that the universe is the ultimate free lunch.

CHAPTER 2

THE COSMIC VISTA FROM ITHACA, NEW YORK

The beginnings of my own exploits in cosmology, which led ultimately to the inflationary universe theory, can be traced to the fall of 1978. Prior to that season, my interest in cosmology had been only casual. I remember, for example, that during my high school years in Highland Park, New Jersey, I had been fascinated by Lincoln Barnett's *The Universe and Dr. Einstein*. As a graduate student at the Massachusetts Institute of Technology (MIT), I had taken the basic courses in general relativity and cosmology. My specialty, however, was the theory of elementary particles—the study of the laws of nature at their most fundamental level. Cosmology seemed very remote in 1978, when I was beginning my second year of a three-year postdoctoral research position with the particle theory group at Cornell. I did not know that during this academic year a few chance events would occur that would completely reshape my career (and even my life).

My position at Cornell, known informally as a "postdoc," was an entry-level academic position. Jobs called "postdocs" exist in many fields of science, but the nuances of the unwritten job description vary from field to field. Postdocs in theoretical physics are certainly not students—they already have their Ph.D. or its equivalent, and they are paid a decent, although by no means luxurious, salary. They are not really apprentices, either, as they do not ordinarily work under the direct supervision of anyone. Rather, postdocs are young researchers turned loose, given the opportunity to choose their own research problems and, if they wish, their own collaborators. While postdocs are sometimes called upon to teach or to help supervise students, for the most part they have free rein to pursue their research. The freedom and the intellectual atmosphere are wonderful, but there is a catch: The positions last typically for two or at most three years, and then the young scientist must seek a new job. Thus a postdoc's career is in the balance, and success depends on carrying out research that will be noticed rapidly by the physics community.

In a highly technical and overcrowded field like particle theory, a successful scientist typically has perhaps two postdoctoral positions before obtaining an assistant professorship. My stint on the postdoctoral circuit had been unusually long. I had begun my first postdoc at Princeton in 1971, without having quite yet completed my Ph.D. thesis at MIT. The thesis, finished during my first year at Princeton, was an exploration of an early theory of how quarks—the hypothetical constituents of protons, neutrons, and other elementary particles—interact with each other. During my stay at Princeton I did more than the usual amount of teaching, but still found some time for research. During this three-year period I produced only a single publication, a very long (forty-two page) paper describing my thesis and some extensions of it. The work was rapidly becoming obsolete, however, with the invention of the now-accepted theory of quark interactions, which is called *quantum chromodynamics*. The paper attracted very little attention, and to this day it has been cited in the particle physics literature only five times.* At the end of the Princeton years, I was in a very marginal situation on the job market.

With some luck, however, late in the season I received an offer from Columbia University. They had apparently offered the position to one or two others before me, but fortunately for me those candidates had chosen to go elsewhere. The two-year position at Columbia was extended for a third year, and during this time my research began to build momentum.

My job search from Columbia was much more encouraging than the previous round, although there were still no offers beyond the postdoctoral level. The Harvard Physics Department had chosen me as a finalist for an assistant professorship opening, although the position was given instead to an old friend from MIT student days. Nonetheless, I felt at least that the person who had edged me out was in the same league as I was—and therefore I must be in the right league. The job that I took at Cornell was billed as a glorified "senior" postdoctoral position, and its previous holder had moved on to an assistant professorship at Harvard.

At this point the focus of my career was very far from cosmology. While I am certainly more of a physicist than a mathematician at heart, I have always been fascinated by the inexplicable success of mathematics in describing the physical world. Until this time I had always sought out clean, mathematically well-defined problems for my research. I had tried to avoid

* Citation information is from the high energy physics database at the Stanford Linear Accelerator Center. Although possibly incomplete, the database is certainly indicative.

Robert Dicke in his office.

problems that were contaminated by the need for crude approximations or the involvement of large amounts of empirical information. Elementary particle theory presented many problems that were just right for my taste, while astrophysics seemed to abound with just those qualities that I preferred to avoid. Cosmology is for the most part a subfield of astrophysics, with the added drawback of being less developed than most branches of science. How much can we know, I asked myself, about the first seconds of the existence of the universe? At the time, cosmology seemed to me to be the kind of subject about which you could say anything you like—how could anyone prove you wrong?

So, when Robert Dicke (rhymes with Mickey) of Princeton University gave a lecture on the big bang at Cornell on Monday, November 13, 1978, I had no compelling reason to attend. If the Nobel Prize had not been awarded the previous month to Penzias and Wilson, cosmology might not have seemed interesting enough to draw me to the lecture. I was battling a case of bronchitis at the time, so I would have missed the lecture if my doctor had scheduled an appointment for Monday afternoon instead of Tuesday morning. If my wife Susan had stubbed her toe that afternoon instead of the week before, I might have taken her to the Tompkins County Hospital for X rays at the time of Dicke's talk. Indeed, if the week had been just a bit more hectic, I might very well have decided to skip this lecture. But it wasn't, and I didn't.

In the lecture Dicke explained something called the *flatness problem*, a peculiarity that he took as an indication that something important is missing in the standard big bang theory. While I will describe the big bang theory in more detail in the next few chapters, for now we need only understand that the universe is expanding. Indications of this expansion were first noticed by astronomers in the 1920s, and the expansion law was codified in 1929 by Edwin Hubble of the Mt. Wilson Observatory. Hubble found that with just a few exceptions, every galaxy that he observed is moving away from us. Furthermore, Hubble found that on average, the speed with which each galaxy is receding from us is proportional to the distance: If one galaxy is twice as far away as another, then the farther galaxy is receding twice as fast as the other. The precise value of the expansion rate, or *Hubble constant*, is still not well-determined, but it is somewhere between 15 to 30 kilometers per second per million light-years.*

If Hubble's law had been discovered in the early 16th century, it would undoubtedly have been taken as further evidence for what people already knew—that we are in the center of the universe. Why else, after all, would all the galaxies be moving radially outward from us, like the spokes on a wheel? Since the time of Copernicus, however, astronomers and physicists have become instinctively skeptical of this kind of inference. Today we believe that neither our planet, nor our galaxy, occupies a special position in the universe.

In the modern big bang theory, Hubble's law is interpreted as evidence that the universe is undergoing *homogeneous expansion*, as illustrated in Figure 2.1. The three frames are intended to show successive images of the same region of the universe. Each frame is approximately a photographic blowup of the previous frame, with all distances enlarged by the same percentage. All galaxies are roughly equivalent, and are spread more or less uniformly throughout all of space. It is further assumed that there is no center and no edge to the distribution of galaxies. As the universe evolves, from diagram (a) to (b) to (c), all intergalactic distances are enlarged. So, regardless of which

* A light-year is the distance that light travels in one year, 5.88 trillion miles or 9.46 trillion kilometers. Astronomers never quote distances in light-years, however, but invariably use the parsec, equal to 3.26 light-years. For larger distances they use the megaparsec, a million parsecs. In these units, the Hubble constant lies between 50 and 100 kilometers per second per megaparsec.

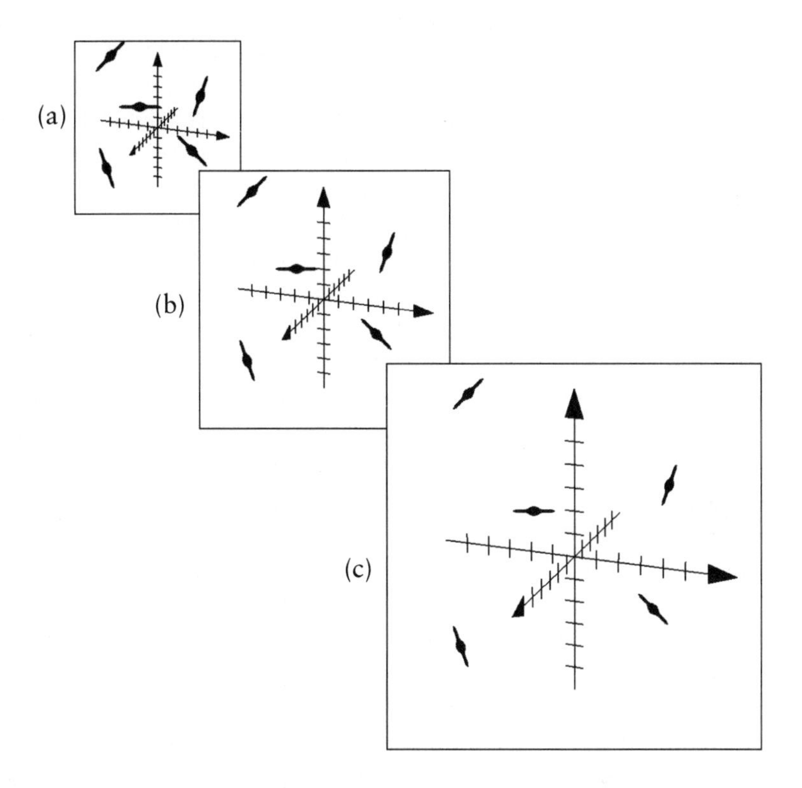

Figure 2.1 **The expanding homogeneous universe.** The three pictures show snapshots of the same region of the universe, taken at three successive times. Each picture is essentially a photographic blowup of the previous picture, with all distances enlarged by 50 percent. An observer on any galaxy would conclude that all the other galaxies are receding from him. Although the distances between the galaxies are growing, the size of the individual galaxies is remaining approximately constant.

galaxy we are living in, we would see all the other galaxies receding from us. Furthermore, this picture leads immediately to the conclusion that the recession velocities obey Hubble's law. Since all distances increase by the same percentage as the universe expands, larger distances increase by larger amounts. The velocity of a galaxy is proportional to the change in distance from one diagram to the next, and so it is proportional to the distance.

The expansion of the universe does not go unchecked, however, since every galaxy is attracting every other galaxy by the force of gravity. This attraction slows the expansion, although the slowing occurs at a rate that is too small for us to measure directly. Even when we take advantage of powerful telescopes that can look out to distances of billions of light-years, and hence billions of years into the past, the signs of the slowing cannot be disentangled from other effects. Nonetheless, the theoretical understanding of the slowing, based on our understanding of the nature of gravity, is seldom doubted. Gravity is created by masses, so the rate of slowing is determined by the amount of mass in the universe. If we only knew the average mass density—the average number of grams of matter per cubic centimeter in the universe—we could calculate the rate at which the expansion of the universe is slowing.

The ultimate fate of the universe, according to the big bang theory, depends on the average mass density of the universe. If this density exceeds a certain critical value that can be calculated from the expansion rate, then gravity will eventually win out, and the expansion of the universe will be reversed. In this case the universe would eventually collapse, in what is sometimes called the Big Crunch. (Although the Big Crunch would certainly be a disaster that would make all previous disasters seem inconsequential, the reader need not be alarmed—if the Big Crunch is to happen at all, it is at least 60 billion years in the future. So you will have to pay off your thirty-year mortgage, but if someone offers to sell you 60-billion-year futures in corn, you might want to discount the value accordingly.) A universe destined for recollapse is called a *closed universe*. On the other hand, if the mass density is less than the critical value, the universe will go on expanding forever—an *open universe*. If the mass density is precisely at the critical density, then the universe is called *flat*. Such a universe would continue to expand without limit, but the rate of expansion would become closer and closer to zero as time goes on.

The value of the critical mass density is believed to lie between 4.5×10^{-30} and 1.8×10^{-29} grams per cubic centimeter, depending on the value for the expansion rate (i.e., the Hubble constant) that one uses in the calculation. By the standards of our everyday experience, this density is astonishingly low. The critical density corresponds to somewhere between 2 and 8 hydrogen atoms per cubic yard, a density that is more than ten million times lower than that of the best vacuum that can be achieved in an earth-bound laboratory!

Cosmologists would like very much to know whether the average mass density of the universe is more or less than this critical value, but, unfortunately, we cannot tell. For the present discussion, however, it suffices to say that the total mass density of the universe is believed to lie somewhere between one tenth of the critical density and twice the critical density.

Cosmologists use the uppercase Greek letter omega (written "Ω") to denote the ratio of the actual mass density of the universe to the critical density. If omega is greater than one, the universe is closed and will eventually recollapse; if omega is less than one, the universe is open and will expand forever. If omega is exactly equal to one, then the universe is flat. Our belief that the actual mass density lies between one tenth and twice the critical density can be expressed succinctly by saying that omega lies between 0.1 and 2.

Since the range of uncertainty for the mass density is so large, one would not expect that any deep conclusions can be drawn from it. Dicke pointed out, however, that the evolution of omega is like a pencil balanced on its point. If the pencil is perfectly balanced, then the laws of classical physics imply that it will stand on its point forever. If the pencil tilts just slightly to the left or right, however, then the tilt will increase rapidly as the pencil falls over, as in Figure 2.2. The situation of perfect balance corresponds to a value of omega equal to one—a mass density precisely equal to the critical density. If omega is exactly one at any time, then it will remain exactly one forever.* However, if omega in the early universe were just slightly less than one, then it would rapidly fall toward zero. Alternatively, if omega in the early universe were just slightly greater than one, then it would rapidly increase without limit.† Figure 2.3 illustrates the behavior of omega for the first 30 seconds of time, for various choices of the value at 1 second. The graph shows that omega will remain one if it begins at exactly one, but a deviation as small as

***Figure* 2.2 A pencil.** The evolution of omega, the ratio of the actual mass density to the critical density, is like a pencil balanced on its point. If the pencil is leaning even slightly in either direction, it will rapidly fall over. Similarly, if omega differs in either direction from one, the deviation will rapidly grow.

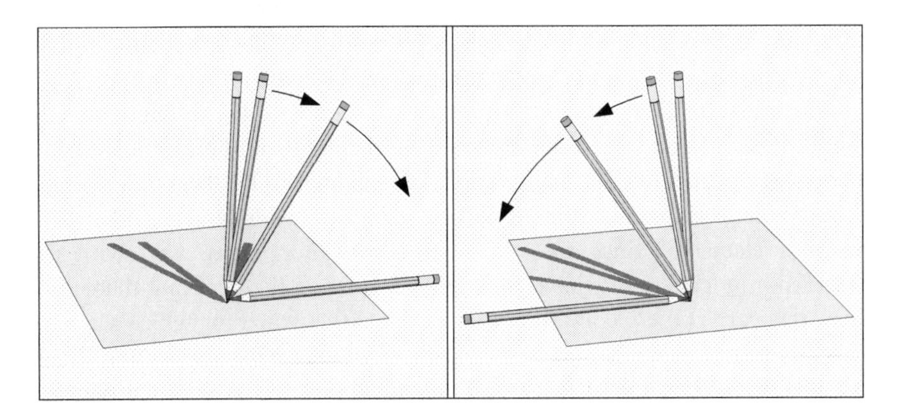

* This statement is actually a tautology—"omega equals one" is the statement that the ultimate fate of the universe lies on the border between eternal expansion and eventual collapse, and clearly the ultimate fate cannot change with time.

† The actual mass density of the universe would of course decrease as the universe expanded, but the critical mass density, which changes with the expansion rate, would decrease even faster. Since omega is the ratio of the two, it would increase.

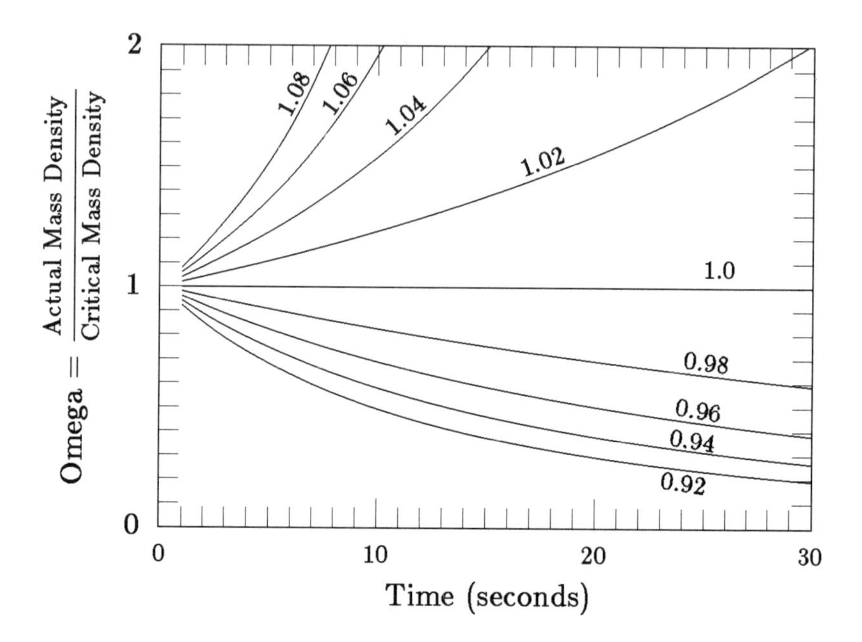

Figure 2.3 **The evolution of omega for the first 30 seconds.** The curves start at one second after the big bang, and each curve represents a different starting value of omega, as indicated by the numbers shown on the graph.

0.02 will become a large deviation within the time period shown. For omega to remain near one for ten billion years or more, any deviation from one in the early universe must have been extraordinarily small.

To illustrate the situation with realistic numbers, Dicke chose to talk about the expansion rate at one second after the big bang. This moment has a special significance in the big bang theory—it marks the beginning of the nuclear reactions that determined the abundance of helium and the other light chemical elements. Since the calculated abundances agree well with the observed abundances, we have good reason to trust the big bang theory for times this early. One second, however, is the earliest time for which we have direct empirical evidence for the accuracy of the big bang description. If we ask what the average mass density of the universe must have been at one second after the big bang, in order for it to be somewhere between a tenth and twice the critical value today, the answer is amazing. Figure 2.4 shows the relationship between the value of omega at one second, and its value at present. As can be seen from the graph, the mass density at one second must have been equal to the critical density to an accuracy of better than one part in 10^{15}. That is, it must have been at least 0.999999999999999 times the critical density, but no more than 1.000000000000001 times the critical density!

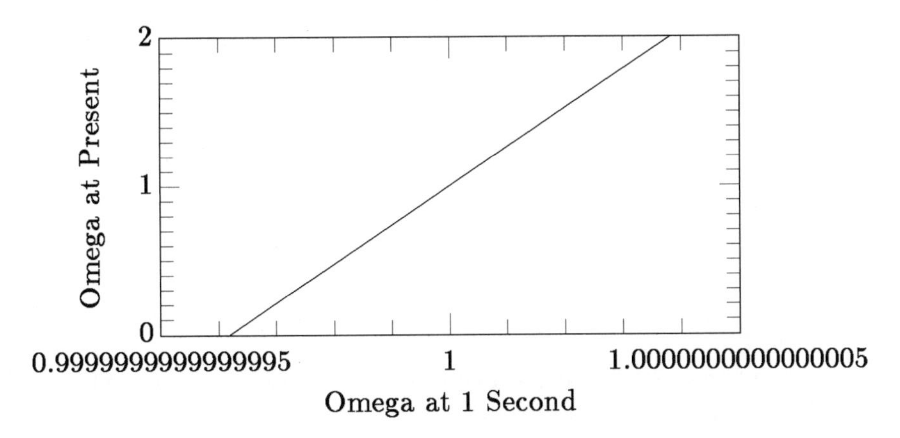

Figure 2.4 **The relation between omega at one second and omega today.** Since the evolution of the universe magnifies any deviation of omega from one, the value of omega today depends very sensitively on the value of omega at one second. The center of the horizontal axis represents omega = 1, and each marked interval on the horizontal axis represents a change in omega of only 10^{-16}, or 0.0000000000000001.

Dicke did not use the argument in this form, exactly, but instead asked what values for the mass density at one second after the big bang would lead to a universe that in any way resembles our own universe. This criterion gives more flexibility, but Dicke nonetheless concluded that the mass density at one second must have equaled the critical density to one part in 10^{14}. If the mass density were less than 0.99999999999999 times the critical value, he argued, then the density would have dwindled to a negligible value so quickly that galaxies would never have had time to form. If the mass density at one second were more than 1.00000000000001 times the critical value, on the other hand, then the universe would have reached its maximum size and collapsed before galaxies had a chance to form.

In the big bang model, the initial value of the mass density is what physicists call a *free parameter.* That is, it is a number that is not determined by the theory, but rather has to be determined by observation. The big bang theory gives an internally consistent description for any value of the mass density at one second after the big bang. If observation requires that the mass density be chosen to lie between 0.999999999999999 and 1.000000000000001 times the critical density, then such a choice can be made. The problem discussed by Dicke is therefore not a failure of the theory in the strict sense—there is no discrepancy between observation and the predictions of the theory. The problem, instead, is one of plausibility. If we do not know what determined the mass density at one second after the big bang, it seems very unlikely that it would lie within one part in 10^{15} of the critical density that puts the universe on the border between eternal expansion and eventual collapse. The suggestion, then, is not that the big bang theory is wrong, but rather that it is incomplete. Since it appears that the initial mass density was spectacularly close to the critical value, most physicists would expect that there must be some explanation. If no explanation can be found in the standard big bang model, then something else is needed.*

At the time I heard Dicke's lecture, my knowledge of basic cosmology was rather rusty, and I would not have been able to reproduce the calculations that led to Dicke's conclusions. But I did understand the gist of what he was saying, and I was struck by it. My diary contains the following notation (with bracketed words added for clarity):

> Colloquium by Bob Dicke on cosmology—fascinating—most outrageous claims: velocity [of Hubble expansion] at 1 sec tuned to 1 part in 10^{14} to allow for star formation.

I tucked the information away in the back of my mind, not knowing that this morsel would be a key piece to a puzzle that I would soon encounter.

Dicke had come to Cornell to deliver a series of lectures in honor of Hans Bethe, the Cornell physicist who had won the Nobel Prize for physics in 1967 for discovering how stars produce their energy. In 1978 Bethe was 72 years old and still an active physicist.† The cosmology lecture on Monday was followed by a lecture on general relativity experiments on Tuesday, and a lecture on solar physics on Thursday. My diary notes describe the second lecture as "very disorganized," and the third lecture as

* Dicke went on to suggest that the problem might be resolved by an oscillating universe theory—a theory in which the universe goes through an infinite succession of big bangs and big crunches. Such a theory is not ruled out by anything that we know, but it currently attracts very little interest among cosmologists. We know that general relativity—the most accurate theory of gravity that we have—does not allow a crunching universe to bounce into a big bang. This is not a fatal objection, however, since general relativity presumably breaks down at the extraordinarily high densities encountered in a big crunch. Nonetheless, since there is no reliable theory that describes how a universe might bounce, the basis of the oscillating universe theory relies solely on speculation.

† As of the writing of this book in 1996, Bethe is 90 years old and still an active physicist. The Cornell Physics Department recently held a symposium to celebrate his sixtieth year at Cornell. One of my fond memories from the Cornell years was an afternoon when Hans came to my office. I was in charge of organizing the weekly particle theory seminar, and Hans came to ask if I would invite a colleague of his from another institution. As a lowly postdoc, I was amazed at the meekness with which a Nobel Prize-winning physicist was approaching *me* with a request. I of course said yes, and Hans then asked, with equal humility, if the seminar would be able to cover the colleague's expenses. It was wonderful to see that at least one great physicist felt no need to remind anyone of his greatness.

Henry Tye in 1996, relaxing in his living room in Ithaca, New York, where he is now Professor of Physics at Cornell University.

"okay." No one can produce all gems, but a single gem can make a month of lectures worthwhile.

On that same Thursday, however, another crucially important slow fuse was set. The protagonist in this case was Henry Tye, a fellow postdoc in particle theory at Cornell. Henry had been born in Shanghai, China, and raised in Hong Kong, but had come to the United States to attend the California Institute of Technology and later graduate school at MIT. He and I had both done our theses under the supervision of Francis Low, although Henry had spent most of his graduate student career working with Gabriele Veneziano and Sergio Fubini.* Despite his foreign birthplace, Henry's English was flawlessly American and his understanding of the U.S. university system was uncanny.

Henry's advice was invaluable to me in many ways. For example, several weeks earlier Henry had suggested that I might benefit by arranging to spend the following year on leave from Cornell. It would probably not be difficult to find a one-year job at a major laboratory, such as the Stanford Linear Accelerator Center (usually known by its acronym, SLAC), where I

* Henry Tye's graduate program had been complicated by the fact that both Veneziano and Fubini left MIT, one after the other, to move to CERN, the giant particle physics laboratory outside Geneva, Switzerland. After Fubini's departure, Henry marched into the lunch room and asked the assembled faculty whether any of them would stay around MIT long enough for him to graduate. Francis answered affirmatively, remarking that even if he wanted to leave the Boston area, his wife would not let him.

would have access to new contacts, fresh perspectives, and additional areas of expertise. The plan would also mean, Henry pointed out, that I would have one more year of security before I had to look for another job. Henry himself had spent the previous year on leave at the Fermi National Accelerator Laboratory near Chicago, so he knew what he was saying. Following Henry's advice, I submitted applications for one-year positions to MIT, SLAC, and the Institute for Advanced Study in Princeton. At the beginning of March, Henry advised me that the time was ripe for follow-up phone calls to SLAC, since they would be making decisions soon. I asked two senior colleagues at Cornell to call, and personally accosted a theorist from SLAC who happened to be visiting Cornell that week. The effort was a success. In the middle of March I received a one-year offer from SLAC, including even the promise to pay one-way moving expenses. Consulting his internal encyclopedia of the ins and outs of academia, Henry suggested that I ask for two-way moving expenses. They might agree if I made the request before I went, but once I was there the regulations would not allow the reimbursement of my expenses to leave. I requested, and they agreed.

Henry's knowledge, however, was by no means limited to the practical side of things, and the idea that he suggested on that particular Thursday was much deeper. It was in fact the key idea that led eventually to the invention of the inflationary universe theory. At lunch that day, Henry suggested that maybe grand unified theories would give rise to magnetic monopoles.

At the time Henry Tye made his suggestion, I knew almost nothing about grand unified theories. Henry, however, had returned from his year at Fermilab with a tremendous enthusiasm for grand unified theories, or GUTs,* as they were usually called. He was convinced—correctly, as it turned out—that grand unified theories would be the exciting new theme in particle theory in the years to follow. Henry explained to me that grand unified theories were first suggested back in 1974, but that real interest in GUTs was just beginning to take off. There were several reasons for the delay, he explained, including the fact that the so-called "standard model of particle physics," which GUTs were intended to improve, was not yet fully accepted in 1974. Furthermore, one of the key predictions of grand unified theories was in conflict with experimental data in 1974, but since then improved experiments had agreed beautifully with the prediction.

* I never like to miss a chance to publicize my ideas, so I mention here that I am trying to persuade my colleagues that the acronym "GUT" is incorrect. The word "theory" should be abbreviated as "TH", since the original Greek root "theoria" ($\theta\epsilon\rho\iota\alpha$) uses the single Greek letter theta (θ) to indicate the "th" sound. I must admit, however, that this improved orthography has not been widely adopted.

Although I knew nothing about grand unified theories, I had spent several years while I was at Columbia working on *magnetic monopoles,* hypothetical particles that produce a special type of magnetic field. The magnetic field of ordinary magnets is caused by the motion of the electrons in the material, and all such magnets have both a north and a south pole of equal strength, as illustrated in Figure 2.5. The lines of the magnetic field, which could be followed by placing compasses near the magnet, extend from the

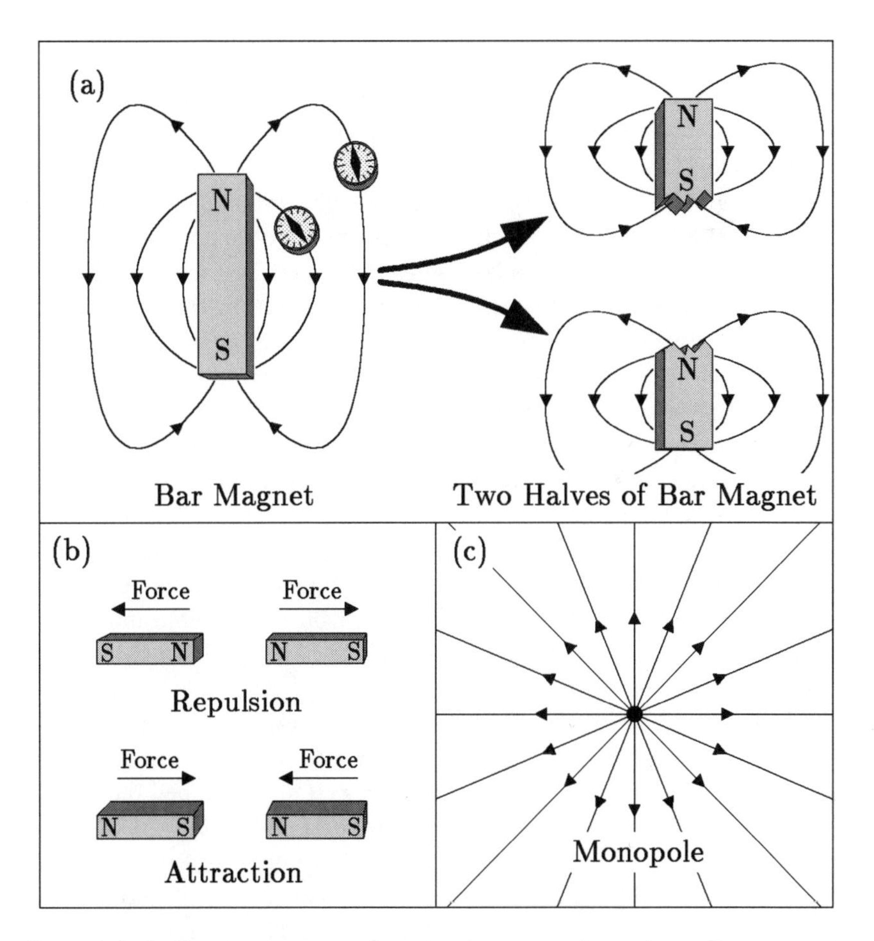

Figure 2.5 **Ordinary magnets and magnetic monopoles.** An ordinary magnet has a north and a south pole of equal strength, as shown in part (a). The curved lines indicate the direction of the magnetic field. Two compasses are shown aligning with the field. If the bar magnet is broken in two, each half will have both a north and south pole. Part (b) illustrates the forces between magnets: poles of the same type repel each other, while poles of the opposite type attract. Part (c) shows the field lines of a magnetic monopole.

north pole to the south. By holding one bar magnet in each hand, one could verify that two north poles repel, while a north and south pole attract. If a bar magnet is broken in two, one does not obtain separate north and south magnets; each piece has its own north and south pole, as illustrated in part (a) of the figure. A monopole, as its name suggests, is an isolated pole, either north or south, as shown in part (c). The magnetic field of the monopole points directly outward from the monopole, just like the electric field of a spherical charge or the gravitational field of a spherical mass, as shown in Figure 1.2.

Like the unicorn, the monopole has continued to fascinate the human mind despite the absence of confirmed observations. The monopole, however, has a much better chance of actually existing. As early as 1894 the French physicist Pierre Curie emphasized that the equations of electromagnetism make only one distinction between electric and magnetic fields: electric charges exist, but magnetic charges (i.e., monopoles) apparently do not. If magnetic monopoles were to be discovered, a perfect symmetry would be established between electricity and magnetism. Driven by the beauty of this symmetry, the British physicist Paul Adrien Maurice Dirac tried to extend the theory of magnetic monopoles to the realm of quantum theory. In 1931 he showed not only that monopoles are consistent with quantum theory, but also that quantum theory gives a unique prediction for the strength of their magnetic charge. Interest in magnetic monopoles was rekindled in 1974, when Gerard 't Hooft in the Netherlands and Alexander Polyakov in the Soviet Union independently discovered that some types of modern particle theories imply the existence of magnetic monopoles, even though no one had previously suspected it. My earlier work on monopoles, and Henry's question about monopoles in grand unified theories, were both based on the work of 't Hooft and Polyakov [1].

Henry's suggestion that grand unified theories might imply the existence of magnetic monopoles seemed intriguing, and also seemed like the kind of problem that I was prepared to attack. If grand unified theories did predict monopoles, then searches for monopoles could help to confirm or refute them. I understood magnetic monopoles quite well, I thought, and I realized that I would need only to learn the bare outline of grand unified theories to tell if they produced monopoles. Under Henry's tutelage, I began a brief dose of grand unified education.

Despite the grandiosity of their name, grand unified theories did not attempt a complete unification of physics, which instead was left as a problem for the future. Nonetheless, grand unified theories did provide a more unified description of the known processes of nature than had previously existed. To achieve this unification, however, it was necessary to extrapolate our understanding of elementary particle interactions to energies that are far beyond those that can be probed experimentally. For reasons that will be discussed in Chapter 8, the most distinctive predictions of the grand unified theories occur at energies in the range of 10^{16} GeV (where 1 GeV = 1 billion electron volts). To explain how large this amount of energy is, I mention first that it would not seem very impressive by the standards of your local power company; it is about what it takes to light a 100 watt bulb for 4½ hours. The novel predictions of grand unified theories, however, describe what happens when that much energy is deposited on a single elementary particle. Such an extraordinary concentration of energy is about 10 trillion (i.e., 10^{13}) times larger than the highest energy beams available from existing particle accelerators.

To appreciate how outrageous such an energy is, suppose we imagine trying to build a GUT-scale accelerator. One can do it with present technology, in principle, by building a very long linear accelerator. The largest existing linear accelerator is the Stanford Linear Accelerator, with a length of 2 miles and a maximum beam energy of 50 GeV. The output energy is proportional to the length, so a simple calculation shows the length required to reach an energy of 10^{16} GeV. The answer: about 70 light-years! Unfortunately, both the U.S. Department of Energy and NASA have been unreceptive to proposals for funding a 70-light-year accelerator. The most important consequences of grand unified theories, alas, seem well out of reach.

Given the inaccessibility of GUT predictions, it was hard for me to believe Henry's claims that GUTs would be the hot topic in particle theory in the coming years. Nonetheless, once I understood the basics of grand unified theories, I was able to answer Henry's original question about magnetic monopoles. The good news was that grand unified theories do contain magnetic monopoles. The bad news, however, was that the magnetic monopoles would be extraordinarily heavy, with masses in the vicinity of 10^{17} GeV, which is about 10^{17} times as heavy as a proton. So, I explained to Henry, the monopoles would be completely uninteresting. There would be no way to produce such behemoths at either present or foreseeable accelerators, so their theoretical existence would have no consequences for the observable world.

While human-made accelerators could not approach the energy needed for monopole production, Henry recognized that there was one legendary "accelerator" which had. "Okay," replied Henry, without hesitation. "So why don't we try to figure out how many monopoles would have been produced in the big bang?"

Henry's question left me cold. If our ignorance of 10^{16} GeV physics was to be combined with our lack of knowledge of the big bang, enlightenment seemed an improbable result.

While I was very reluctant to work on this problem, Henry was extremely enthusiastic. In the coming months, Henry's persistence slowly began to have an effect.

CHAPTER 3

THE BIRTH OF MODERN COSMOLOGY

I recently had the experience, while visiting Haverford College in Pennsylvania, of leafing through an original 1687 edition of Isaac Newton's *Philosophiae Naturalis Principia Mathematica.* Exposed to this kind of monumental work, a student of physics like myself can be daunted, and even a bit demoralized. It is hard to imagine how Newton invented and so thoroughly mastered such entirely new fields of both mathematics and physics. By handling the original book I was at least reassured that the work was constructed from paper and ink, and had not exploded through some ethereal process directly into our intellectual heritage. The achievements of the *Principia* can be exhilarating, but sometimes it is hard to remain enthusiastic while feeling so inadequate compared to the great masters of the past.

Against this awe-inspiring backdrop, the history of modern cosmology is a refreshing antidote. Maybe there were too few researchers working in cosmology for the usual scientific point/counterpoint to keep things on track. Maybe the field of cosmology is just too primitive to bring out the best in scientific creativity. Or maybe the questions of cosmology are too close to those of religion and philosophy for scientists to maintain a clear head. For whatever reason, cosmology has a history that can help us all to restore our equanimity. It is a field in which the giants of science, including Newton and Einstein, fell prey to mistaken prejudices and even technical incompetence. The greats of the field, such as Edwin Hubble, committed errors that held back progress for decades. Meanwhile the underdog foot soldiers of the field, such as Alexander Friedmann, Georges Lemaître, and later Ralph Alpher and Robert Herman, pushed forward the frontiers in ways that were not fully appreciated by the scientific community until many years later.

Before I could begin to work with Henry Tye on the question of how many magnetic monopoles would have been produced by the big bang—a study that would lead eventually to the inflationary theory of the universe—I had to relearn, or in some respects learn for the first time, the theory of the hot big bang. I acquired this knowledge in bits and pieces, but I will try to summarize it here as a coherent narrative. My approach will be mainly historical, since I want to emphasize that science is not merely a collection of facts, but is instead an ongoing detective story, in which scientists passionately search for clues in the hope of unraveling the mysteries of the universe. While reading these chapters on the big bang, the reader should keep in mind that this theory is now generally accepted by almost all scientists actively working in cosmology. The inflationary universe theory derives from the big bang theory and adds to it, but does not replace it. In Chapter 6, once the necessary background is established, I will return to the story of how my work with Henry led to a new theory of the origin of essentially all the matter and energy in the universe.

It is hard to know where to begin the story of cosmology, since it is among the most ancient of sciences. From observations carried out with the human eye alone, Greek philosophers such as Eudoxus, Aristotle, Aristarchus, and Ptolemy constructed complicated theories of rotating spheres on spheres to describe the motion of the earth, sun, moon, planets, and stars. Progress accelerated in the sixteenth century, when Nicolaus Copernicus reasserted Aristarchus' theory that the earth revolved around the sun.* With the work of Tycho Brahe, Johannes Kepler, Galileo Galilei, and Isaac Newton, the presently accepted picture of the solar system emerged. In 1750 Thomas Wright suggested that the visible stars are distributed in a flat slab, which we now know as our galaxy, the Milky Way. Shortly afterward, Immanuel Kant and then William Herschel suggested that some of the wispy nebulae that are sprinkled across the night sky are in fact other island universes, or galaxies, like our own. It was not, however, until 1923 that Edwin Hubble, using the newly completed 100-inch telescope at Mt. Wilson, resolved the stars of the Andromeda nebula to demonstrate conclusively that our galaxy is not alone in the cosmic void. It was soon recognized that the universe is filled with similar galaxies, extending in all directions as far as our telescopes can see. While these achievements and

* Edward Harrison has said that "Those who think that nothing happened in cosmology between Ptolemy and Copernicus badly need a course of reading in Pierre Duhem." For those who wish to fill in this gap, an abridged edition of Duhem's work has been edited and translated into English by Roger Ariew, under the title *Medieval Cosmology* (University of Chicago Press, 1985).

many others can each make a fascinating story, for the sake of brevity I will attempt to pick up the story of cosmology with the work of Albert Einstein at the beginning of the twentieth century.

In 1916 Einstein published his celebrated theory of general relativity, eleven years after the publication of his equally celebrated theory of special relativity. The terminology here is likely to be confusing, since the words "special" and "general" are used in common English in a wide variety of ways. An army general ranks far above a specialist, but a special delivery letter is almost guaranteed to reach its destination well before a general delivery letter.

In the case of relativity, the special theory represents the first of Einstein's major revisions of our ideas of space and time. Prior to Einstein, physicists thought that space was permeated by the *luminiferous ether* (or *aether*), a substance which was assumed to be weightless, transparent, and frictionless. (The word ether comes from the same root as ethereal, and is not related to the organic compound widely used as an anesthetic.) The theoretical purpose of the ether was to provide a medium for the transmission of light. It was assumed that light travels at a fixed speed, usually denoted by the letter c, relative to the ether, just as sound waves travel at a fixed speed relative to the air. (The value of c is 2.998×10^{10} centimeters per second or 186,283 miles per second.) Einstein, however, discarded the ether hypothesis and adopted instead the premise that the speed of light would be measured at the standard value c by all observers who move at a constant speed in a fixed direction, even if those observers are in motion relative to each other. This premise was supported by the Michelson-Morley experiment of 1887, and is by now very well established.

This basic premise of special relativity, however, seems at first to be nonsensical. Suppose, for example, that I am at rest, and a light beam passes me. The speed would have to be c, the standard speed of light. Suppose, now, that I take off in a spaceship to chase the light beam at $\frac{2}{3}$ of the speed of light. Common sense (or Newtonian physics) implies that I would then see the light pulse receding from me at only $\frac{1}{3}$ of c. The premise of special relativity, however, holds that I would still measure the recession speed as c. No matter how hard I might try to catch a light beam, I will always see the beam recede at speed c.

In spite of this conflict with common sense, Einstein showed that the premise of special relativity *is* consistent, provided that we are willing to give up some of our preconceived notions about space and time. In particular, we must give up our belief that measurements of space and time give the same results for all observers. Measurements instead become *relative*, in that they depend on the motion of the observer and the object observed. In particular, suppose a spaceship flies by us at a significant fraction of the speed of light. According to Einstein, the spaceship will appear to us to be shorter than the length that would be measured by observers inside the ship, although the width would be unchanged. The clocks on the ship would appear to us to be running slowly, and the clocks at the back of the ship would look like they were set to a later time than the clocks at the front of the ship, even though they would look synchronized to the crew.

Einstein gave a specific set of rules to describe how the measurements of one observer would be related to those of another, and showed that with these rules the apparent contradictions associated with the premise of special relativity would disappear. Although specific measurements would depend on the observer, the laws of physics, including the laws that describe light transmission, would appear the same to all observers (as long as the observers are not accelerating). It was no longer necessary to have one set of laws for observers at rest with the ether, and a more complicated set for moving observers. As a by-product of this reasoning, Einstein concluded that mass and energy are equivalent, and that no information or matter can travel faster than light.

The theory of special relativity is "special" in the sense that it is restricted to a special case: the case in which gravitational fields are either absent or negligible. Einstein noticed as soon as he invented special relativity that it is inconsistent with Newton's theory of gravity, so he immediately set to work on a new theory of gravity. The eventual result was the theory of general relativity, which is "general" in the sense of "universal"—it holds under all circumstances, whether the effects of gravity are negligible or not.

Given that special relativity already existed, it is fair to say that the theory of general relativity is nothing more nor less than a theory of gravity. Gravity is described in a surprising way, which is not at all similar to the way that electric or magnetic forces are treated. Gravity is represented as a bending, twisting, or stretching of the geometry of space and time. The picture of space as a rigid background governed by Euclidean geometry is abandoned. Instead space becomes plastic, distorted by the presence of the matter that it contains.

Immediately after completing his paper on the theory of general relativity, Einstein set out to study the consequences of the new theory of gravity for the universe as a whole. In less than a year he finished his classic paper on cosmology, titled "Cosmological Considerations on the General Theory of Relativity." Since this publication, almost all theoretical work in cosmology has been carried out in the context of general relativity.

In pursuing these studies, Einstein discovered something that surprised him a great deal: It is impossible to construct a mathematical model of a static universe consistent with general relativity. Einstein was perplexed by this fact. Like his predecessors, he had looked into the sky, had seen that the stars appear fixed, and had erroneously concluded that the universe is effectively motionless. He did not question this assumption in any way, but merely stated it as a matter of fact [1]:

> The most important fact that we draw from experience as to the distribution of matter is that the relative velocities of the stars are very small as compared with the velocity of light. So I think that for the present we may base our reasoning upon the following approximative assumption. There is a system of reference relative to which matter may be looked upon as being permanently at rest.

The same problem that Einstein discovered in the context of general relativity also existed in Newtonian mechanics, although it had not been fully understood until the work of Einstein. The problem is nonetheless fairly simple to understand: if masses were distributed uniformly and statically throughout space, then everything would attract everything else and the entire configuration would collapse, contracting without limit.

Newton himself had wrestled with this problem, and had realized that any *finite* distribution of mass would collapse by gravitational attraction. He went on, however, to erroneously conclude that the collapse would be avoided if the universe were infinite and filled with matter throughout. In that case, Newton reasoned, there would be no center at which the mass might collect. The fallacy of Newton's logic, which I discuss in Appendix B, provides an interesting example of the subtleties that arise in the discussion of infinite spaces.

Having discovered that he could not construct a mathematical model of a static universe consistent with general relativity, Einstein nonetheless remained convinced that the universe is static. He therefore modified his equations of general relativity, adding what he called a *cosmological term*—a kind of universal repulsion that prevents the uniform distribution of matter from collapsing under the normal force of gravity. (The coefficient of this term—the number that determines how large an effect the term has—is called the *cosmological constant.*) Of course, if one is constructing a theory of gravity, one does not want to introduce a repulsive force that nullifies the attractive force completely. The cosmological force, however, depends on distance differently than the attractive force. In the Newtonian approximation, the attractive force between two particles becomes four times as weak whenever the distance between them is doubled, while the cosmological force becomes twice as strong. The effects of the cosmological term, therefore, are most important for objects that are widely separated. The cosmological force can stabilize the universe against collapse, while its effect on the solar system or smaller structures is negligible. Einstein found that the cosmological term fits neatly into the equations of general relativity—it is completely consistent with all the fundamental ideas on which the theory was constructed.

By adjusting the value of the cosmological constant to provide a repulsive force that exactly cancels the gravitational attraction caused by the mass density, which Einstein approximated as a constant, Einstein was able to construct the static solution that he sought. The solution is *homogeneous*, which means that it looks exactly the same from any point in space. It is also *isotropic*, which means that if one stands at any point in space, the model universe looks exactly the same in all directions. The hypothesis that the universe is both homogeneous and isotropic is called the *Cosmological Principle*, a term that was introduced in 1933 by the British astrophysicist Edward Arthur Milne.

The spatial geometry of Einstein's construction is what we now call a *closed universe*, a striking example of the non-Euclidean geometry introduced by general relativity. Although Einstein's static universe is no longer acceptable now that we know the universe is expanding, the closed universe geometry remains a possibility. The closed universe is finite in volume, but it has no edge or boundary. If a rocket ship were to cruise for a sufficiently long time in what would appear to be a straight line, it would eventually find itself back at its starting point.

A very good analogy for a closed universe is the two-dimensional surface of a sphere in three Euclidean dimensions. Three illustrations of such a universe are shown in Figure 3.1. You must try to imagine a two-dimensional creature who lives on the surface of the sphere—a non-Euclidean cousin of the *Flatland* creatures of Edwin A. Abbott. To make the analogy precise, however, you must insist to yourself that the radial direction (i.e., the direction perpendicular to the surface of the sphere) has no reality. To the *Flatland* creatures, the only possible motion—the only *conceivable* motion—is within the two-dimensional surface of the sphere. The radial direction is included in the picture only to make it possible for incorrigibly three-dimensional creatures like ourselves to visualize the situation. A *Flatland* creature who surveys its universe would discover that the area is finite, yet no matter how far it wanders it will never encounter an edge.

To complete the analogy, everything must be elevated by one dimension. You must imagine a four-dimensional Euclidean space, and then imagine a sphere in the four-dimensional space. The three-dimensional surface of the sphere is precisely the geometry of Einstein's cosmology. (If you have difficulty visualizing a sphere in four Euclidean dimensions, rest assured that you have a lot of company, including the author. Luckily, the analogy with the sphere in three Euclidean dimensions is enough to understand most of the properties of closed universes.)

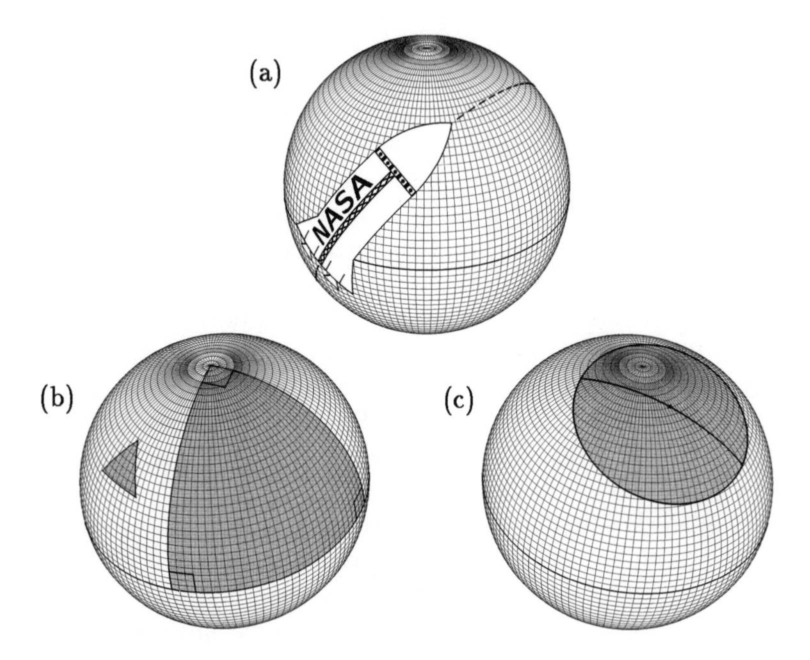

Figure 3.1 **Three views of a *Flatland* closed universe.** In part (a) a *Flatland* rocket ship travels along what we would call a great circle, returning eventually to its point of origin. A *Flatland* resident views the rocket path as a straight line, since it does not bend to either the left or right. Part (b) shows two triangles of different sizes. Their sides are straight lines (in *Flatland* language), which appear to us as arcs of great circles. The total of the angles in either triangle is more than the usual 180°. The excess is larger for the bigger triangle, which contains three right angles, so the total is exactly 270°. In part (c) a circle is drawn on the surface of the sphere; the diameter of this circle is a little longer than the circumference divided by π (3.14159. . .), which is the value that would describe a Euclidean circle. All these effects are characteristic of closed universes.

Einstein's theory of a static universe was an important start for cosmology based on general relativity, although it is no longer accepted. The theory is now known to have two serious flaws, the more obvious of which is the fact that the real universe is expanding. While the expansion was not known in 1917, Einstein could in principle have noticed the second flaw: the model universe is unstable.

Recall that the theory depends on choosing the value of the cosmological constant to give a perfect balance between the normal, attractive force of gravity and the repulsive cosmological force. Whenever a balance of this sort is required, the modern physicist has been trained to ask what would happen if the balance were upset by a small amount. Suppose, for example, that due to small fluctuations in the random motion of galaxies, the universe came to be ever so slightly larger than the size for which the perfect balance had been established. Would the universe return to its balanced state, or would it head off to oblivion? If the universe became slightly larger, then the average distance between particles would increase. With an increase in distance, the normal attractive force of gravity becomes weaker, and the repulsive cosmological force becomes stronger. The particles would therefore continue to fly apart, and the size of the universe would increase without limit. Conversely, if a fluctuation caused a slight decrease in the size of the universe, it would proceed into catastrophic collapse. Since small fluctuations are viewed as inevitable, the universe could not conceivably remain approximately static for eternity, as Einstein had proposed.

This criticism of Einstein's theory was discussed in 1930 by the British astronomer Arthur S. Eddington [2], and was also emphasized in 1934 by Richard C. Tolman in his classic book, *Relativity, Thermodynamics, and Cosmology*. By this time the observed expansion of the universe had already discredited the static universe theory, but the discovery of the instability was nonetheless an important accomplishment. Previously the Einstein static universe was a theoretical possibility, and only by observation could we learn that we are not living in such a universe. Now, as Eddington emphasized, we also had a compelling theoretical reason for believing that we *could not* be living in a static universe.

In the same year as Einstein's cosmology paper, 1917, the Dutch astronomer Willem de Sitter constructed an alternative theory, which was also based on general relativity. De Sitter shared Einstein's conviction that the universe is static, but unlike Einstein he suspected that the mass density of the universe was too low to be relevant. Keeping Einstein's cosmological constant, de Sitter sought and found a static solution which contained no matter at all.

Einstein was disappointed to learn of the de Sitter solution, for he had hoped that the general theory of relativity would determine cosmology uniquely. He looked for ways to rule it out, but with no success.

The understanding of de Sitter's work took a surprising twist in 1923, when it was discovered that the universe described by de Sitter's equations was not static after all. It may seem shocking that such a basic feature of the theory could have been misconstrued by both de Sitter and Einstein, but the episode shows how easy it is to misinterpret the equations of general relativity. (General relativists today are equipped with an arsenal of mathematical techniques to help prevent this kind of misconception.) The confusion arose mainly because de Sitter's universe contained no matter, and therefore no markers to make the expansion visible. However, when Arthur Eddington and the German-born mathematician and theoretical physicist Hermann Weyl considered what would happen if the empty de Sitter universe were sprinkled with small particles, they discovered that the particles flew apart.

There was considerable interest in the de Sitter theory through the 1920s, but in hindsight it appears that the interest was misguided. The solution was initially considered appealing because it was thought to be static, but it was later found to be expanding. The expansion is no longer a problem, since we know that the real universe is expanding, but now the de Sitter theory has another problem. The theory assumes that the mass density is negligible, while the mass density of the real universe is too high for the de Sitter solution to be even a rough approximation. Nevertheless, elegant equations have a mysterious habit of reappearing in other contexts. The de Sitter solution re-emerged in 1980 to play a major role in the inflationary universe theory.

While Einstein was disappointed to learn from de Sitter that his equations had two possible solutions instead of one, he was further disappointed by a large class of solutions that were discovered by an obscure Russian meteorologist and mathematician, Alexander Friedmann. These solutions were published in the well-known German journal *Zeitschrift für Physik* in 1922 and 1924 [3], but were widely ignored until much later.

Friedmann [4] was born in St. Petersburg in 1888. His father was a composer and his mother was a piano teacher, but Friedmann claimed that he inherited none of his parents' musical talent. He studied mathematics at the University of St. Petersburg and then meteorology at the Pavlov Aerological Observatory. At the outbreak of World War I, Friedmann joined a volunteer air corps, making a number of flights over friendly territory to test some very successful bombing procedures which he had worked out. In 1916–17 he headed the air-navigation service on all fronts, and co-directed the computation of ballistic tables. After the revolution of 1917 he became a full professor of mechanics at Perm' University, near the Ural Mountains about 750 miles from Moscow. In 1920 he returned to St. Petersburg to work at the Academy of Sciences, where he became interested in quantum theory, relativity, and other topics in theoretical physics. Friedmann died in 1925 at the age of 37, a death that was attributed by official accounts to typhoid. However, George Gamow, a student of Friedmann's who remained in the Soviet Union until 1933, reports that Friedmann died of pneumonia which was contracted following a chill that he caught while flying a meteorological balloon.

Friedmann's papers laid the foundation for cosmology based on general relativity, deriving the key equations that now appear in all cosmology textbooks. (The textbook derivations, however, do not follow the methods of Friedmann, which were actually rather awkward. The modern approach is due mainly to later work by Howard Percy Robertson and Arthur G. Walker.) Friedmann made what now appear to be all the right choices. He dropped Einstein's and de Sitter's assumption that the universe should be static; he assumed, correctly, that there was no real evidence to support this prejudice. He maintained, however, the assumption that the universe is homogenous (the same at all places) and isotropic (the same in all directions). Friedmann did not need the cosmological constant, since his solutions were not static, but he nonetheless chose to explore solutions both with and without it.

Friedmann found that the solutions without a cosmological constant fall into three classes. First, there are the closed universe models. These are mathematical models describing an expanding universe in which the mass density is so high that the gravitational field will eventually halt the expansion. Such a universe would reach a maximum size, and then contract. If we choose two arbitrary points, or galaxies, to follow, we would find that their separation starts from zero, reaches a maximum value, and then decreases again to zero, as shown in Figure 3.2. The mass density causes the space to curve back on itself, so at any instant the space is identical to the closed space of the static Einstein model. Thus, Friedmann found that if the universe is closed in time (i.e., if the universe recollapses), then it is also closed in space (i.e., it has finite volume). Like the separation between two arbitrary galaxies, the circumference of the universe starts from zero, reaches a maximum value, and then decreases again to zero.*

A second class of solutions, which Friedmann discussed in his second paper (published in 1924), are called *open universe* models. These are expanding models in which the mass density is low, so the gravitational field is too weak to halt the expansion. The distance between two arbitrarily chosen galaxies starts from zero, and then grows without limit, as shown in Figure 3.2. As time increases the velocity of separation between the two galaxies levels off at a constant value.

* Since the circumference (and hence the radius) of the Friedmann universe changes with time, it is quite possible that I have left my readers confused about the earlier statements which said that the radial direction in Figure 3.1 has no physical meaning. It is still true that the radial direction has no physical meaning, in that physical objects cannot move in the radial direction. However, the circumference of the sphere is a physically meaningful property, since it can be measured without leaving the surface of the sphere. In this case the circumference changes with time. If we visualize the surface in a three-dimensional Euclidean space, as in Figure 3.1, then the radius must change with time along with the circumference.

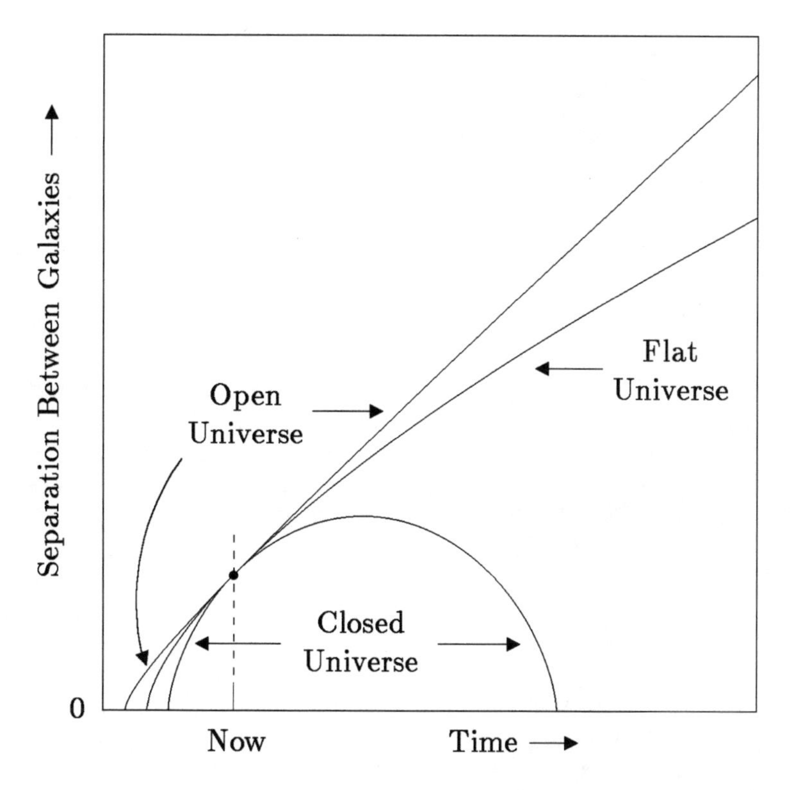

Figure 3.2 **The evolution of Friedmann universes.** If the cosmological constant is assumed to vanish, models of the universe fall into three classes. A mass density high enough to reverse the expansion leads to a closed universe. In a low density universe, called open, the expansion continues forever, and the velocity of any given galaxy levels off at a constant value. The mass density at the borderline between these two cases is called the critical density, and the corresponding universe is called flat. A flat universe will expand forever, but the velocity of any given galaxy will become smaller and smaller as time goes on. The curves are drawn so that they coincide at the present time, labeled "Now," but they begin at different times in the past.

The geometry of an open universe is non-Euclidean, but it differs from Euclid in the opposite way from the closed universe. While the space of the closed universe curves back on itself to produce a finite space, the space of the open universe curves in some sense away from itself, producing an infinite space. While the closed space can be illustrated by the surface of a sphere, as in Figure 3.3(a), the open space can be illustrated by a saddle shape, shown in Figure 3.3(b). The sphere and the saddle, however, are not really on an equal footing. The surface of a sphere is an exact representation of the geometry of the closed universe, while the saddle shape is only an

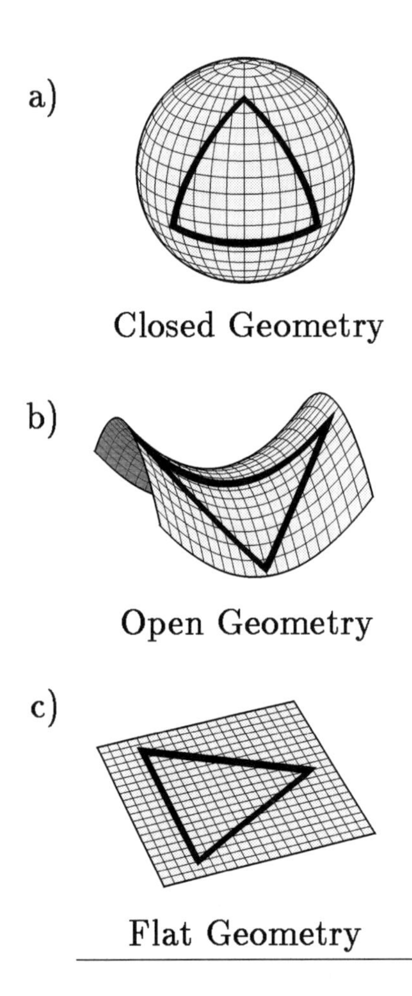

Figure 3.3 **The geometry of Friedmann universes.** If the universe is assumed to be homogeneous (i.e., the same at all locations) and isotropic (i.e., the same in all directions), there are exactly three types of possible geometries. A closed geometry is the three-dimensional analogue of the surface of a sphere—a triangle in this geometry contains more than 180°, and the circumference of a circle is a little shorter than π times the diameter. An open geometry is the analogue of a saddle-shaped surface—a triangle in this case has less than 180°, and the circumference of a circle is a little longer than π times the diameter. The third possibility is flat geometry, the Euclidean geometry with which we are familiar, in which each triangle contains exactly 180°, and the circumference of a circle is exactly π times the diameter. If Einstein's cosmological constant vanishes, then each type of geometry occurs only in combination with the time-evolution shown in Figure 3.2 with the corresponding name. If the cosmological constant is nonzero, however, then any type of geometry can occur with any type of time evolution.

approximation that is valid near the center. Thus, if the cosmological model is open in time (i.e., if it does not recollapse), then the Friedmann equations imply that it is also open in space (i.e., has infinite volume).

Finally, there is a case that is precisely the borderline between the open and closed models of the universe. Friedmann did not actually discuss this case explicitly, but it can be obtained as a limiting case of either the closed or open universe solutions. That is, the borderline case can be obtained by pushing the mass density of the closed model to its lowest possible value, or the mass density of the open model to its highest possible value. The mass density that puts the universe just at the boundary between eternal expansion and eventual collapse is called the *critical density.* The space is neither closed nor open—it is Euclidean. For this reason such universes are called *flat*. The spatial volume is then infinite, like the open universe. The time evolution is also similar to the open models, in that the universe starts from zero size and grows without bound. The difference, however, lies in the behavior of the velocity of separation between two galaxies as time increases. In the open case this velocity levels off at some nonzero value, while in the flat case the velocity becomes closer and closer to zero, but never quite reaches it, as time goes on. The time evolution of the flat models is shown in Figure 3.2, and the flat (i.e., Euclidean) geometry is included for completeness in the illustration of Figure 3.3.

In all of the models without a cosmological constant, the distance between any two arbitrarily chosen galaxies starts at zero and then grows. Friedmann did not discuss the implications of this mathematical curiosity—it means that all of the matter began from a state of infinite compression. In the 1940s Fred Hoyle would coin the term "big bang" to describe these theories to the audience of a BBC radio series. Friedmann, however, offered no rubric whatever to label this dramatic hypothesis about the origin of the universe.

How did Einstein react to all this? Einstein was already convinced that de Sitter had found one more solution than the equations of general relativity really ought to have, so Friedmann's solutions had to be wrong. Einstein looked over Friedmann's first paper and decided that the solution was incorrect, because it violated the conservation of energy. To prevent any confusion from arising, Einstein immediately published a short comment in the *Zeitschrift für Physik* (the journal where Friedmann had published), announcing Friedmann's mistake [5].

But Einstein was wrong, as anyone with a basic knowledge of general relativity would have been able to tell. It would probably have taken days for anyone to check Friedmann's complicated calculations, but Einstein's objection was specific enough that a relatively quick calculation would have revealed that he was off base. Eight months later, after a visit by a friend of Friedmann's, Yuri A. Krutkov, Einstein had no recourse but to admit his error, submitting a retraction to the *Zeitschrift für Physik* [6]. The text of the retraction read in its entirety:*

> I have in an earlier note criticized the cited work. My objection rested however—as Mr. Krutkov in person and a letter from Mr. Friedmann convinced me—on a calculational error. I am convinced that Mr.

* George Gamow's posthumously published autobiography, *My World Line* (The Viking Press, New York, 1970) recounts a frequently retold story that Friedmann found a crucial error in Einstein's 1917 paper. Friedmann showed, the story goes, that Einstein's alleged proof that the universe must be static was based on a mathematical error involving a division by zero. Although this story is delightful, as far as I can tell it has no basis in fact. In particular, Einstein's 1917 paper takes the static nature of the universe as an observational fact, not the result of a mathematical derivation. Perhaps Gamow's recollections were based on Friedmann's discovery of the error in the reasoning behind Einstein's criticism of the Friedmann solutions.

> Friedmann's results are both correct and clarifying. They show that in addition to the static solutions to the field equations there are time varying solutions with a spatially symmetric structure.

In spite of this comment, Einstein still believed that the Friedmann solutions were of no relevance, as he remained confident that the universe is static. He had chosen his words deliberately in admitting the existence of time-varying "solutions," while avoiding the word "universe." Einstein's original handwritten draft of this comment included the crossed-out words "a physical significance can hardly be ascribed," [7] which no doubt reflected Einstein's feelings at the time. It is interesting, however, that something prevented Einstein from including these words in the publication; perhaps he realized that they had no purely rational justification.

For several years more Einstein maintained his belief in the static universe. This belief was abandoned, however, when Edwin Hubble of the Mt. Wilson Observatory announced in 1929 that the universe is demonstrably expanding.

Hubble's initial result on the expansion of the universe was based on the observation of twenty-four galaxies. For each galaxy it was necessary to measure the velocity and the distance, so that the dependence of velocity on distance could be determined.

The velocities were determined by careful measurements of the spectra of the light from the galaxies. When light from a star or galaxy is broken into its components by a prism or diffraction grating, it is found that the pattern includes sharp lines produced by particular elements, such as hydrogen, helium, calcium, and iron. If the source is moving, the frequency of these lines is altered by an effect known as the *Doppler shift*, predicted in 1842 by the Prague mathematician, Johann Christian Doppler. The Doppler shift in starlight had been measured as early as 1868 by Sir William Huggins, and the Doppler shifts for a few bright galaxies had been measured between 1910 and 1920 by Vesto Melvin Slipher of the Lowell Observatory.

The Doppler shift occurs for both light and sound, and is an effect that is familiar to most of us. If you stand by the highway and listen to a car going by with its horn blowing, you will hear the pitch shift to a lower frequency as the car passes.

To understand the Doppler effect, we first need to visualize a wave. Both light and sound are waves, which means that they can be described as a periodic variation that moves through space. An illustration of a sound or light wave is shown in Figure 3.4. If the wave is sound, the dark and light regions represent high and low pressures in the air. If the wave is light, the

Figure 3.4 **A wave moving through space.** In the case of sound waves, the dark and light regions represent high and low values of the air pressure. For light waves, the dark and light regions represent strong and weak electric and magnetic fields. In either case, the entire pattern moves to the right with a fixed speed.

---

dark and light regions represent stronger and weaker electric and magnetic fields. In either case, the dark regions will be called the crests of the waves. The distance between crests is called the *wavelength*. The number of crests per second that pass a given point is called the *frequency*, and the time between crests is called the *period*. When the source is moving toward the observer, the frequency of arrival of the crests is increased, as explained in Figure 3.5. This effect is called a *blueshift*, because blue lies at the high-frequency end of the visible spectrum. If the source is moving away from the observer, then the frequency is lowered and the effect is called a *redshift*.

While the velocities of distant galaxies can be measured accurately by their redshifts, the distances are much more problematic. A variety of methods are used, each appropriate to a different range of distances. The various methods have overlapping regimes of validity, so each method is used to calibrate the next, in a sequence of steps called the *cosmic distance ladder* [8]. The uncertainties tend to compound with distance, which is part of the reason why they are large and difficult to estimate.

Using this distance ladder, Hubble measured the distances of twenty-four extragalactic nebulae. The plot of this data, exactly as it appeared in Hubble's paper [9], is shown as Figure 3.6. Hubble concluded from this data that the recession velocity of a distant galaxy (as measured by the redshift) is proportional to the distance. Now known as Hubble's law, this statement forms the cornerstone of modern cosmology. After centuries of erroneous theorizing, it was now clear that the universe is not static—it is expanding!

In his original paper, Hubble estimated the expansion rate at about 150 kilometers per second per million light-years of distance. He was not explicit about the estimated uncertainty in this number, but the discussion in the

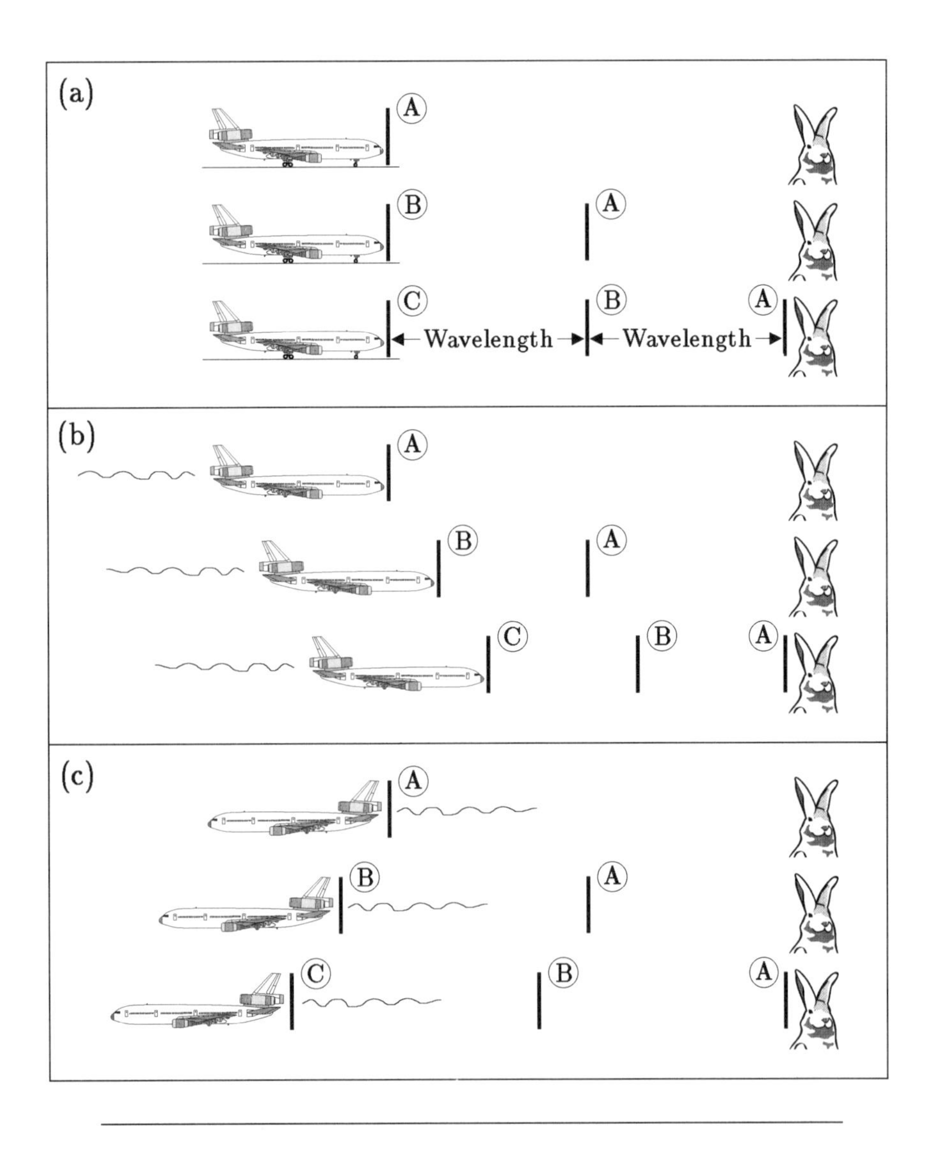

paper suggests an uncertainty in the range of 10–15%. Two years later Hubble and Milton L. Humason published a more extensive study, extending to about eighteen times the distance of Hubble's original data. They found an expansion rate of 170 kilometers per second per million light-years, with an uncertainty explicitly stated as 10%. The expansion rate (recession velocity per distance) is now known to astronomers as the Hubble constant. It should be emphasized, however, that the word "constant" is misleading. It is a constant in the sense that it is independent of distance, and in the sense that it changes very little over the lifetime of an astronomer. It no doubt var-

Figure 3.5 **(facing page) The Doppler effect for sound waves.** The top sequence of three pictures, (a), shows sound waves emitted by a stationary jet airplane. The time interval between the pictures is exactly equal to the separation between the wave crests, which is called the period of the wave. The top picture shows the emission of the first wave crest, labeled *A*, and the second and third pictures show the emission of the next two wave crests, *B* and *C*. Between pictures each wave travels a distance equal to the wave speed times the wave period, which is called the wavelength. In the third picture the rabbit's ears receive wave crest *A*, and one period later they will receive wave crest *B*. The crests arrive with the same time separation with which they were emitted, so the frequency is unchanged. The second set of three pictures, (b), which are also separated by time intervals of one period, show the jet approaching the rabbit at one quarter the speed of sound. Between successive pictures each wave crest travels one wavelength, as before, but this time the jet also moves, by a quarter of a wavelength. Wave crests *B* and *C* are therefore emitted closer to the rabbit. The wave crests bunch together, and arrive at the rabbit with a higher frequency. For the case of light waves, this effect would be called a blueshift. The third set of three pictures, (c), show that if the jet is moving away from the rabbit, the separation between the wave crests is stretched. The received frequency is lower, which for light waves is called a redshift.

---

ies, however, over the lifetime of the universe, as the force of gravity acts to persistently slow the Hubble expansion.

Many have argued that the straight line which Hubble drew through the points of Figure 3.6 was not well-justified, considering the scatter of the points. The magnitude of this scatter is not difficult to understand from a modern perspective, since Hubble's law is really only an approximation. In addition to the cosmic expansion, galaxies are subject to the tug of the local gravitational fields created by neighboring concentrations of galaxies. Since Hubble's original study included only nearby galaxies, these locally induced velocities were large enough to compete with the cosmic expansion. Modern studies, extending to galaxies over a hundred times more distant, have shown that Hubble's law is valid, on average, beyond any reasonable doubt. An example of a more modern data set is shown as Figure 3.7.

There is also no doubt, however, that Hubble very badly mismeasured the distance scale of the universe, and therefore obtained a wildly inaccurate value for the Hubble constant. The errors were not corrected quickly. It was not until key papers were published by Walter Baade in 1952 and Allan Sandage in 1958 that the accepted value of the Hubble constant was dropped to within its currently accepted range, which is 15–30 kilometers per second per million light-years—smaller than Hubble's original value by a factor of 5 to 10!

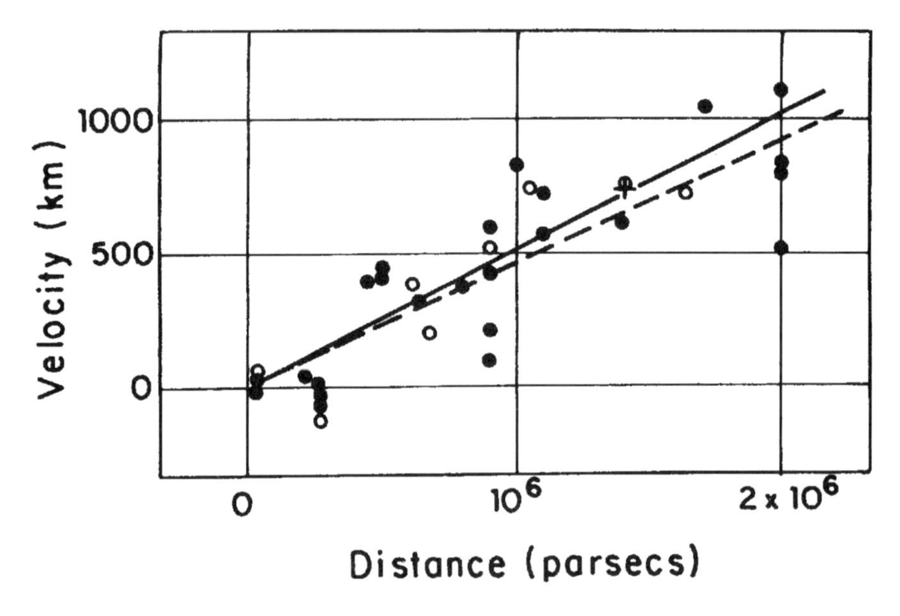

Figure 3.6 **Hubble's original data.** The horizontal axis shows the estimated distance to the galaxies, in units of parsecs, where 1 parsec = 3.26 light-years. The vertical axis shows the recession velocity, in kilometers per second. (Each black dot represents a galaxy, and the solid line shows the best fit to these points. Each open circle represents a group of these galaxies, selected by their proximity in direction and distance; the broken line is the best fit to these points. The cross shows a statistical analysis of 22 galaxies for which individual distance measurements were not available.)

Hubble's error had important consequences, since the Hubble constant sets the entire time scale of evolution for the big bang theory. Using Hubble's original value for the expansion rate, the estimated age of the universe was in the range of two billion years or less. Even in the 1930s there was geological evidence that the earth is much older than this, so the big bang theory was in serious observational trouble from the start.

The connection between the Hubble constant and the age of the universe can be seen in Figure 3.2. An enlarged version is shown as Figure 3.8, which also includes some additional annotation. To consider the most straightforward case first, suppose that there was no gravity at all. In that case all galaxies would move at constant speed, so the expansion of the universe would be described by the broken line labeled "zero-gravity approximation." To find the age of such a universe, choose one galaxy to consider. Since the galaxy is receding, it must have been closer in the past, and if we think back far enough, it would have been on top of us. Since the zero-gravity approximation means that the galaxy has been moving at constant speed, we can find

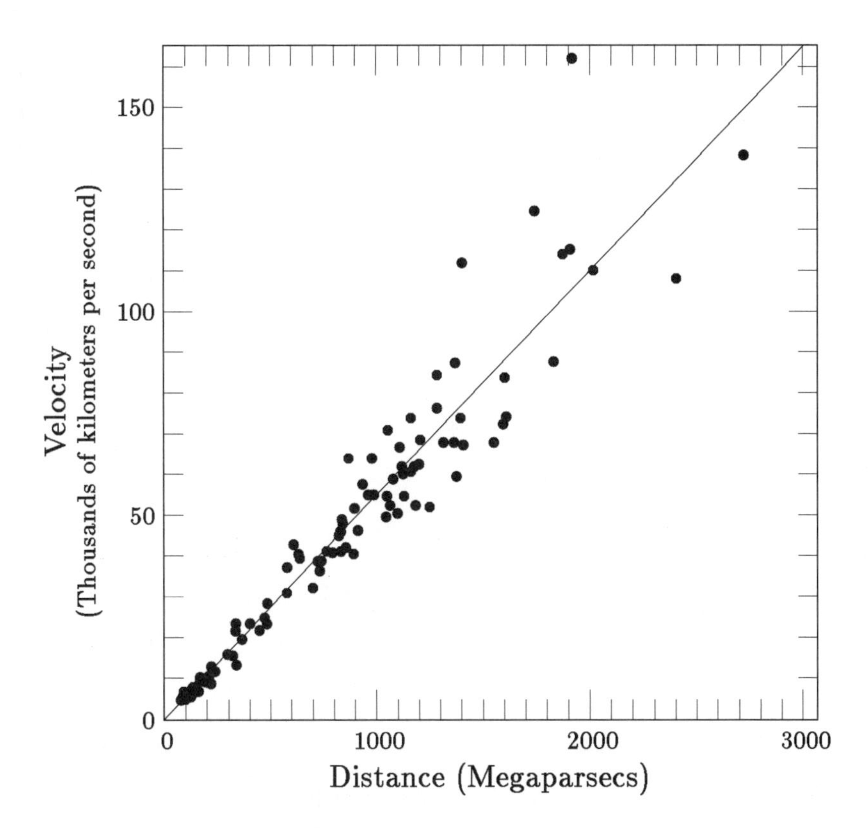

Figure 3.7 **Modern example of the Hubble diagram.** This graph is adapted from the data of Jerome Kristian, Allan Sandage, and James A. Westphal [10]. The authors measured the distances only on a relative scale, but for clarity I have introduced an approximate scale in megaparsecs (1 megaparsec = 3.26 million light-years). The largest distances are hard to measure so the scatter is still large, but the left half of the graph shows a very convincing straight line.

when it was on top of us by dividing its distance by its velocity. But the velocity is Hubble's constant times the distance, so the distance cancels out of the calculation. We conclude that *every* galaxy was on top of us at a time $1/H$ in the past, where H is the Hubble constant. $1/H$ is called the *Hubble time*, and in the zero-gravity approximation it is the age of the universe. For Hubble's original value, the Hubble time was only about 2 billion years,* so this was the predicted age of the universe in the zero-gravity approximation.

* To express $1/H$ in seconds, it is necessary to convert H from kilometers per second per million light years to kilometers per second per kilometer. To carry out the conversion, use 1 million light-years = 9.46×10^{18} kilometers. To relate seconds to years, use 1 year = 3.16×10^{7} seconds. See Appendix D for more information.

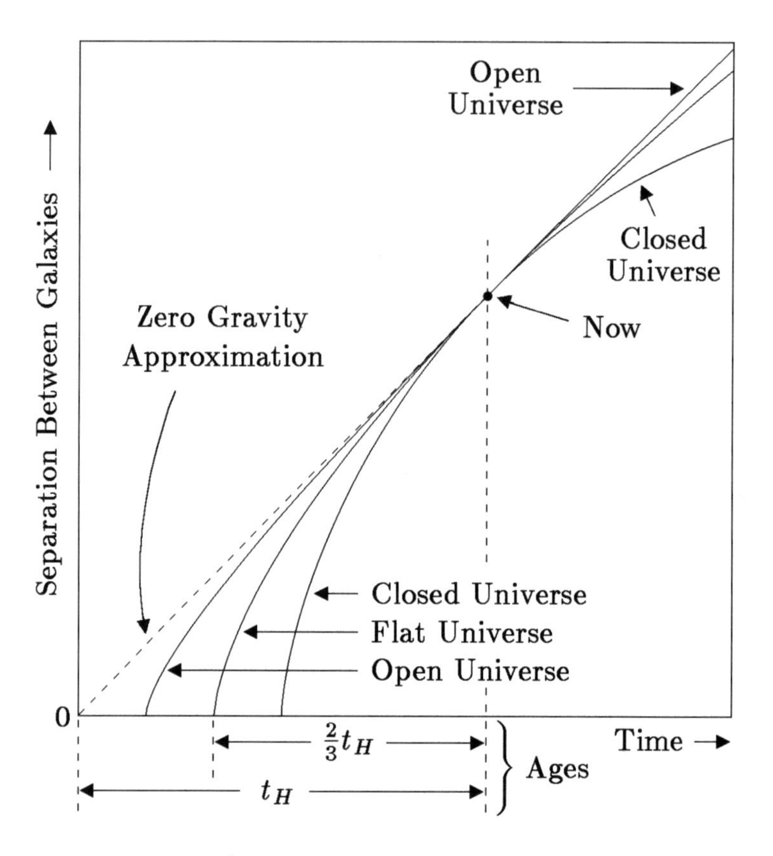

Figure 3.8 **Friedmann models and the age of the universe.** The diagram shows the evolution of Friedmann models without a cosmological constant. The curves are drawn to coincide at the present time, indicated by the vertical broken line. The various models begin at different times in the past. The ages are expressed in terms of the Hubble time t_H, which is defined to be $1/H$, where H is the present value of the Hubble constant. If gravity were absent, the separation between galaxies would grow at a constant rate, and the age of the universe would be exactly t_H. The effect of gravity is to slow the expansion, implying a faster expansion in the past and hence a younger universe.

If gravity is added to the calculation, the problem only gets worse. The effect is to slow the expansion as time goes on. It follows that the galaxies were moving faster in the past, so the universe is even younger! For open universes, the age can be anywhere from two-thirds of the Hubble time up to the Hubble time, depending on the mass density. For a flat universe the age is exactly two-thirds of the Hubble time, and for a closed universe it can be any number less than two-thirds of the Hubble time.

The age discrepancy was a serious problem for the fledgling big bang hypothesis, and it was no doubt part of the reason why some scientists worked hard to find alternatives, such as the steady state theory. The whole field was clouded, and research on the big bang proceeded very slowly. Despite this conflict, Hubble's estimate of the Hubble constant was accepted for several decades without question. According to J. D. North [11],

> There was a lack of scepticism shown towards the findings in extra-Galactic astronomy in the 1930's which can perhaps be explained in part by the fact that there was only one telescope in the world really capable of being used to this end. Whatever the explanation, criticism of the empirical value of the Hubble factor was not often forthcoming, and it appears at times that astronomers were more concerned with bringing the rest of astrophysics into line with the parameter in Hubble's Law than the other way about.

For currently accepted values, on the other hand, the Hubble time lies between 10 and 20 billion years, so the large discrepancy between the Hubble time and other estimates of the age of the universe has melted away. (The issue has not completely disappeared, however, as some astronomers today argue that the Hubble time is probably only about 12 billion years. The inflationary model predicts a flat universe, so the age would be two-thirds of the Hubble time, or 8 billion years. This number conflicts with estimates of stellar ages, which suggest a universe at least 11 or 12 billion years old. The estimates of both the Hubble constant and the stellar ages remain somewhat uncertain, however, so it is not clear if the current "age crisis" will be any more meaningful than the age crisis of the 1930s and 40s.)

Since Friedmann's papers were hardly noticed and Friedmann was no longer alive, the work on time-varying cosmologies had to be reinvented. Enter Georges Lemaître. While Friedmann's background as a mathematician and meteorologist seemed unlikely for the role, Lemaître's was perhaps even more so. He was a civil engineer and a priest. Born in 1894 in Belgium, Lemaître served as an artillery officer in the Belgian army during World War I. After the war he studied briefly at the University of Louvain, then entered a seminary, and was ordained a Roman Catholic priest in 1923. The next year Lemaître studied at the solar physics laboratory at the University of Cambridge, where he worked with Arthur Eddington. Traveling to the United States, Lemaître spent two years studying at the Massachusetts Institute of Technology. Here he learned the latest findings of Edwin Hubble and also Harlow Shapley, the Harvard College Observatory astronomer who played a key role in the establishment of the extragalactic distance scale.

Lemaître returned to Europe in 1927, when he became a professor of astrophysics at the University of Louvain. In the same year he wrote his famous paper on cosmological solutions, published in the *Annales de la Société Scientifique de Bruxelles*. The paper discussed only closed universes. Friedmann's 1922 paper was included in the list of references, but there is curiously no evidence that Lemaître was familiar with its contents. Lemaître wrote down the general evolution equations for a closed universe containing arbitrary amounts of matter, electromagnetic radiation, and a cosmological constant. The paper discussed the redshift relation, which Friedmann's papers had not done, and Lemaître "adopted" (without any reference) a value for the Hubble constant of 190 kilometers per second per million light-years. It is not known, but it seems likely that Lemaître had received some inside information from Hubble or his colleagues at the Mt. Wilson Observatory.

The paper went on to propose a particular model of the universe, a model that is now of mainly historical interest. Here Lemaître apparently became the first theorist to fall victim to Hubble's mismeasurement of the Hubble constant, two years before Hubble even published it! Lemaître considered a variety of possible solutions to his equations, but noted that the models typically had beginnings about a billion years in the past—"i.e., quite recently for stellar evolution," he pointed out. Apparently Lemaître had strong confidence in the value of the Hubble constant that he "adopted," because he allowed it to control his conclusions. By judicious use of the cosmological constant, Lemaître was able to construct a model [12] with an age much larger than the Hubble time, $1/H$.

After the announcement of Hubble's results in 1929, Eddington became very enthusiastic about Lemaître's work. In 1931 he arranged to have Lemaître's article translated into English and republished in the widely read *Monthly Notices of the Royal Astronomical Society* [13]. Lemaître's models rapidly became the standard tool for discussing redshift data and the Hubble expansion.

Today, however, Lemaître's models have been essentially forgotten, since the motivation for these complicated constructions disappeared with the radical decrease in the accepted value of the Hubble constant during the 1950s. Lemaître's models can be viewed as an example of what goes wrong when theorists blindly accept unreliable observations and force their models to fit them. Perhaps cosmology would have advanced more quickly if Lemaître had taken the attitude of his mentor Arthur Eddington [14]:

> Observation and theory get on best when they are mixed together, both helping one another in the pursuit of truth. It is a good rule not to put overmuch confidence in a theory until it has been confirmed by

> observation. I hope I shall not shock the experimental physicists too much if I add that it is also a good rule not to put overmuch confidence in the observational results that are put forward *until they have been confirmed by theory.*

As a leading cosmologist who was also a priest, Lemaître was often asked how he reconciled his faith in the Bible with the discoveries of modern science. "There is no conflict," he would invariably reply [15]. "Once you realize that the Bible does not purport to be a textbook of science, the old controversy between religion and science vanishes. . . . There is no reason to abandon the Bible because we now believe that it took perhaps ten thousand million years to create what we think is the universe. Genesis is simply trying to teach us that one day in seven should be devoted to rest, worship and reverence—all necessary to salvation."

Lemaître became probably the first person to take seriously the physics of the big bang model. He attempted to investigate what the universe might have been like at the beginning when all the matter was compressed to enormous density, and coined the phrase "primeval atom" to describe it. At the time, however, very little was known about how matter would behave at high densities, so the serious study of the early universe would have to wait for the next generation of big bang theorists.

CHAPTER 4

ECHOES OF A SCORCHING PAST

While the basic features of big bang cosmology fell into place during the 1920s with the work of Friedmann, Lemaître, and Hubble, progress during the subsequent decades was very slow. At this point the only observational evidence for the big bang was the relationship between velocity and distance that was first observed by Hubble. The mismeasurement of the expansion rate, however, made it very difficult to reconcile the big bang picture with age estimates based on geological or stellar evolution. The big bang theory had been invented, but its health was marginal. Cosmology in general was still not taken very seriously by most scientists.

In 1948 a reasonable alternative to the big bang was proposed by Hermann Bondi, Thomas Gold, and Fred Hoyle. Known as the *steady state theory,* this hypothesis pushed the cosmological principle one step further. The cosmological principle holds that the universe looks roughly the same, no matter where the observer is located and no matter what direction he or she is looking. The steady state theory introduced the *perfect cosmological principle,* which adds the assumption that it does not matter when the observations are made—the universe is eternal and looks about the same at all times. Since the galaxies were known to be moving apart, the theory requires that matter be continually created, in order to produce new galaxies to fill in the expanding gaps between the old galaxies. While such matter production is at odds with conventional physics, the steady state theorists were confident that they were on safe ground observationally. The required rate of matter production was only about two hydrogen atoms per cubic meter per billion years, so there was no way that any feasible experiment could rule out the possibility of such a rare event. Furthermore, the steady state theorists argued, if the advocates of the big bang could assume that all the matter in the universe was created in a single flash, why should they balk at the possibility of continuous matter creation?

With two rival cosmologies in the theoretical marketplace, it became all the more important to find observational tools to address these questions. Curiously, the most compelling piece of evidence was discovered more or less by accident, by two young radio astronomers who had no intention of investigating the origin of the universe. In the spring of 1964, Arno A. Penzias and Robert W. Wilson discovered the first signs of a hissing radio noise that is now believed to fill the universe—a radio noise that we now view as the faint echo of the violent explosion of the big bang.

Penzias and Wilson were working at a satellite communications center set up by Bell Telephone Laboratories at Crawford Hill, a few miles from the major Bell Labs installation at Holmdel, New Jersey. They were using a low-noise 20-foot horn-reflector radio antenna that had been built for communications research with the *Echo I* satellite, which had been launched in August 1960. Only three years after the first *Sputnik* satellite, this early venture into satellite communications consisted of a large Mylar balloon, about 100 feet in diameter, with a thin coating of aluminum. The satellite was inflated in orbit and used as a reflector for radio signals beamed between Bell Laboratories, the Jet Propulsion Laboratory in California, and several other locations. (Modern communications satellites, by contrast, actively receive the signals from earth and rebroadcast them with amplification.) The signals reflected by the *Echo* satellite were very weak, so the Crawford Hill receiver was constructed to be extremely sensitive. By 1963 the research interest in the *Echo* satellite had ended. *Telstar*, the first actively rebroadcasting communications satellite, had been launched in 1962 and had become the focus of satellite communications research at Bell Labs. The horn antenna had been modified to serve as an auxiliary monitor for *Telstar*, but its use in this capacity was not very important. In any case, the Bell Labs program in satellite communication was winding down, since the United States Communications Satellite Act of 1962 had given exclusive rights to United States interests in international satellite communications to the newly created COMSAT corporation. Since Bell Labs had a strong tradition of devoting part of its resources to basic research, Penzias and Wilson were authorized to refurbish the radio antenna for use in radio astronomy.

Penzias and Wilson were both just out of graduate school, having received their Ph.D.'s in 1961 and 1962, respectively. As one might expect, the radio astronomy program that they began at Bell Labs was designed as an extension of the research that each had done in earning his degree. The first project would use the precision of the 20-foot horn to measure to 2% accuracy the intensity of *Cassiopeia A*, a supernova remnant in our galaxy,

Arno Penzias and Robert Wilson in front of the 20-foot horn-reflector radio antenna that they used in 1964 to discover the cosmic background radiation.

which due to its brightness is frequently used as a calibration source by radio astronomers. Penzias and Wilson then expected to use their highly sensitive antenna to explore the radio emissions of our galaxy at a wavelength of 21 centimeters, corresponding to an emission line of atomic hydrogen. The intended program, however, never went beyond the *Cassiopeia A* project, as the pair of astronomers were soon caught up in what is now viewed as one of the most important astronomical discoveries of the century.

To clarify some frequently used terminology, I mention that both radio and light are examples of an *electromagnetic wave,* a pattern of electric and magnetic fields that moves through space. Different types of electromagnetic radiation are distinguished by their wavelength, the distance between one crest of the wave and the next. Electromagnetic waves with a wavelength of 21 centimeters fall within a category called microwaves, a term applied to electromagnetic waves with wavelengths ranging from one millimeter up to 30 centimeters or sometimes one meter. Shorter waves are called infrared radiation, and longer waves are called radio waves. Visible light has a wavelength shorter than infrared.*

To detect feeble radio signals emitted from natural sources in space, Penzias and Wilson needed to understand all other possible sources of electrical signals. One important source of background noise is the random thermal motion of electrons within the circuitry that detects and amplifies the signals. To correct for this effect, Penzias and Wilson needed to know what level of noise their amplifier would emit if no signal whatsoever were fed in from the antenna. Why not simply disconnect the antenna? It is not so simple, because a dangling antenna lead would pick up an intolerably large amount of random radio noise. The antenna lead could be shielded, but the random motion of the electrons in the shield would again produce far too much radio noise.

A logically simple solution would be to cool the shield to the coldest possible temperature, absolute zero†, at which all random motion ceases and no radiation is emitted. Absolute zero is impossible to reach, but Penzias and Wilson came close by using a device called a *cold load,* consisting of a dewar cooled by liquid helium at its liquification temperature, 4.2°C above absolute zero. Even at this temperature the microwave emissions cannot be ignored, but Penzias and Wilson were able to calculate the intensity and account for it in their data analysis. Manually switching the receiver between the cold load

* Even more information for the scientific terminology buffs: The microwave region encompasses at least part of the UHF band (Ultra High Frequency), which extends from wavelengths of 10 centimeters to 1 meter. The UHF band includes the broadcast frequencies for television channels 14 and above. Shorter wavelengths in the microwave region are used for radar, satellite communication, and microwave communication links, while microwave ovens generally operate at a wavelength of 12.2 centimeters. For comparison, standard television and FM radio are broadcast in the VHF band (Very High Frequency), with wavelengths from 1 to 10 meters, while AM radio is broadcast at wavelengths ranging from 180–560 meters. The wavelength of visible light ranges from 0.4 to 0.8 microns (1 micron = 10^{-6} meter).

† Absolute zero is 273 degrees below zero centigrade, or 460 degrees below zero Fahrenheit.

and the antenna at intervals of about 30 seconds, they used the difference between the two signals to determine the true signal from the antenna.*

To confirm their estimates of the background noise in the system, Penzias and Wilson ran a series of observations at the relatively short wavelength of 7.35 centimeters, where radio signals from galactic sources were believed to be negligible. They expected that when the known sources of electrical background were subtracted from the measured antenna output, the result would be equal to zero, at least to within some small experimental uncertainties.

When Penzias and Wilson began measuring the antenna output in the spring of 1964, however, they discovered that the background subtraction did not cancel the measured power. Instead, the system seemed to be picking up some unforeseen source of microwave signal. The strength of this signal did not change when they pointed the antenna in different directions, it did not change with the time of day, and as the seasons progressed it became clear that it did not vary with season. The unremitting constancy of the signal suggested very strongly that it was noise within the system, but the cold-load comparison is supposed to eliminate that possibility. Could there be some other source of noise in the antenna that had been overlooked? Was it possible, for example, that the signals were related to the pigeons who enjoyed sauntering through the heated part of the horn? After ejecting the pigeons, the scientists put some effort into cleaning off the "white dielectric material," as Penzias described it, with which the pigeons had coated the throat of the antenna. All this, however, had no significant effect on the mysterious hiss. Penzias and Wilson remained thoroughly puzzled, but they were determined to track down the source of the enigmatic signal.

Meanwhile, less than 30 miles away in Princeton, New Jersey, there was a small group of physicists who had a pretty clear idea about what such a signal might mean. At the center of this group was Robert H.

* The background noise problem was really a bit more complicated, because there were other sources of power that could come through the antenna. The largest was the emission of radio waves from the thermal agitation of molecules in the Earth's atmosphere. The atmosphere could not be eliminated, but the amount of atmospheric contamination could be varied by pointing the antenna in different directions. The line of sight intercepts a longer stretch of atmosphere when the antenna is pointed at the horizon than when it is pointed straight up, so Penzias and Wilson estimated the atmospheric contribution by comparing the signals in the two cases. Another problem was the generation of electrical noise within the antenna itself. Whenever an electric current passes through a material, electrical resistance converts some of the power to heat. An inverse effect, also proportional to the electrical resistance, converts some of the energy of thermal molecular agitation into a fluctuating electrical signal. Penzias and Wilson therefore strove to minimize the resistance of the antenna, carefully sealing all the seams with aluminum tape to ensure a clean electrical connection.

Dicke, the same physicist who fifteen years later would teach me about the flatness problem in a lecture at Cornell. Dicke is an extremely versatile physicist who has pursued a large number of interests. He is known mainly as an experimentalist, but he is also responsible for some important theoretical developments. During World War II he worked on radar, developing a sensitive microwave detector known as the Dicke radiometer. Dicke later became interested in gravitational physics, and in 1961 he collaborated with Carl Brans to develop a new theory of gravity, now called the Brans-Dicke theory. Although Einstein's general relativity continues to be the accepted theory of gravity, most books on general relativity discuss the Brans-Dicke theory as a viable alternative which should not be ignored. Dicke is also well-known for a 1964 experiment that confirmed the fact that any two bodies in a gravitational field will experience the same acceleration, independent of their mass and composition. According to legend, Galileo demonstrated this fact by dropping masses from the leaning tower of Pisa, but Galileo's result did not come close to the one part in 10^{11} accuracy achieved by Dicke.

In the early 1960s, Dicke spearheaded a very active research group in gravitational physics at Princeton University. The group would get together once a week for an evening seminar or discussion, after which they would usually retire to a nearby restaurant for beer and pizza. These meetings were sometimes held on Friday nights, despite the objections of outraged spouses. One of the participants in that group, P. James E. Peebles, tells the story [1] of a very hot night in the summer of 1964 when the group met in the "ridiculously hot" attic of the Palmer Laboratory with a smaller-than-average number of people.

Dicke described to the group some thoughts he had on the history of the universe. Dicke had decided sometime earlier that he could not accept the idea that all the matter in the universe was created in one flash of a big bang. He preferred an alternative idea that was frequently discussed at the time: Perhaps the universe is oscillating, with successive phases of expansion, contraction, and then expansion again. Dicke was aware, however, that astronomical evidence showed that the earliest stars had condensed from hydrogen and helium only. Elements heavier than helium are synthesized in stars, but they were not present, at least not in significant amounts, when the earliest stars were formed. What, then, has happened to the heavy elements that were synthesized in the previous cycle of cosmic existence? Dicke could invent only one possible answer: At the time of the bounce, when the contraction ended and the expansion began, the universe must have been so hot that the nuclei of heavy elements were shattered into their constituent protons and neutrons. In this way the heavy-element debris from the previous cycle can be cleaned up, so each cycle of the universe can start off fresh. If that were true, then there must be a radiation background—an afterglow of this intense heat—that would continue to permeate the universe.

Jim Peebles

Although Dicke's idea of an oscillating universe is no longer favored, the prediction of a radiation background remains relevant. An oscillating universe can be viewed as a succession of big bangs, each of which is nearly indistinguishable from a single-episode big bang. If each bang of the oscillating theory would produce a radiation background, then the unique bang of the big bang theory would have the same effect.*

* Important ideas in science are sometimes invented and reinvented, and the cosmic background radiation is a striking example. Dicke would soon learn that he was not the first to understand that a big bang origin would leave the universe filled with radiation. The same ideas had been examined in the 1940s and 1950s by George Gamow, Ralph Alpher, and Robert Herman. As Dicke's group was beginning its work on the background radiation, closely related calculations were being completed at Cambridge University by Fred Hoyle and Roger Taylor. Like many physicists at the time, however, Dicke was unaware of the previous work. I will discuss the earlier work in the following chapter.

Dicke thought it would be fun to look for the background radiation, so he persuaded two young researchers, Peter G. Roll and David T. Wilkinson, to set up the experiment. Then he turned to Peebles and said, "Why don't you go and think about the theoretical consequences?" Soon Roll and Wilkinson were busily setting up an antenna on the roof of the geology building. The experiment used a Dicke radiometer tuned to microwaves with a wavelength of 3.2 centimeters. The system included a liquid helium cold load similar to the Crawford Hill experiment, but the horn itself was much smaller—only a foot across.

Following Dicke's suggestion, Peebles enthusiastically worked out the consequences of a hot early phase in the history of the universe. By early 1965 Peebles had written an article on the cosmic radiation. He found, as Dicke had expected, that the universe today should be uniformly bathed with a background of electromagnetic radiation which is a remnant from the big bang. He was also able to predict the *spectrum* of the radiation—the way in which the energy density varies with the frequency or wavelength. It would be what physicists call a thermal, or *blackbody* spectrum, at a temperature of 10 degrees centigrade above absolute zero.*

The concept of blackbody radiation is well-known to physicists, since it arises in such a wide variety of situations. Experimentally, it is found that any object at any temperature other than absolute zero emits a glow of electromagnetic radiation. We are all familiar with the glow of the sun, or the coals in a hot barbecue grill. Objects at room temperature do not appear to glow, but that is only because the radiation is predominantly in the infrared part of the spectrum, to which our eyes are not sensitive. If the light from a hot glowing object is broken into its spectrum, by a prism or a diffraction grating, one finds that the light contains characteristic peaks in intensity at certain wavelengths, which can be used to identify the glowing material. These peaks, known as spectral lines, are used by chemists to analyze the composition of substances on earth, and by astronomers to probe the composition of the sun and stars. Shifts in the wavelengths of these lines, usually toward the red, are used to measure the velocities of distant galaxies.

Suppose, however, that we construct a closed box from some material, and heat it to a uniform temperature. On the outside the box will emit radiation with the usual spectrum characteristic of the material. On the inside, however, the radiation will be emitted and absorbed, and in a very short time a steady-state situation, or equilibrium, will be established. If we mea-

* Physicists and chemists often measure temperature in centigrade units above absolute zero, a system known as the Kelvin scale. Thus, 10 degrees centigrade above absolute zero can be called 10 degrees Kelvin, with the abbreviation 10°K.

sure the spectrum of this equilibrium radiation, we will find that the spectral lines have disappeared. The walls of the box will continue to emit light most strongly at the wavelengths of the spectral lines, but the material will also *absorb* light most strongly at precisely these same wavelengths. The equilibrium spectrum will have no lines, but instead will have a universal form which is completely independent of the material that is used to construct the box. This is the spectrum known as thermal, or blackbody, radiation—the spectrum predicted by Peebles' calculations for the cosmic radiation. If the temperature is 10°K, then the spectrum will look exactly like Figure 4.1. The underlying physics that determines the shape of the spectrum is described qualitatively in Appendix C.

Why, you might ask, is this type of radiation called blackbody? Why is the phrase "black body" not reserved for an object that emits no radiation at all? The reason is that there is no such thing as a material that emits no

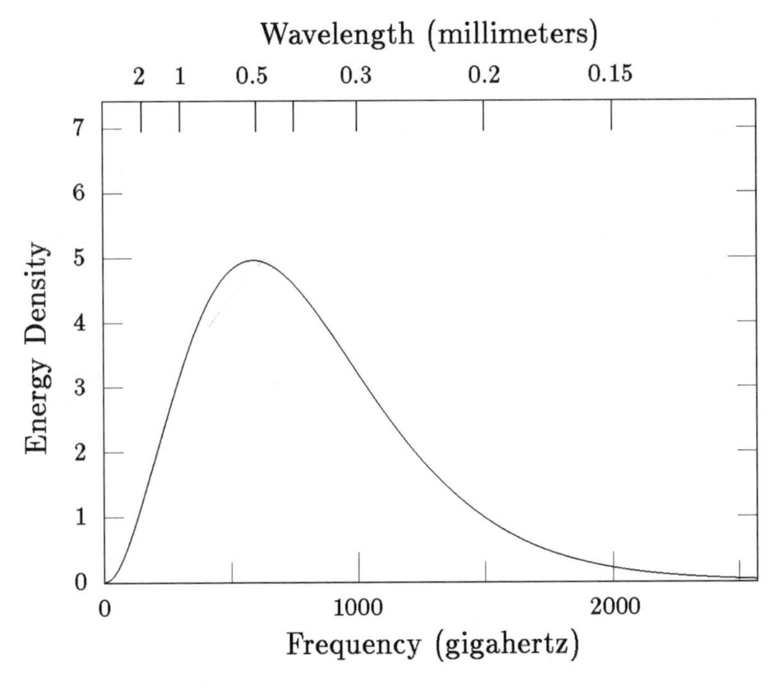

Figure 4.1 **The spectrum of blackbody radiation at 10 degrees centigrade above absolute zero, or 10°K.** The graph shows how the energy density depends on the frequency, for the cosmic background radiation predicted by Peebles in 1965. The frequency is shown on the lower horizontal axis, in gigahertz, or billions of cycles per second. The corresponding wavelength is shown on the upper horizontal axis. (The energy density is measured in units of 10,000 electron volts per cubic meter for each one gigahertz interval of frequency.)

radiation at all. If such a material did exist, Penzias and Wilson would have used it for their cold load. For any material at a temperature other than absolute zero, there is nothing that can prevent the random motion of the molecules on the surface of the material from converting energy into electromagnetic radiation. In general, when an electromagnetic wave, such as light, strikes the surface of a material, three things can happen: some of the light can be transmitted, as usually happens with glass; some can be reflected, as from a mirror; and some can be absorbed, as when light hits an asphalt pavement. A physicist calls an object *black* if it absorbs all electromagnetic radiation that hits it. Such an object does not transmit or reflect any radiation, but it will nonetheless emit radiation thermally, due to the random motion of its molecules. If the object is truly black by this definition, then the radiation that is emitted will have precisely the blackbody spectrum discussed above.

Now we return to the activities of Jim Peebles, who submitted his paper on the theory of the cosmic background radiation to the *Physical Review* in March 1965. The paper was rejected by the journal, as was a revised version that Peebles wrote in response to the referee report on the first submission. The rejections apparently hinged on the issue of credit given to prior work, particularly the work of Gamow and his collaborators [2]. Letters went back and forth, but agreement was never reached and the paper was never published.

Peebles, however, was very excited about the project. Even before the first submission, Peebles accepted an invitation from Johns Hopkins University, in Baltimore, Maryland, to give a colloquium about the work. He presented the colloquium on February 19, 1965 (he still has the notes!), and the subsequent events represent a spectacular success story of the informal communication network that exists in the scientific community. In the audience at Peebles' talk at Johns Hopkins was a radio astronomer from the Carnegie Institution of Washington, D.C., Kenneth Turner. Turner had been an old friend of Peebles' from graduate student days at Princeton, and in fact Peebles and his wife and two young daughters were all staying with the Turners during their visit. Fascinated by Peebles' prediction of a background radiation filling the universe, Turner mentioned the colloquium to a fellow radio astronomer, Bernard Burke, also at the Carnegie Institution (although now at MIT). Arno Penzias happened to be a friend of Burke's, and at the end of a phone conversation about other matters, Burke asked Penzias how the measurements with the Crawford Hill horn were coming. Penzias told Burke about the unexplained signals, and Burke suggested that Penzias might learn something very interesting from the group at Princeton.

Dicke received a phone call from Penzias, and the Princeton crew was soon on the road to Crawford Hill. When Dicke and his collaborators saw

David Wilkinson with his 3 cm radiometer on the roof of Guyot Hall, Princeton University, in the summer of 1965.

the results that Penzias and Wilson were obtaining, they were quickly convinced that the Bell Labs team had made the crucial discovery—the echo of the big bang had been found.

The two groups decided to submit separate papers to the *Astrophysical Journal*, to be published back-to-back. The Bell Labs paper would describe the observations, while the Princeton paper would describe the theoretical interpretation. Penzias and Wilson remained, however, a little skeptical of the far-reaching cosmological interpretation. Penzias was partial to the big bang theory, but Wilson had a fondness for the steady-state view, a predilection that began in his graduate student days while studying cosmology with Fred Hoyle at the California Institute of Technology. Wilson told me, "I was not particularly reluctant to accept a big bang interpretation because of that philosophical bias, but I rather didn't want to immediately accept the big bang explanation without some period of time to see if the Steady Staters or other theorists would come up with an alternate explanation." The Bell Labs physicists chose the title of their paper with caution and precision: "A Measurement of Excess Antenna Temperature at 4080 Mc/s [3]." There is no way that this team would be accused of hyperbole. (Note that 4080 megacycles per second, or million cycles per second, is the frequency corresponding to the 7.35 centimeter wavelength of their observations.) The paper did not mention any cosmological relevance of the result, except to say that "A possible explanation for the observed excess noise temperature is the one given by Dicke, Peebles, Roll, and Wilkinson in a companion letter in this issue." Shortly after the original submission, however, Penzias and Wilson sent the journal a short addendum that discussed the relevance of earlier observations by another group. While still making no explicit reference to cosmology, they concluded that "This clearly eliminates the possibility that the radiation we observe is due to radio sources of types known to exist."

Contrasting the guarded attitude of the Bell Labs team, the Princeton physicists had been thinking about the big bang for some time, and were prepared to write about it. Their paper came straight to the point with the title "Cosmic Black-Body Radiation [4]." The conclusions, however, were stated with some caution, since everyone recognized that observations at a single wavelength are not enough to tell if the radiation is blackbody. "While all the data are not yet in hand," the Princeton group wrote, "we propose to present here the possible conclusions to be drawn if we tentatively assume that the measurements of Penzias and Wilson do indicate blackbody radiation . . . ." Such radiation, they explain, is exactly what would be predicted by the big bang theory: "The presence of thermal radiation remaining from the fireball is to be expected if we can trace the expansion of the universe back to a time when the temperature was of the order of 10^{10}°K."

Even after submitting the article, Penzias and Wilson remained concerned about the reliability of their results. They were cautious experimenters, and they understood how difficult it is to prove that the constant hiss was a real astronomical signal—that it was not caused by some imperfection in the detection system. As the article awaited publication, they made one further check of the noise pickup from the ground, reassuring themselves that it was not the source of their signal. They had eliminated every source of noise that they could imagine, but was there any source that they had not imagined? To satisfy themselves that the signal was real, they took over a small horn detector near the main building at Holmdel, attempting to repeat the measurement on a totally new piece of apparatus. Only the cold load was carried down from Crawford Hill, as it would be too difficult to rebuild from scratch. After several months the second horn began observations at the same wavelength as before, 7.35 centimeters. Penzias and Wilson were relieved to find that it also showed a uniform background of radiation. The uncertainties were larger than with the original antenna, but the measured intensity of the background radiation was reassuringly consistent with the earlier measurement.

Bob Wilson says that when the article was submitted, he had not yet fully appreciated the impact of the discovery. He was rather surprised when Walter Sullivan of *The New York Times* found out about the work and wrote an article under the headline "Signals Imply a 'Big Bang' Universe [5]." The opening sentence of the article announced to the world that "Scientists at the Bell Telephone Laboratories have observed what a group at Princeton University believes may be remnants of an explosion that gave birth to the universe." Wilson's father was visiting from Texas on the day that the article appeared, and he went out early to buy the newspaper. "When my father returned with the newspaper showing a picture of *our* horn antenna on the front page," Wilson explained, "it was then that I began to realize that the rest of the world was taking this seriously."

Penzias and Wilson had measured their mysterious signal at only one wavelength, so they had no way of knowing whether the spectrum in any way resembled that of a blackbody. If they assumed, however, that it was a blackbody spectrum, then the intensity measured at one wavelength was sufficient to determine the temperature, which was found to be 3.5°K; more precisely, Penzias and Wilson concluded that it was higher than 2.5°K and lower than 4.5°K.* (In a recalibration carried out shortly afterward, they reduced their estimate to 3.1°K, again with an uncertainty of 1°K in either direction.)

Although the calculation by Peebles had predicted a background temperature of 10°K, the Princeton physicists were not shaken by the discrepancy. The calculation depended on a number of estimations, so the possibility for error was significant. One of the uncertain numbers needed for the calculation is the mass density of the universe. Peebles had used the value of 7×10^{-31} grams per cubic centimeter, a value that had been estimated in 1958 by Jan H. Oort for the mass density in ordinary galaxies. This number is somewhere between 4% and 15% of the critical density that would be needed to close the universe (as discussed in Chapter 3). To change the prediction from 10°K to 3.5°K, one would have to assume that the mass density was about twenty times lower than this.† The Princeton team concluded, however, that the Oort estimate was probably not reliable enough to rule out such a low density universe.

* In hindsight, evidence of the microwave signal discovered by Penzias and Wilson can be found in several earlier publications. DeGrasse, Hogg, Ohm, and Scovil (*Proceedings of the National Electronics Conference*, vol. 15, p. 370 (1959)) carried out noise measurements on a prototype horn-reflector system, and found the noise slightly higher than expected. E.A. Ohm (*Bell System Technical Journal*, vol. 40, p. 1065 (1961)) measured the noise on the same 20-foot horn reflector later used by Penzias and Wilson, finding the level 3.3°K higher than expected. He concluded, however, that the known sources of noise must be a little more intense than estimated. After the 20-foot horn was fitted with the 7.35 centimeter receiver later used by Penzias and Wilson, the noise level measured by W.C. Jakes, Jr. (*Bell System Technical Journal*, vol. 42, p. 1421 (1963)) exceeded the sum of known contributions by 2.5°K. In this case the author did not even mention the excess noise, but the astute reader who did his own arithmetic could have spotted the discrepancy. None of these experiments used a cold-load comparison, so the excess signal was never firmly established.

† The predicted temperature is proportional to the cube root of the mass density.

While a discrepancy by a factor of 3 in the temperature or a factor of 20 in the mass density sounds very significant, one must remember that basic cosmological numbers such as the Hubble constant or the mass density are very difficult to measure. The article in *The New York Times* described the 10°K prediction, and then added without further comment that the 3.5°K measurement "was considered quite close to the prediction." When I asked Jim Peebles if the discrepancy in temperatures or mass densities was secretly a cause for worry at the time, he replied:

> I remember gathering from a very nice review article by Oort (in a Solvay conference) that there was considerable uncertainty in the measured mean mass density. In short, I don't remember being worried about the mass density, but I do now wish I had kept a diary!

David Wilkinson was even more emphatic about the lack of concern:

> I wasn't the least bit worried that Jim's 10°K prediction was too high. The whole [theoretical] story seemed very fishy at the time, and I didn't take primordial nucleosynthesis seriously.* It seemed like a second order problem compared to showing that the cosmic microwave radiation existed. Also, I probably didn't understand it.

Over the next year Peebles refined his calculations, submitting a detailed paper to the *Astrophysical Journal.* This paper *was* published, and it established the standard techniques that have been used ever since in papers examining the origin of the cosmic background radiation. Most importantly, the refinements eliminated the discrepancy between the predicted and observed values of the temperature. The observed value of 3°K to 3.5°K for the cosmic background temperature was now found to be consistent with the Oort value for the mass density of the universe! Peebles concluded that even a critical density of matter, which he estimated as 25 times larger than the Oort value, would be just barely consistent. The issues involved in this analysis will be discussed in the next chapter.

Meanwhile, the major experimental challenge was to find out whether the radiation has the predicted blackbody spectrum. About six months after submission of the Penzias and Wilson paper, Roll and Wilkinson [6] completed their own measurements at 3.2 centimeters. They again found a uniform background of radiation that could not be attributed to any conventional astronomical source. The temperature they found was 3.0°K, with an estimated error of 0.5°K in either direction—a value com-

* The background for this sentence is discussed in the next chapter.

pletely consistent with the measurement by Penzias and Wilson, and completely consistent with the expected blackbody spectrum. Meanwhile, Penzias and Wilson proceeded with their original plans to adapt the horn antenna for 21 centimeter radiation, but now with a new goal in mind. They dropped their original program to measure radiation from the galaxy, but instead made another measurement of the cosmic background. They were pleased to find that the background radiation at this longer wavelength was again consistent with blackbody radiation at a temperature near 3°K [7].

Excitement grew, and a number of radio astronomy groups joined the enterprise. Within a few years, the cosmic background radiation was measured at 75 centimeters, 50 centimeters, 1.5 centimeters, 9.2 millimeters, 8.6 millimeters, 8.2 millimeters, and 3.3 millimeters. Most of these data points are shown on Figure 4.2.

The points all fit the expected blackbody spectrum very well, but so far there were no points to confirm the sharp decrease of the energy density expected at wavelengths shorter than about 2 millimeters. Measurements at such wavelengths are very difficult because of the emissions of ozone, oxygen, and water vapor in the atmosphere. In addition, detectors operating in the millimeter range were just being developed.

The first measurements [8] in the vicinity of the peak of the blackbody spectrum were made indirectly, using a carbon-nitrogen molecule called cyanogen as a kind of cosmic thermometer. The technique, suggested independently by George B. Field of the University of California at Berkeley and Neville J. Woolf of the University of Texas, relies on the fact that cyanogen molecules in interstellar gas clouds are bathed in the cosmic background radiation. Astronomers observe visible-wavelength starlight that passes through the gas clouds, and from the absorption lines in the spectrum they can infer the temperature of the microwave radiation absorbed by the cyanogen. This technique not only allowed the measurement of the background radiation at wavelengths that were then otherwise inaccessible, but it also provided direct evidence that the 3°K radiation is found not only in our own solar system, but also at many locations elsewhere in the galaxy.*

* Cyanogen had been used to measure temperatures by Andrew McKellar of the Dominion Observatory in Canada as early as 1941, but the connection with cosmology was not recognized at the time. In 1960 George Field, who was then at Princeton, concluded that the behavior of interstellar cyanogen could be explained only by assuming that it was bathed with radiation at about 3°K, although again no connection was made to cosmology. A senior colleague, Lyman Spitzer, convinced Field that it was too speculative to publish. Field buried the draft in his desk drawer, and continued to overlook the implications even several years later, as he watched his colleagues Roll and Wilkinson setting up a microwave detector on the roof of a nearby building. Shortly after moving to Berkeley in 1965, Field learned about the discovery of the cosmic background radiation from a phone call, he thinks from the legendary microwave background telephone caller, Bernard Burke. This time he realized the connection immediately, and his paper was soon off to the journal.

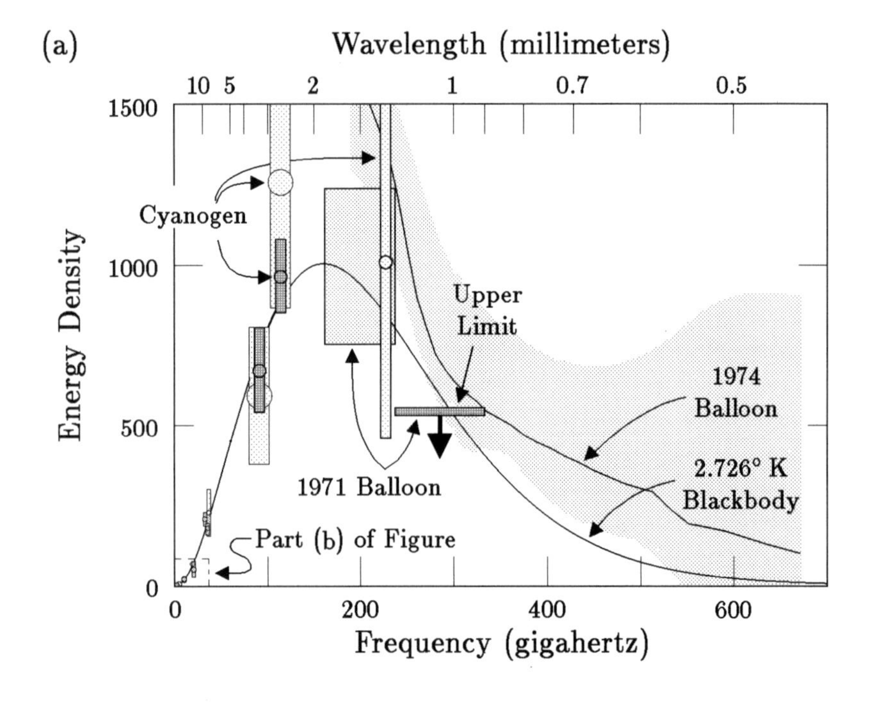
(a)
Wavelength (millimeters)
10 5 2 1 0.7 0.5
1500
1000
500
0
Energy Density
Cyanogen
Upper Limit
1974 Balloon
2.726° K Blackbody
1971 Balloon
Part (b) of Figure
0 200 400 600
Frequency (gigahertz)

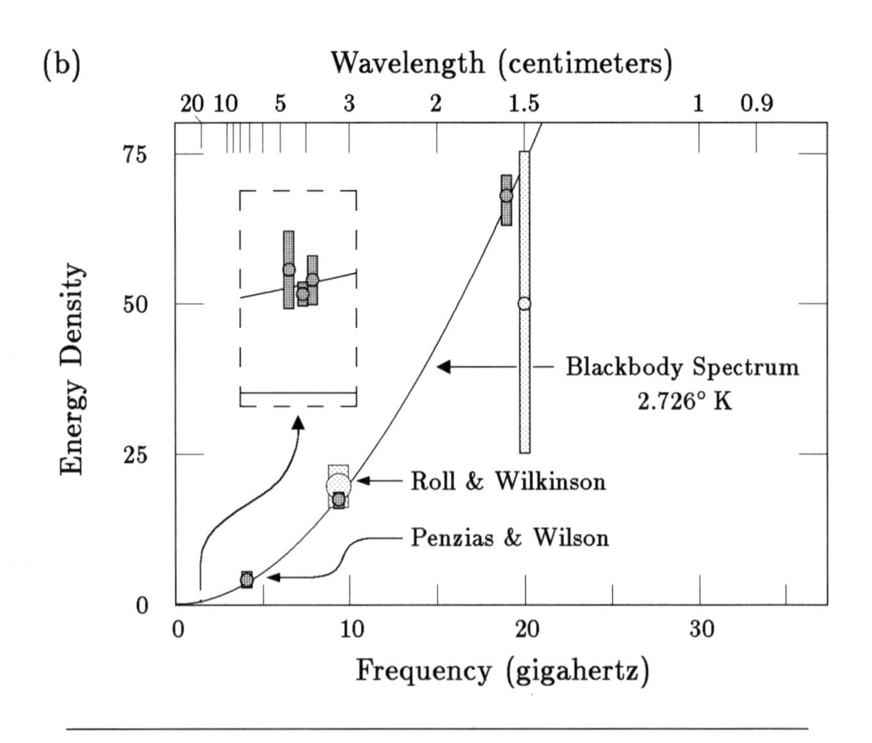
(b)
Wavelength (centimeters)
20 10 5 3 2 1.5 1 0.9
75
50
25
0
Energy Density
Blackbody Spectrum
2.726° K
Roll & Wilkinson
Penzias & Wilson
0 10 20 30
Frequency (gigahertz)

A direct measurement of the cosmic background spectrum just above the peak was accomplished in 1971 by Dirk Muehlner and Rainer Weiss [9] of the Massachusetts Institute of Technology. Using a liquid-helium-cooled detector lofted to an altitude of 24 miles by a football-field-sized balloon that had originally been developed for military reconnaissance, they determined a data point near the peak of the blackbody curve, indicated as "1971 Balloon" on part (a) of Figure 4.2. They also showed that the energy density, at a frequency just above the peak, was no higher than the marker on the diagram labeled "Upper Limit."

In 1974 a broadband balloon measurement of the cosmic background radiation, extending well beyond the peak, was carried out by David P. Woody, John C. Mather, Norman S. Nishioka, and Paul L. Richards [10], all from the University of California at Berkeley (1974 Balloon in Figure 4.2). The uncertainties in the experiment were large enough so that a blackbody spectrum was not excluded, but the data strongly suggested that there was excess radiation. Early rocket experiments and another balloon experiment in 1981 gave further evidence of excess radiation, leading to discussions of a "Woody-Richards distortion" of the cosmic background spectrum.

While the early measurements of the high frequency component of the cosmic background radiation were inconclusive, they at least showed that the spectrum was not too far from a blackbody. In addition, these investigations pioneered valuable experimental techniques that would be used more effectively later.

The efforts to pin down the spectrum of the cosmic background radiation have continued to the present, and at the end of this chapter I will fill in some exciting recent highlights. First, however, I would like to

---

Figure 4.2 **(facing page) Data on the cosmic background radiation spectrum as of 1975.** The graphs show measurements of the energy density in the cosmic background radiation at different frequencies (or wavelengths). The lower horizontal axis shows the frequency in gigahertz (10^9 cycles per second), and the upper horizontal axis shows the corresponding wavelength. The solid line is the expected blackbody distribution, shown for the best current determination of the temperature, 2.726°K. Part (a) includes the full range of interesting frequencies, while part (b) shows a magnified view of the low frequency measurements. The result of each measurement is marked by a circle, with a vertical bar to indicate the range of the estimated uncertainty. Measurements with small uncertainties are emphasized by dark shading. A high-frequency broadband measurement is shown on part (a) as a solid line labeled "1974 Balloon," with the estimated uncertainty indicated by gray shading. (The energy density on both graphs is measured in electron volts per cubic meter per gigahertz.)

discuss some frequently asked questions about the cosmic background radiation.

I am often asked why it is that we are still surrounded by the radiation from the heat of the big bang. Suppose, the questioner elaborates, we consider the explosion of a stick of dynamite or a nuclear bomb. In either case, light is emitted and matter is strewn outward. But since the matter moves much more slowly than the light, after a short time the light has moved well beyond the matter. An observer riding on the matter would not be able to see the radiation surviving from the fireball of the original explosion. Why, then, can we see the remnant radiation from the fireball of the big bang?

The answer to this question lies in understanding the geometry of the big bang. One must realize that the big bang theory is not a description of a localized explosion, like a stick of dynamite or a nuclear bomb. On the contrary, the models of Friedmann and Lemaître were constructed to be completely homogeneous. This means that the matter is assumed to have uniformly filled all of space at all times, right back to the instant of the big bang. There is no edge and no center to the distribution of matter. The expansion of the universe happens homogeneously, with every galaxy (on average) moving away from every other galaxy, as the entire space expands. Since the matter fills all of space, it is impossible for the radiation to leave the matter-filled region, as it would for a stick of dynamite.

If the questioner is very persistent, he or she might go on to ask why we should believe in this homogeneous model, since the image of a localized explosion—one which has a center and an edge—seems so much more plausible. After all, every explosion that we have ever seen on earth has been a localized explosion. It is hard to imagine what could possibly cause an explosion to happen simultaneously throughout the universe, as the Friedmann-Lemaître models assume.

Here I freely admit that the questioner has a good point—the idea of a simultaneous explosion throughout the universe does seem hard to accept. In 1962 a localized explosion model was in fact proposed, by Oskar Klein and Hannes Alfvén [11]. They suggested that the observed Hubble expansion could have been caused by an explosion due to the annihilation of matter and antimatter. Such a model sounds very plausible, but it appears to be incompatible with observations. The problem is that a localized explosion cannot explain the observed uniformity in the cosmic background radiation. The earliest measurements of the background radiation showed that it was isotropic (that is, uniform in all directions) to an accuracy of about one half of a percent. More recent measurements (which will be discussed in Chapter 14) show that the background radiation is uniform to about one part in 100,000! If there were a localized explosion that occurred in some particular direction in the sky, on the other hand, then one would expect that this direction would be clearly visible as a hot spot in the background radiation. The idea of a localized explosion would work only if we happened to be living right at the center of the explosion. Since it seems very unlikely that we should be so close to the center, the possibility of a localized explosion is not given much serious consideration.

If the reader still finds it difficult to accept the assumption of a simultaneous explosion throughout the universe, he or she should read on. One of the attractive features of the inflationary universe theory is a new understanding of the homogeneity of the big bang. The inflationary universe can explain why the big bang explosion occurred homogeneously throughout the observed universe, while at the same time the theory suggests that the observed universe is only a minute part of a vastly larger space that is far from homogeneous.

Continuing the discussion of the standard big bang model, I will assume for now that the big bang was a uniform explosion filling all of space. We must still understand how the intense heat of that explosion has been reduced to the faint reverberation at 3°K that fills the universe today.

The answer to this question hinges on the Doppler effect, which was discussed in Chapter 3. The application to an expanding universe in the context of general relativity is in some ways more complicated than the earlier discussion, but in the end the result is very simple. Proceeding one step at a time, let us begin by considering what happens when a light pulse emitted by some distant galaxy is received on Earth. Since Hubble's law tells us that the distant galaxy would be receding, we expect that the light pulse would be redshifted when it reaches the earth. As long as the velocity of the distant galaxy is small compared to the speed of light, nothing more needs to be said. The redshift could be accurately calculated by using the logic that accompanied the jet airplane and rabbit illustration of Figure 3.5 (see page 48). For more distant galaxies, however, the discussion would have to include relativity, since the velocities could be comparable to the speed of light. Furthermore, at large distances it will no longer be safe to neglect the gravitational field caused by the intervening mass. This gravitational field, according to general relativity, will also contribute to the Doppler shift: the frequency of an electromagnetic wave is decreased if it travels uphill in a gravitational field, and increased if it travels downhill. Despite these complexities, the result is remarkably simple: The wavelength of light is simply stretched with the expanding universe. If the separation between galaxies increases by a factor of 3 while a light pulse travels from one galaxy to another, then the wavelength of the pulse will be increased by exactly the same factor. This simple rule can be understood by comparing a light wave moving through an expanding universe with a line of bugs marching along a stretching rubber band, as shown in Figure 4.3 and explained in the caption.

Knowing that the wavelength of light is stretched with the cosmic expansion, we next ask what happens to blackbody radiation that fills the universe. The answer depends on a remarkable property of the blackbody spectrum that can be seen by comparing Figure 4.1 with Figure 4.2(a). The first graph shows a blackbody spectrum for 10°K, while the second graph is for 2.726°K. Nonetheless, if you rip out a page to place one graph on top of the other, you will see that they coincide exactly. This does not mean that the two spectra are identical. On the contrary, if you look at the scales you will see that they are different. It does mean, however, that a change in temperature can be completely compensated by a judiciously chosen change in the scales, without any change in the shape of the graph itself. If the temperature is doubled, for example, then the new graph can be obtained from the old by multiplying the frequencies on the horizontal axis by 2, and the energy density values on the vertical axis by 8 (where $8 = 2^3$). The argument is a little too technical to include here, but it can be shown that this scaling property of the blackbody spectrum leads to a simple result: As the universe expands, the blackbody spectrum is preserved exactly. The only thing that changes is the temperature, which falls as the universe expands. During any time interval in which the distance between galaxies doubles, the temperature of the cosmic background radiation falls to half its initial value. Thus, we can assume that the cosmic background radiation was created with a blackbody spectrum at a very high temperature early in the history of the universe. As the universe expanded, the spectrum remained blackbody, but the radiation was cooled by redshifting until now the temperature has reached the impressively cold reading of approximately three degrees above absolute zero.

Measurements of the cosmic background spectrum continued at a high level of activity through the 1970s and 1980s, and for the most part the data continued to conform to a blackbody spectrum at 2.7°K. Minor discrepancies were found, but invariably disappeared when further experiments were performed. Nonetheless, information about the spectrum for frequencies near the peak of the blackbody spectrum and beyond remained very meager.

In 1982 planning began for a joint Japanese-American project to measure the high frequency cosmic background spectrum from a rocket that would carry the detector far above the Earth's atmosphere. The Japanese (led by Toshio Matsumoto of Nagoya University) would supply the rocket, while Paul Richards and Andrew Lange, from the University of California at Berkeley, would supply the instrumentation.

After a 1985 flight was ruined by an instrument cover that failed to open, a 1987 launch carried the detector flawlessly to an altitude of 200 miles over

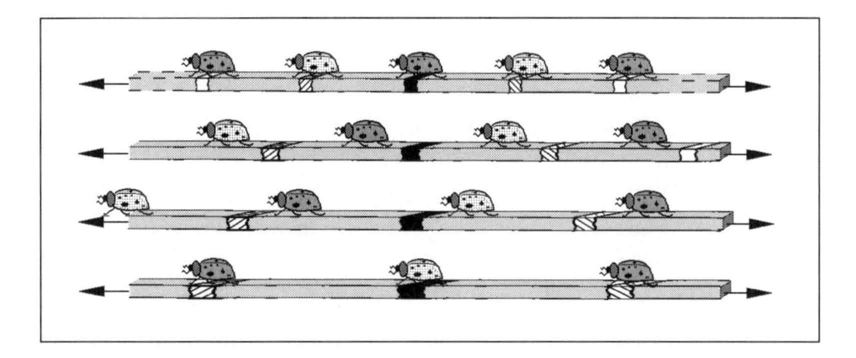

Figure 4.3 **Bugs marching on a stretching rubber band.** A sequence of four snapshots illustrates a line of bugs marching along a uniformly stretching rubber band—a situation closely analogous to a light wave moving through an expanding universe. The bugs are evenly spaced, and all the bugs move at the same, fixed speed relative to the rubber under their feet. Suppose the rubber band is painted with evenly spaced markers, which in the top frame line up with the bugs. By the time the center bug moves one marker ahead, which happens in the fourth frame, each of the other bugs has also moved one marker ahead. The bugs will always remain exactly one marker apart, so their distance in centimeters will increase in direct proportion to the stretching of the rubber band. If each bug is taken to represent the crest of a light wave, we can see that the wavelength stretches with the expanding space.

the southern tip of Japan. The measurements of the cosmic background radiation intensity, at three frequencies well above the peak of the spectrum, stunned the cosmological community. The data points, shown in Figure 4.4 [12], indicated a background radiation level far above the blackbody curve, with extremely small estimated uncertainties. Point (2) differed from the expected blackbody curve by an amount 12 times larger than the estimated uncertainty, and point (3) differed by an amount that was 16 times larger! The energy indicated by this excess radiation amounts to about 10-15% of the total energy of the cosmic background radiation. The Berkeley-Nagoya team recognized that such a radical result should be viewed with caution, but they nonetheless explained that every conceivable source of error had been carefully studied, and nothing that could account for such a large discrepancy had been found.

Theorists attempted to explain the excess radiation by hypothesizing a pregalactic generation of stars, but in the end no plausible explanation could be found. Yet the experiment was carried out by some of the most respected cosmic background researchers in the world, everything had functioned perfectly, and the observed discrepancies were as much as 16 times larger than the estimated uncertainty.

The Berkeley-Nagoya group launched another rocket in July 1989 to check their results, but this time an error in wiring contaminated their data with electrical noise. Attempting to salvage the data by computer enhancement, they found results that were highly uncertain, but consistent with a blackbody spectrum and not with the previously measured excess. The team was divided on what to do. Should the previous result be retracted on the basis of such flawed data? Knowing that another experiment was about to answer the question, the Berkeley-Nagoya collaboration decided to wait.

In November of 1989, just four months after the rocket launch, the National Aeronautics and Space Administration (NASA) launched the *Cosmic Background Explorer*, also known by its acronym, COBE (rhymes with

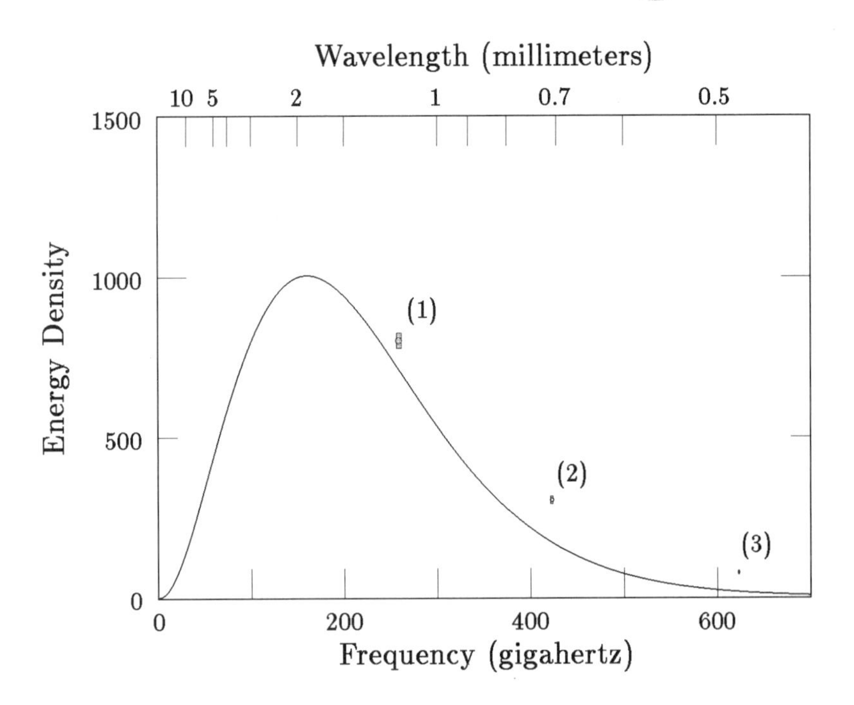

Figure 4.4 **Cosmic background measurements by the Berkeley–Nagoya rocket experiment.** A rocket flight in 1987 measured the intensity of the cosmic background radiation at three frequencies well beyond the peak of the blackbody spectrum. The intensities were found to be far larger than the blackbody prediction, especially for the points labeled (2) and (3). The estimated uncertainty is shown by the height of the bars; for point (3) the estimated uncertainty is almost invisibly small. (The energy density is measured in electron volts per cubic meter per gigahertz.)

Moby of *Moby Dick*), shown in Figure 4.5. Although the Soviet satellite *Relikt-I* had searched in 1983-84 for nonuniformities in the cosmic background radiation, COBE was the first satellite to take full advantage of a space platform for probing the cosmic background radiation. Its launch was the fruition of fifteen years of effort by over 1500 scientists, engineers, technicians, computer specialists, and administrators.

The project had been initiated in 1974 by three proposals, the most extensive of which was led by John Mather, a 28-year-old former graduate student of Paul Richards. Mather and six other scientists proposed a *Cosmological Background Radiation Satellite* that included four separate instruments. The two competing proposals each included only a single instrument to measure the nonuniformities in the cosmic background radiation [13]. Two years later, NASA Headquarters invited Mather to lead a study group to prepare a more detailed proposal that would incorporate the ideas of all three groups.

The resulting proposal was called COBE, and included three of the four instruments that had been proposed by Mather and his collaborators. The measurement of the spectrum of the cosmic background radiation would be

John Mather

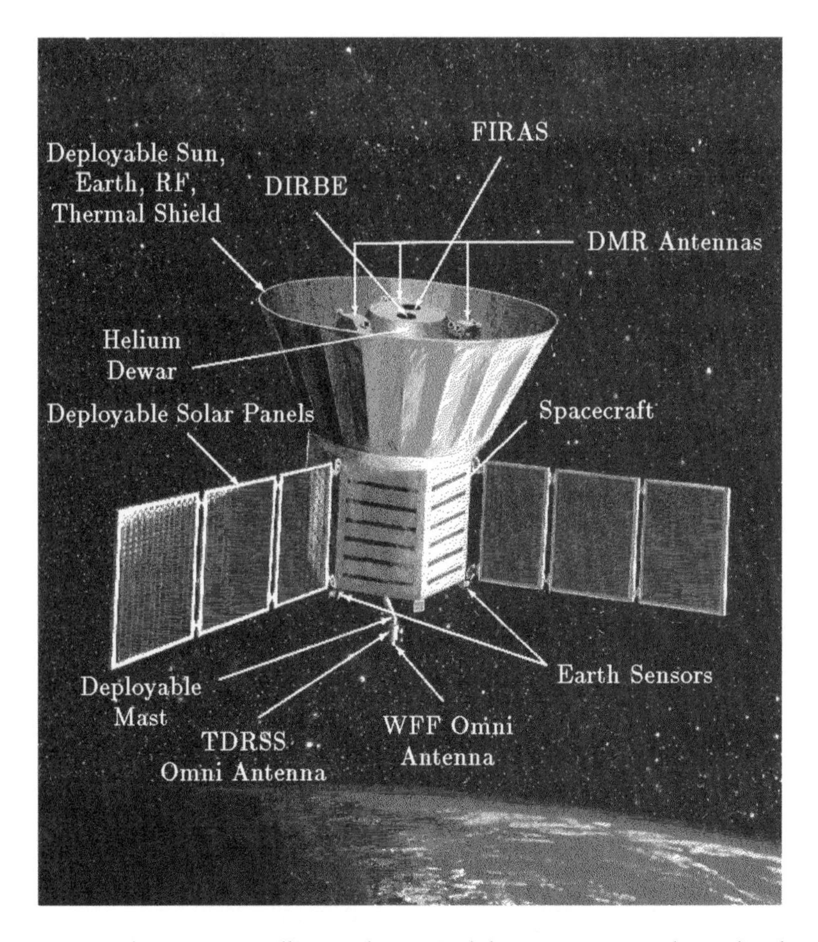

Figure 4.5 **The COBE satellite.** The 2,270 kilogram structure has a height of 5.5 meters, and a diameter of 2.4 meters when the Sun-Earth shield and solar panels are folded, or 8.5 meters when they are extended. The shield protects the instruments from sunlight and light reflected from Earth, and also from radio frequency (RF) interference originating from Earth and from the spacecraft's own transmitting antennas. The antennas are mounted on a mast that was deployed after the satellite was in orbit. One antenna communicates with the Tracking and Data Relay Satellite System (TDRSS), and the other communicates directly with the Wallops Flight Facility (WFF) on the ground. The 170 (U.S.) gallons of liquid helium in the dewar lasted for 10 months, during which time the FIRAS surveyed the sky 1.6 times. The DIRBE and the DMR completed their four-year mission in December 1993. The drawing is courtesy of the NASA Goddard Space Flight Center and the COBE Science Working Group.

carried out by the *Far Infrared Absolute Spectrophotometer*, known as FIRAS. The nonuniformities of the cosmic background radiation would be probed by the *Differential Microwave Radiometer* (DMR), and the diffuse background at infrared frequencies, relevant to such questions as galaxy and star formation, would be measured by the *Diffuse Infrared Background Experiment* (DIRBE) [14].

The COBE project was originally planned for seven years, but major projects are always prone to delay. The satellite was initially designed for launch by a *Delta* rocket, but after several years NASA's commitment to the space shuttle program led to a redesign for a shuttle launch. Full-scale funding of the project was not approved until 1982. Despite the slow start, by 1986 the satellite was being assembled for launch. In January of 1986, however, the *Challenger* space shuttle exploded tragically on take-off, and the American space program was placed on hold. It soon became clear that a shuttle launching for COBE was not going to happen. Shuttle operations would eventually resume, but the stricter safety standards, the reduced demand for military satellite launches, and the shortage of money would mean that the Vandenberg Air Force Base in California would no longer be maintained as a shuttle launch site. In order to obtain the full-sky coverage that was needed for mapping the cosmic background radiation, COBE would have to be placed in an orbit that takes it over the Earth's north and south poles. Since such an orbit can be achieved only from Vandenberg, the entire launch plan had to be changed.

Engineers at the Goddard Space Flight Center in Greenbelt, Maryland, began a round-the-clock effort to redesign COBE for launch by a *Delta* rocket, as had been originally planned. At this stage, however, the change required that the satellite be reduced to 5000 pounds, approximately half of its previous weight. Part of this reduction was achieved by dispensing with a hydrazine propulsion system that was no longer needed to carry the satellite to its final altitude. Fortunately no large changes were needed in the instruments or associated electronics, but the sunshield, solar panels, antennas, battery system, and mechanical structure had to be significantly redesigned. Despite the upheavals, COBE was ready for launch on November 18, 1989. The *Delta I* rocket successfully carried the satellite into orbit at an altitude of 560 miles.

In January of 1990, less than two months after launch, the COBE team announced their first results at the annual meeting of the American Astronomical Society, in Washington, D.C. Reports were given by all three instrument teams, but excitement centered on the announcement of the first data on the cosmic background spectrum from FIRAS. In front of a crowded hall, John Mather reached the climax of his talk by placing on the transparency projector the graph that is shown here as Figure 4.6 [15]. Measurements of the cosmic background radiation spectrum were shown at 67 different wavelengths, ranging from well below the blackbody peak to far above it. The data points were shown as small squares, with a size

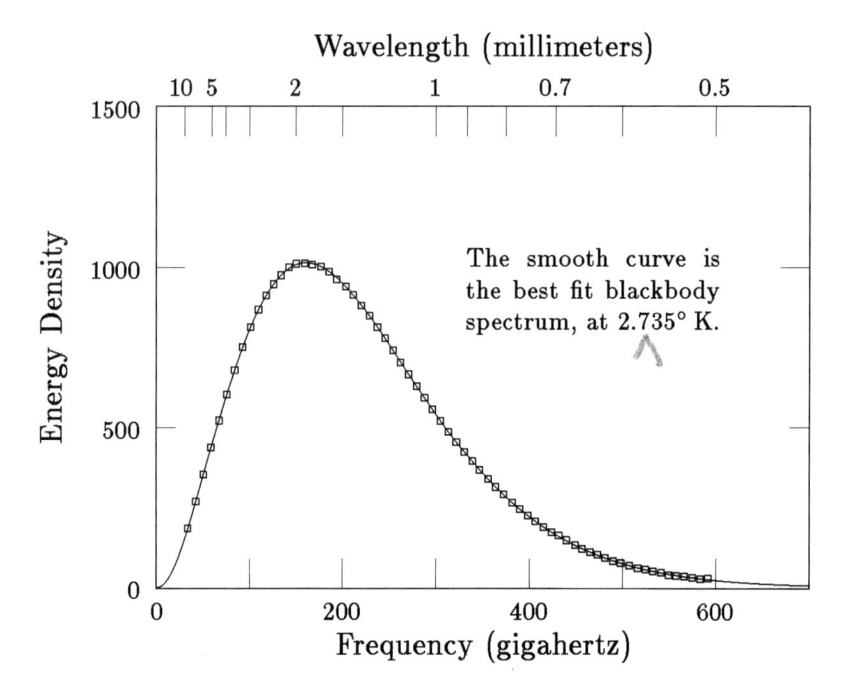

Figure 4.6 **Cosmic background measurements by the COBE satellite.** In January 1990, the COBE team released the first spectrum of the cosmic background radiation measured from a satellite. The size of the boxes indicates estimated uncertainty, taken to be 1% of the peak energy density. (The energy density is measured in electron volts per cubic meter per gigahertz.)

that indicated the estimated uncertainty: 1% of the peak energy density. The data fell on the theoretical blackbody curve with a precision that was totally mind-shattering to cosmologists who had learned their trade by staring at graphs such as those shown in Figures 4.2 and 4.4. With one graph, all of the previous data, with all the complications and uncertainties, could be forgotten. The excess radiation seen by the Berkeley-Nagoya rocket flight was evidently some kind of an instrumental glitch, although the cause remained unknown.* With one graph, the description of the cosmic background radiation spectrum was reduced to one simple statement: It is a

* The members of the Berkeley-Nagoya group agreed without hesitation that the COBE results were more reliable than their rocket data. They now believe that the error originated from a failure of communication. The rocket engineers in Japan were not aware of the electrical grounding requirements of the experiment, and the physicists in Berkeley were not aware of the grounding procedures used in the Japanese rocket. The result was a measurement of electrical interference that was confused with the cosmic background radiation.

blackbody. The temperature is 2.735°K, with an uncertainty of only 0.06°K.

NASA had been responsible for a major scientific triumph. The audience at John Mather's talk rose to give him and the COBE team a standing ovation.

In January of 1993, John Mather again addressed the American Astronomical Society. The first graph had come from just nine minutes of data, but now the team had analyzed the data from the entire mission [16]. The error boxes were shrunk beyond visibility to only 0.03%, and the background spectrum was still perfectly blackbody, just as the big bang theory predicted. The new measurement of the temperature was just a little colder, 2.726°K, with an uncertainty of less than 0.01°K.

The perfection of the spectrum means that the big bang must have been very uncomplicated. The COBE team estimated that no more than 0.03% of the energy in the background radiation could have been released anytime after the first year. Theories that predict energy release from the decay of turbulent motions or exotic elementary particles, from a generation of exploding or massive stars preceding those already known, or from dozens of other interesting hypothetical objects, were all excluded at once. Although a few advocates of the steady state universe have not yet given up, the COBE team announced that the theory is ruled out. A nearly perfect blackbody spectrum can be achieved in the steady state theory only by a thick fog of objects that could absorb and re-emit the microwave radiation, allowing the radiation to come to a uniform temperature. Steady state proponents have in the past suggested that interstellar space might be filled by a thin dust of iron whiskers that could create such a fog. However, a fog that is thick enough to explain the new data would be so opaque that distant sources would not be visible.

In the light of data so precise, it is difficult to imagine any reasonable alternative to the big bang.

CHAPTER 5

CONDENSATION OF THE PRIMORDIAL SOUP

Anyone who has ever felt a bicycle tire that was just pumped up knows that the temperature of a gas increases when the gas is compressed. Conversely, air conditioners and refrigerators function on the principle that a gas cools when allowed to expand. The big bang theory, resulting from the work of Friedmann, Lemaître, Hubble, and others, describes the universe as a giant refrigerator, forever cooling itself by its continued expansion. The theory implies that the universe began as a scorching cauldron, with temperatures well in excess of a trillion degrees. The intense heat blasted the matter into its elementary constituents. Even atomic nuclei could not exist at these temperatures, but would have been ripped apart into the protons and neutrons from which they are made. Today we live amid the embers of this inferno, the ash that remains after billions of years of expansion and cooling. Yet we are still surrounded by the signs of the primeval fireball. As discussed in the last chapter, the heat of the big bang survives today as an afterglow of 2.7°K radiation that permeates the universe. In addition, the light chemical elements in today's universe are the product of nuclear reactions that took place during the first few minutes—the condensation of the primordial soup. By comparing calculations of these nuclear reactions with the observed composition of the universe, cosmologists have been able to test and refine their ideas about the big bang.

To understand big-bang nucleosynthesis, one must first recognize that the big bang theory is much more than a cartoon image of a big explosion that hatched a universe. The big bang is actually a very detailed theory. The matter in the early universe is described as a uniform, hot gas of particles undergoing uniform expansion. The gas is characterized by its temperature, and some simple but plausible assumptions about its composition. The cooling caused by the expansion is understood by using conventional thermal physics, and the slowing of the expansion due to gravity is described by the equations of Friedmann and Lemaître, based on Einstein's theory of general relativity. Putting together this rather compact set of ideas, cosmologists can calculate how fast the early universe was expanding at any given time, and also how the temperature, density, and pressure varied with time. By combining this information with knowledge of nuclear physics, they can go on to calculate the expected abundances of the various nuclei produced in the big bang.

Using the big bang equations to extrapolate back to 100,000 years after the big bang, we find that the temperature of the universe was about equal to the present temperature on the surface of the sun, 5,800°K. The mass density was only about 10^{-19} times the density of water (which is 1 gram per cubic centimeter). This density was much higher than the present value, but still astonishingly low. Extrapolating further back to just a year after the big bang, the temperature was two million degrees Kelvin, about as hot as the solar corona today. The mass density was still less than a billionth that of water, but the pressure of the extremely hot gas was almost a million times larger than that of the Earth's atmosphere. At the end of the first seven days, the universe was at 17 million degrees, a million degrees hotter than the inferno at the center of the sun. The density was about a millionth that of water, but the pressure was more than a billion atmospheres. Going all the way back to one second after the big bang, cosmologists estimate a temperature of ten billion degrees, comparable to the core of a supernova explosion—the highest temperature known to exist in the universe today. The mass density was very high, half a million times that of water, and the pressure was an unfathomable 10^{21} atmospheres. To make contact with the grand unified theories that Henry Tye was trying to persuade me to work on, one would have to trust the extrapolation all the way back to 10^{-39} seconds after the big bang, when the temperature was 10^{29}°K. At that temperature the average energy per particle would be about 10^{16} GeV (1 GeV = one billion electron volts), the energy at which the new effects predicted by grand unified theories become significant. The mass density under these extraordinary conditions would be roughly 10^{84} times higher than water, the same density as a trillion suns jammed into the volume of a proton!

If one continues the extrapolation backwards in time, one comes to a point of infinite density, infinite pressure, and infinite temperature—the instant of the big bang explosion itself, the time that in the laconic language of cosmologists is usually called "$t = 0$." It is also frequently called a *singularity*, a mathematical word that refers to the infinite values of the density, pressure, and temperature. It is often said—in both popular-level books and in textbooks—that this singularity marks the beginning of the universe, the beginning of time itself. Perhaps this is so, but any honest cosmologist would admit that our knowledge here is very shaky. The extrapolation to arbitrarily high temperatures takes us far beyond the physics that we understand, so there is no good reason to trust it. The true history of the universe, going back to "$t = 0$," remains a mystery that we are probably still far from unraveling.*

How far back, then, should we trust the equations of the big bang theory? One might start to answer this question by comparing the temperatures of the early universe with those of laboratory experiments. For example, the Tokamak Fusion Test Reactor, a controlled fusion facility at the Princeton Plasma Physics Laboratory, has reached temperatures of 200,000,000°K, equal to the calculated temperature of the universe an hour after the big bang. Although this temperature is spectacularly high, more than ten times hotter than the center of the sun, it is still far from the highest temperatures that physicists can probe. To learn about higher temperatures, physicists turn to accelerators, which can produce beams of particles at energies that are much higher still. Thus, the extraordinary temperatures of the early universe have led to a peculiar marriage of disciplines. Cosmology, the study of the largest objects known, has become intimately linked to particle and nuclear physics, the study of the smallest objects known.

Table 5.1 shows a sampling of particle accelerators and how they have helped us to delve closer and closer to the instant of the big bang. Even the very first circular particle accelerator, the cyclotron built by Earnest O. Lawrence in 1930, reached an energy per particle four times higher than that of the Princeton Tokamak. The highest energy particle accelerator in operation in 1996, and presumably through the end of the twentieth century, is the Tevatron proton accelerator at the Fermi National Accelerator Laboratory (Fermilab). The beam energy corresponds to the temperature of the universe at just 2×10^{-13} seconds after the big bang. If construction proceeds according to plan, at the beginning of the next century the Large Hadron Collider at CERN will allow us to study the conditions in the universe at just 5×10^{-15} seconds after the big bang.

The accelerator experiments provide crucial information, but a high energy beam is not exactly the same as the hot gas that we believe filled the early universe. The particles in an accelerator beam all have essentially the

* If the extrapolation to $t = 0$ is not trustworthy, then a description such as "one second after the big bang" also becomes ambiguous. This ambiguity is similar to that experienced by a person who measures dates in the standard A.D. (*anno Domini*) system, even though he might not be confident in his knowledge of the true birthdate of Jesus Christ. The time measurements nonetheless have meaning relative to each other, even though the significance of time zero is uncertain.

Accelerator	*Date*	*Beam Energy*	*Equivalent Temperature*	*Time*	*Comment*
First Cyclotron	1930	80,000 eV	0.93 billion°K	200 sec	Built by Ernest O. Lawrence
Cosmotron	1952	3 GeV	35 trillion °K	3×10^{-8} sec	At Brookhaven National Laboratory, on Long Island, New York
Tevatron	1987	1000 GeV	1.2×10^{16}°K	2×10^{-13} sec	At Fermi National Accelerator Laboratory, outside Chicago
Large Hadron Collider	2002	7000 GeV	8.1×10^{16}°K	5×10^{-15} sec	Planned at CERN,* near Geneva in Switzerland and France

* European Laboratory for Particle Physics, formerly Conseil Européen pour la Recherche Nucléaire

Table 5.1 **Particle accelerators and the early universe**

same energy, while the particles of a gas have a range of energies. In addition, the particles in an accelerator beam are much further apart, on average, than the particles of the early universe. Accelerator experiments show us what happens when two particles collide at high energy, but we must rely on theoretical calculations to infer how the temperature and pressure of a hot dense gas will vary as the gas expands. In most cases these calculations are believed to be very reliable, so we can use use experimental particle physics to probe the early universe.

So, once again, how far back should we trust the equations of the big bang theory? There is of course no definitive answer. If you asked a dozen physicists, you would probably get fourteen different answers along with five or six abstentions. Later in the book I will be discussing extrapolations all the way back to the era of grand unified theories, at roughly 10^{-39} seconds. This takes us far beyond the reach of any accelerator, but I will nonetheless argue that such extrapolations are not crazy, so long as one is interested in qualitative features that do not depend on having all the details right.

The rest of this chapter will focus on the role of nuclear reactions that are believed to have taken place between about one second and three to four minutes after the big bang. The underlying principles of this low-energy nuclear physics have been studied since the 1930s and 1940s, so the foundation for these studies seems very secure. Through these reactions—known as *big-bang nucleosynthesis*—the light chemical elements were forged from the hot soup of particles that emerged from the big bang. It was the study of these reactions that allowed Jim Peebles, and others earlier, to predict the existence and approximate temperature of the cosmic background radiation.

The reader may well be surprised that scientists dare to study processes that took place so early in the history of the universe. On the basis of present observations, in a universe that is some 10 to 20 billion years old, cosmologists are claiming that they can extrapolate backward in time to learn the conditions in the universe just one second after the beginning! If cosmologists are so smart, you might ask, why can't they predict the weather? The answer, I would argue, is not that cosmologists are so smart, but that the early universe is much simpler than the weather!

The outlandish simplicity of the early universe is seen most clearly in the cosmic background radiation, the afterglow of the primeval heat. Physicists have sensitively measured the radiation coming from different directions, looking for differences in the temperature from one direction to another. Indeed they have found "hot spots" and "cold spots," but the temperatures differ from the average by only about one part in 100,000. Thus the temperature of the early universe, and presumably the density and pressure as well, was uniform to this extraordinary accuracy. The complicated distribution of matter that we see in the universe today arose much later, as the gravitational attraction of the matter caused it to clump. If the temperature on Earth were as consistent as in the early universe, our weather reports would be ludicrously dull: "It was scorching today in Dalol, Ethiopia, where the mercury soared to 45.012°F. Meanwhile, at Plateau Station Antarctica, there were reports of frigid temperatures as low as 45.001°F." If the temperature on Earth were this monotonous, we *could* predict the weather—very accurately. Although we do not necessarily understand why the early universe was so uniform, the cosmic background radiation provides direct evidence that it was. (One of the great successes of the inflationary theory is a possible explanation for this simplicity.) For most purposes one can approximate the early universe as being exactly uniform, greatly simplifying the calculation of its evolution.

Since the uniformity of the early universe makes the calculations easy, and since the underlying physics of hot gases and nuclear reactions seems to be well-understood, it may not be so surprising that cosmologists claim to have plausible theories describing the universe at one second after the big bang. In any case, we will find that the predictions obtained for abundances of the light chemical elements agree well with observation. These predictions are an important success of the big bang theory, strongly suggesting that the extrapolation is valid all the way back to just one second after the big bang.

Big-bang nucleosynthesis is believed to be responsible for the formation of the lightest chemical elements in the universe: hydrogen, helium, and lithium. All heavier elements were produced much later, in the interior of stars, and then strewn into space by a variety of processes. The life cycle of a heavy star is believed to end with a supernova explosion, spewing most of the mass into the surrounding space. Less massive stars go through a period of expansion called the red giant phase, and then blow off their outer shells as they contract to become compact stable objects known as white dwarfs. Astronomers estimate that perhaps half of the material in the thin gas between stars in our galaxy has been processed in this way. Our solar system is believed to have condensed from this interstellar gas about 5 billion years ago, at a time when the gas was already sprinkled with heavy elements from earlier generations of stars.

Although the big bang did not produce the large variety of elements that we see around us, the nuclear processing in the big bang cannot be lightly dismissed. Even though elements heavier than helium are found all around us—in our bodies, in the air we breathe, and in the floor we walk on—they are actually a very small component of the cosmic inventory. The distribution of chemical elements on earth is very atypical, since most of the hydrogen and helium has escaped from the earth's weak gravitational field. While big-bang nucleosynthesis is responsible for only a small fraction of the chemical elements, it is nonetheless the origin of more than 98% of the known matter in the cosmos.

While only three elements (hydrogen, helium, and lithium) were produced primarily in the big bang, there are several different forms of each. The number of protons in a nucleus determines which element it is: one proton for hydrogen, two protons for helium, and three protons for lithium. The number of neutrons for a given element, however, can vary. A helium nucleus, for example, can have either one or two neutrons, so one says that helium has two isotopes. The isotopes are labeled by the total number of protons and neutrons, so a helium nucleus with one neutron is called helium-3 (two protons plus one neutron), and a helium nucleus with two neutrons is called helium-4. The nuclei that are relevant to big-bang nucleosynthesis are shown in Figure 5.1.

The description of nucleosynthesis can be started at 0.1 seconds after the big bang, when the temperature was 31.5 billion degrees Kelvin. This

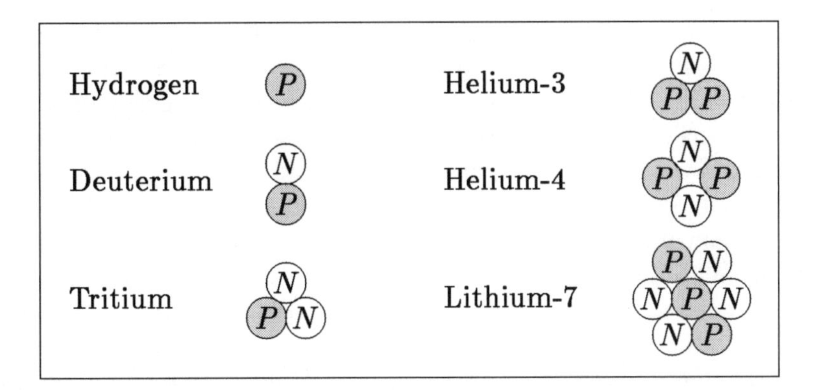

Figure 5.1 **Nuclei produced in the big bang.** The chart shows the six different types of nuclei that were produced in detectable amounts by big-bang nucleosynthesis. The isotopes of hydrogen are given special names: hydrogen-2 is called deuterium, and hydrogen-3 is called tritium. Tritium is unstable, decaying to helium-3 with a half-life of 12.5 years. The other nuclei shown here are stable, and their abundance in the universe today can be used to test the big bang theory.

temperature is so high that neither atoms nor even nuclei would have been stable—the random motion of the intense heat would have torn them apart, setting free the electrons, protons, and neutrons to stream through space as free particles. (The protons and neutrons themselves are stable at these temperatures, although at temperatures higher than about 10^{13}°K they too would be broken into their constituents, called *quarks*.) For every neutron there were 1.61 protons. The density of protons and neutrons was very modest, only about one and a half times that of water. However, although the protons and neutrons are important to us (we are made of them!), they were only a light seasoning in a very dense cosmic soup that was thickened by other ingredients. The extreme heat produced a blaze of blackbody radiation so intense that its energy dwarfed that of the protons and neutrons. According to relativity, the energy density of the radiation is equivalent to a mass density, which was 5 million times larger than the mass density of protons and neutrons. Quantum theory implies that the energy of electromagnetic radiation is not spread smoothly in space, but instead is concentrated in bundles called *photons*, which are essentially particles of light. Although we will not need quantum theory to describe the radiation of the early universe, the language of quantum physics is nonetheless very convenient—we can talk about photons in the same way that we talk about protons or electrons. At this time there were about 1.7 billion photons for every neutron.

The cosmic inventory also included electrons and their antiparticles, called *positrons*, both of which were produced in large numbers by the energetic collisions of other particles. Collisions also produced a thick soup of neutrinos and antineutrinos. For every eight photons there were six electrons, six positrons, nine neutrinos, and nine antineutrinos. There was also one electron for each proton, so that the mix was electrically neutral.

The last paragraph contains a surprising amount of information, considering that all this happened 10 to 20 billion years ago, and none of us were there to watch it. This level of detail is made possible by a wonderfully simplifying phenomenon called *thermal equilibrium*. Consider for example the ratio of protons to neutrons. We have no way of knowing the initial value of this ratio, but fortunately we have no need to care. Suppose, for example, that the universe started out with only neutrons and no protons. This situation would not last long, as some of the neutrons would convert to protons by colliding with neutrinos or positrons in the hot cosmic soup, as illustrated on the left side of Figure 5.2. The density of protons would increase and the density of neutrons would drop. As the density of protons builds up, so will the frequency of collisions between protons and other particles. Some of these collisions will convert the protons back to neutrons, as shown on the

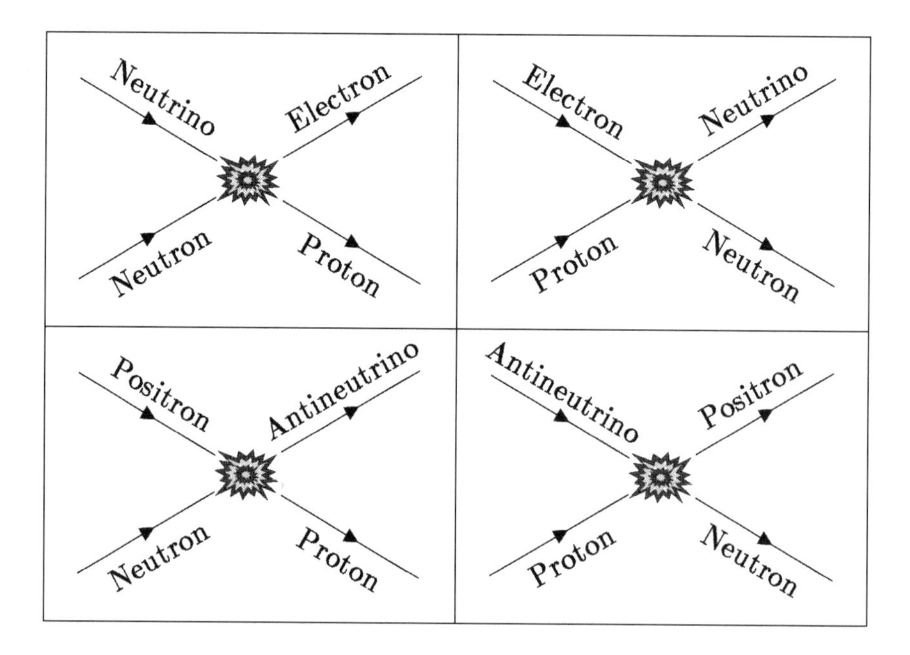

***Figure 5.2* Interconversion of protons and neutrons.** In the hot gas of the early universe, neutrons can be converted to protons by the processes shown on the left. The processes on the right can convert protons into neutrons.

right side of Figure 5.2. If enough time is allowed, then the density of protons will grow until the rate of the proton-destroying reactions, on the right half of Figure 5.2, is equal to the rate of the proton-creating reactions shown on the left half. The density at which these rates match is called the *thermal equilibrium density.* The thermal equilibrium ratio of protons to neutrons at 31.5 billion degrees is 1.61, no matter where or when this situation might occur [1]. Thermal equilibrium leads to fewer neutrons than protons, because neutrons are a tenth of a percent more massive. One must check, of course, that there is enough time for thermal equilibrium to be established. Although the temperature in the early universe plummeted rapidly, the reactions shown in Figure 5.2 are rapid enough to assure thermal equilibrium at 0.1 seconds after the big bang.

By one second after the big bang, the temperature had fallen to 10 billion degrees, and two important changes began to take place. First, the rates for the reactions of Figure 5.2 slowed, with the decreasing temperature, to the point that thermal equilibrium could no longer be maintained. While the thermal equilibrium ratio of protons to neutrons had risen to 4.49:1, calculations show that the actual ratio lagged behind, at 3.18:1. The reactions shown in Figure 5.2 became slower and slower from this time onward, and after about fifteen seconds their effect can almost be ignored. Afterward neutrons continued to convert to protons, but very slowly, by the decay process shown in Figure 5.3. Since neutrons are essential in building any

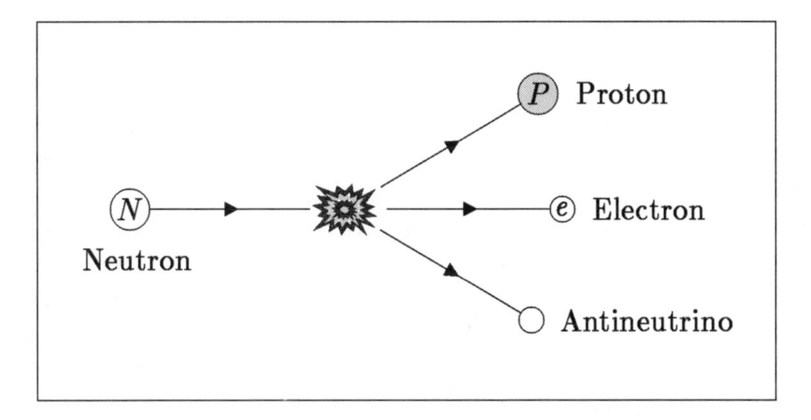

***Figure 5.3* The decay of a free neutron.** The half-life for this decay is difficult to measure. When Gamow began his calculations in 1946 it was believed to be about an hour, but in the next several years the estimate was reduced to 20.8 minutes and then 12.8 minutes. Today the half-life is known to be about 10.3 minutes.

element beyond hydrogen, their shifting abundance is a crucial factor in the element production that will follow.

The second important change concerned the electron-positron pairs, which at this time accounted for about a third of the mass density of the universe. By now, however, the universe had cooled enough so that collisions energetic enough to produce electron-positron pairs were becoming rare. Since electrons and positrons could annihilate when they collided, the electron-positron pairs began to disappear. The disappearance was a slow one, however, so it was not until 30 seconds that half the pairs had vanished. The other particles present in great abundance at this time—photons, neutrinos, and antineutrinos—are believed to have survived to the present day. The photons constitute the cosmic background radiation that we see today. The neutrinos and antineutrinos are believed to form a cosmic neutrino background which also fills the present universe. The neutrinos are predicted to be a little colder than the photons, just 1.95°K, because the photons were heated when the electron-positron pairs annihilated. The neutrinos, on the other hand, interact so weakly that they failed to absorb any of the electron-positron heat. While high energy neutrinos are now routinely detected in accelerator experiments, the interactions of neutrinos become much weaker at low energy; consequently, no one has any idea how to detect the feeble neutrinos of the cosmic background, even though we believe that there are over 300 of them in every cubic centimeter. The detection of these neutrinos, if it ever happens, would be a spectacular confirmation of the hot big bang picture of the universe.

Nucleosynthesis started at one second in the sense that the proton to neutron ratio began to deviate from its thermal equilibrium value—a fact that is crucial in determining the abundances of the chemical elements that are later produced. The actual formation of nuclei, however, did not begin until much later. The main nuclear reactions are shown in Figure 5.4. By about 15 seconds the universe had cooled enough so that helium-4 nuclei could have held together if they were formed, but helium-4 is not at the beginning of the chain. The chain starts with the production of deuterium, so the formation of the elements could not get underway until the universe cooled enough for deuterium to be stable. Since deuterium is an extremely fragile nucleus, this "deuterium bottleneck" delayed nucleosynthesis until about three and a half minutes after the big bang.

The earliest nucleosynthesis calculations were performed by George Gamow of George Washington University, his collaborators Ralph A. Alpher and Robert Herman, and the Japanese physicist Chushiro Hayashi. Born in Odessa, Russia, in 1904, Gamow [2] was educated at the University of Leningrad, and then spent three years abroad at the University of Göttingen in Germany, the University of Copenhagen, and Cambridge University.

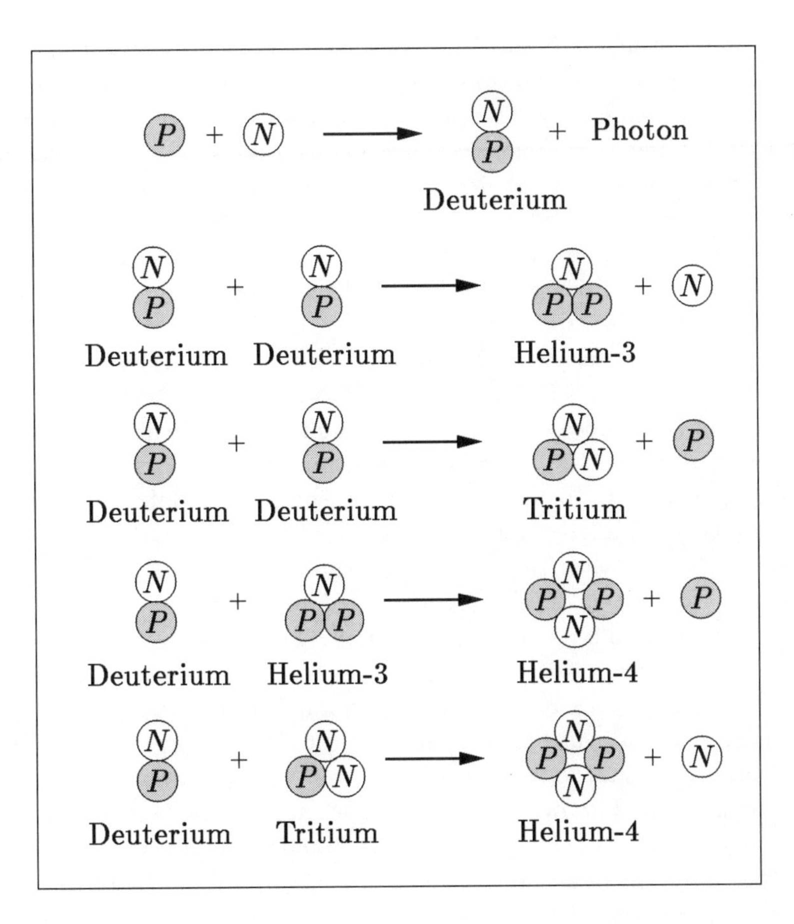

Figure 5.4 **Nuclear reactions for helium production.** The diagram shows the major nuclear processes contributing to modern calculations of big-bang nucleosynthesis. In addition, a trace amount of lithium-7 is produced by other processes.

Returning to Russia in 1931, he found that the earlier spirit of international cooperation in science had been replaced by an atmosphere of rigid dialectical materialism, in which it had become a crime for Russian scientists to fraternize with scientists of the capitalistic countries. Determined to escape this situation, in the summer of 1932 he and his wife Rho* attempted to paddle 170 miles in an inflatable boat across the Black Sea from the

* Her full name was Lyubov Vokhminzeva, but while a physics student she took the nickname from the Greek letter.

Crimean peninsula to Turkey, but were driven back by a storm. Later Niels Bohr began to worry about his old friend, so he enlisted the help of Paul Langevin, a well-known French physicist with Communist connections, who was serving as the chairman on the French side of the Franco-Russian Scientific Cooperation Committee. Although Gamow did not know it at the time, Langevin contacted Moscow and asked the government to appoint Gamow as a delegate to the XIIIth Solvay Congress in Brussels. Gamow was initially promised a passport for his wife as well, but then the second passport was denied. In a spirit of brash defiance, he refused to go to the conference without his wife, and at the last minute his request was for some reason granted. In October, 1933, George and Rho boarded a train in Leningrad for Helsinki, on the way to Copenhagen and Brussels. It was not until after the conference that Gamow and his wife learned how their exit from Russia had been arranged. To their horror, they learned that Bohr had intended only a visit, and had given a personal guarantee to Langevin that they would return to Russia when the conference was over. After a brief panic, the Gamows' freedom was rescued by Madame Marie Curie, who spoke to Langevin and arranged for them to remain in the West.

Gamow, often described as the father of the big bang, was at the very least the driving force behind big bang research in the 1940s and 1950s. Energetic and fun-loving, he was perhaps the only scientist ever described in the American Physical Society magazine *Physics Today* as "swashbuckling [3]." He was also a well-known popularizer of science, author of such classics as *One, Two, Three . . . Infinity** and the *Mr. Tompkins* series (——— *in Wonderland*, ——— *Explores the Atom*, ——— *Learns the Facts of Life*, and ——— *Inside Himself*).

In 1946 Gamow [4] suggested that the early universe was filled with a hot gas of free neutrons, a substance that appeared to him to be the simplest possible starting point. Gamow proposed that the decay of these neutrons and the subsequent nuclear reactions would form all the nuclei that we see in the universe today, from hydrogen to uranium. Although this approach is at odds with the modern view that elements heavier than lithium were produced mainly in stars, the idea was perfectly reasonable at the time.

In a sequence of papers over many years, Gamow and his collaborators developed a description of the early universe which increasingly resembled the

* While most of Gamow's limericks are described by his friends as "unprintable," the following verse from the title page of *One, Two, Three . . . Infinity* gives an example of "Gamowian" humor:

> There was a young fellow from Trinity
> Who took the square root of infinity
> But the number of digits
> Gave him the fidgets;
> He dropped Math and took up Divinity.

presently accepted picture.* In 1948 Gamow [5] realized that his hot neutrons would produce a blackbody spectrum of photons, and Alpher and Herman soon pointed out that these photons would survive as a radiation background in the present universe. They estimated the temperature today to be "of the order of 5°K," a speculation that proved to be surprisingly accurate [6].† They did not, however, recognize that this vestige of the big bang could be identified by its unique spectrum, which would survive to the present day; instead they suggested incorrectly that the cosmic photons would mix with the much hotter starlight to produce blackbody radiation at some intermediate temperature. Alpher and Herman apparently discussed this prediction with a number of radar experts [7], but were told that a signal at such a low temperature could not be detected with equipment then available.

As Gamow, Alpher, and Herman refined their calculations, they gradually became more aware of a crucial problem: Big bang nucleosynthesis essentially stops with the nucleus helium-4, consisting of two protons and two neutrons. The problem is that there is no stable nucleus with five particles, so the reactions have to produce an improbable jump from a four-particle nucleus to one with six or more particles. In 1952, Gamow wrote in *The Creation of the Universe* that the 5-particle gap remained an unsolved problem, but he was still hopeful that the problem would disappear when more careful calculations were carried out.

* Ralph Alpher began his cosmological career as a part-time graduate student working under Gamow at George Washington University. He was working in the evening division to complete his doctorate while he also had a full-time job at the Applied Physics Laboratory of Johns Hopkins University. Robert Herman, also working at the Applied Physics Laboratory, soon joined the collaboration. When it came time to publish Alpher's thesis work, Gamow's impish mind became fascinated with the notion that the names of the two authors so closely resembled two of the first three letters of the Greek alphabet (α alpha, β beta, and γ gamma). Since the author list could plausibly have included Hans Bethe (pronounced "beta"), a Cornell physicist who later won the Nobel Prize for his work on nuclear reactions in stars, Gamow took the liberty of adding his name to the submitted manuscript, with the words "in absentia" included parenthetically alongside. The editors at the *Physical Review* naturally had no idea what to make of this peculiar designation, so the words "in absentia" were dropped, and the paper was published under the names Alpher, Bethe, and Gamow, coincidentally on the date of April 1, 1948.

† Alpher and Herman's 5°K estimate of the cosmic temperature today depended on knowing the mass density of the universe, a very uncertain number which they took from a 1937 paper by Hubble. It also depended on their theory of nucleosynthesis, which attempted incorrectly to attribute the production of all elements to the big bang. Given the crudeness of this input, one might wonder how they obtained a result so close to the correct value of 2.7°K. In part, it was good luck. Their description of the calculation starts with the statement that "the temperature during the element [formation] process must have been of the order of 10^8–10^{10}°K," allowing a full two orders of magnitude of uncertainty. Then "for purposes of simplicity" they choose a value of 6×10^8°K, which gives a nice round number for the equivalent mass density, exactly one gram per cubic centimeter. With this choice, they obtained 5°K. Their original range of temperatures, however, would have given a temperature today somewhere between 0.8°K and 80°K. I mention this not to detract from the importance of the work of Alpher and Herman, but to highlight the important uncertainties in big-bang nucleosynthesis that existed in this period.

The pinnacle of the efforts by the Gamow group was the 1953 paper by Alpher, Herman, and James W. Follin, Jr. [8]. Incorporating a number of important refinements that were introduced by Hayashi [9] in 1950, they introduced a few refinements of their own and employed the latest measurements of nuclear physics. They discussed the shifting proton-to-neutron ratio at a level of detail that closely resembles the best modern treatments. They stopped short, however, of completing the description of element formation, since they realized that the 5-particle gap would prevent them from going beyond helium-4. They expressed hope that this problem would go away if all possible reactions were included in the calculation, but they apparently gave up in despair. While cosmologists eventually came to accept big-bang nucleosynthesis as a description of light element formation only, the Gamow group seems to have abandoned their efforts when it became clear that it would not be possible to explain the abundances of *all* the elements by this approach.

The 5-particle gap was not only a problem for big bang theorists—it was also an obstacle for theorists trying to explain how elements might be synthesized in stars. Hans Bethe, in his classic 1939 paper on the energy production in stars [10], concluded that "under present conditions, no elements heavier than helium could be built up to any appreciable extent" in stellar interiors. Nevertheless, Martin and Barbara Schwarzschild (a husband and wife team) showed [11] in 1950 that the abundance of heavy chemical elements is significantly higher in older stars than in younger stars. In light of this strong observational evidence for element production in stars, the 5-particle gap was reconsidered. By this time more was known about the variety of stages in stellar evolution, and in 1952 Edwin E. Salpeter, a young theorist at Cornell and frequent collaborator of Bethe's, discovered [12] that the 5-particle gap can be bridged in the extremely hot dense core of a dying star in its explosive red giant phase. Salpeter found that under these conditions an unstable intermediate state could survive just long enough for the reactions to proceed. In 1957 a very thorough study of the synthesis of heavy elements in stars, particularly in supernovas, was carried out by Geoffrey and Margaret Burbidge (another husband and wife team), William Fowler, and Fred Hoyle [13].

By the end of the 1950s, astrophysicists had come to believe that all the elements were synthesized in stars, and the work on big-bang nucleosynthesis by Gamow and his collaborators was nearly forgotten. Admitting defeat in his own inimitable style, Gamow wrote a version of *Genesis*, which is included in his autobiography, *My World Line*. Starting with "In the beginning God created radiation and ylem," where *ylem* was Gamow's term for the primordial matter of the big bang, the passage goes on to describe how God called for all of the elements, but

> In the excitement of counting, He missed calling for mass five and so, naturally no heavier elements could have been formed.

> God was very much disappointed, and wanted first to contract the Universe again, and to start all over from the beginning. But it would be much too simple. Thus, being almighty, God decided to correct His mistake in a most impossible way.
>
> And God said: "Let there be Hoyle." And there was Hoyle. And God looked at Hoyle . . . And told him to make heavy elements in any way he pleased.
>
> And Hoyle decided to make heavy elements in stars, and to spread them around by supernova explosions.

Oddly enough, it was Hoyle himself, one of the chief architects of both the steady-state universe and stellar nucleosynthesis, who put an end to the notion that ordinary stars can be solely responsible for element production. In 1964 Hoyle and Roger J. Tayler, both at Cambridge University, wrote a classic paper titled "The Mystery of the Cosmic Helium Abundance [14]." If all the observed helium were synthesized from hydrogen during the lifetime of the galaxy, the authors explained, the energy released by the synthesis would make the galaxy about ten times brighter than it actually is. Their conclusion was stated plainly: "helium production in ordinary stars is inadequate to explain" the observed abundance. They then went on to consider a "radiation origin" of the universe, the theory that we know as the big bang. (While Hoyle himself had coined the term "big bang" in radio broadcasts in the 1940s, at this time neither he nor anyone else would use such a silly phrase in a dignified scientific publication.) Referring back to the papers of the Gamow group, Hoyle and Tayler attempted to extend the calculations of Alpher, Follin, and Herman through the period of nucleosynthesis. Settling for approximate calculations which they admitted "must be repeated more accurately," they concluded that 36% of the mass of the universe would have been transformed into helium in the big bang. Had they carried out the calculations more carefully they might have used the observed helium abundances, typically about 25%, to predict the temperature of the cosmic background radiation. The cosmic background radiation was not their concern, however, and they made no comment whatever about what would happen to the radiation fireball after the period of nucleosynthesis.

Our story has now reached the eve of Penzias and Wilson's discovery, so the next episode in the theoretical development is the work of Jim Peebles that was discussed in the previous chapter. When Peebles submitted his first paper on big-bang nucleosynthesis in March 1965, he was essentially unaware of the earlier work by Gamow and his collaborators, although he did cite a review article by Alpher and Herman. Peebles is not sure when he first learned of the Gamow work, but he thinks that after submitting his first paper he heard about the Hoyle-Tayler paper from a friend. This paper contained references to the earlier work, and he remembers "feeling keen disappointment on seeing that this was not a new idea [15]."

By the beginning of May, 1965, the Penzias and Wilson paper announcing the discovery of the cosmic background radiation was off to the *Astrophysical Journal.* At the same time the Princeton group—Dicke, Peebles, Roll, and Wilkinson—submitted their paper on the theoretical interpretation, which summarized the results in Peebles' unpublished paper. Despite statements in a number of books to the contrary, the paper by Dicke, Peebles, Roll, and Wilkinson *did* reference the earlier work by Gamow and his collaborators. In summarizing the nucleosynthesis calculations used to estimate the expected temperature, the Princeton authors explained that "This was the type of process envisioned by Gamow, Alpher, Herman, and others," with citations to several of the key papers [16]. This understated reference was not enough to satisfy Gamow, however, who wrote an angry letter to Arno Penzias citing the places in which he and his collaborators had "predicted" the existence and temperature of the background radiation, concluding with the sentence: "Thus, you see the world did not start with almighty Dicke."

While Peebles was not the first to study big-bang nucleosynthesis, he was the first to complete the calculation of helium production with the same level of detail with which Alpher, Follin, and Herman had begun it. Unlike his predecessors, he knew from the beginning that heavier elements can be synthesized in stars, so he was not disheartened to learn that element formation stops at helium. Beginning with a calculation of the shifting neutron-proton balance that closely paralleled the work of Alpher, Follin, and Herman, Peebles extended the calculation to include the nucleosynthesis reactions shown in Figure 5.4.

Peebles also seems to have been the first person to trace the evolution of the blackbody radiation to the present. (The Gamow group had computed the redshifting of the radiation, but they had not examined the interaction of the radiation with the matter of the universe.) Peebles found that until the radiation temperature fell to 3000°K, the hydrogen gas was so hot that atoms would not form. The gas remained ionized, meaning that the electrons and protons moved independently through space. Since photons interact strongly with charged particles, especially charged particles with a small mass, the photons were constantly scattered by collisions with the electrons. The frequent collisions assured that the matter and radiation stayed at the same temperature, cooling together as the universe expanded.

After about 300,000 years the universe cooled enough for the ionized gas to convert to neutral atoms. Peebles called this process "recombination," a term still standard in cosmological literature. The prefix "re-" always seemed out of place, however, since according to the big bang theory the electrons and protons were combining for the first time ever. I asked Peebles if the term was a vestige of Dicke's belief in an oscillating universe, but he said no. "Recombination" is the word used by physicists who study ionized gases (also called plasmas) under laboratory conditions, so naturally the name was transported to cosmology.

A gas of electrically neutral atoms is very transparent to photons, so a typical photon in the cosmic background radiation has traveled on a straight line from 300,000 years after the big bang until the present. The blackbody spectrum would be maintained as the radiation redshifted, with the temperature falling as the universe expands. Starlight would not interact with the blackbody radiation; photons would be added at much higher frequencies, but the spectrum in the microwave region would be almost completely unaffected. The cosmic background radiation, therefore, gives us effectively a snapshot of the universe just 300,000 years after the big bang.

To appreciate the significance of a time as early as 300,000 years, consider an analogy that was used in a 1967 *Scientific American* article [17] by Peebles and Wilkinson. Compare our observations of the evolving universe with the view downward from the observation platform of the Empire State Building. Street level corresponds to the instant of the big bang. Updating the numbers from the original, the most distant galaxies discovered so far correspond to a view down to the 10th floor, and the most distant quasars are at about the 7th floor. The cosmic background radiation is equivalent to a glimpse of something just half an inch above the street!*

The calculations of big-bang nucleosynthesis reached a new level of sophistication in 1967, when Robert V. Wagoner, Fowler, and Hoyle [18] wrote an intricate computer code that incorporated 144 reactions involving all nuclei up to and including sodium-23 and magnesium-23. They found agreement with Peebles for helium-4 and deuterium production, and they obtained for the first time predictions for helium-3 and lithium-7 production.

Nucleosynthesis remains to this day an important topic in cosmology, with the predictions changing slowly with time as the measurements of the reaction rates are refined. A large part of the recent work on nucleosynthesis has been carried out by David N. Schramm of the University of Chicago, Michael S. Turner of the Fermi National Accelerator Laboratory and the University of Chicago, Gary Steigman of Ohio State University, and several other collaborators.

* The authors attributed the analogy to their colleague John A. Wheeler, a well-known general relativist whose other contributions to scientific imagery include the coining of the phrase "black hole." The analogy also helps to illustrate the spectacular progress in astronomy since 1967, when the most distant galaxies were up at the 60th floor, and the most distant quasars were on the 20th.

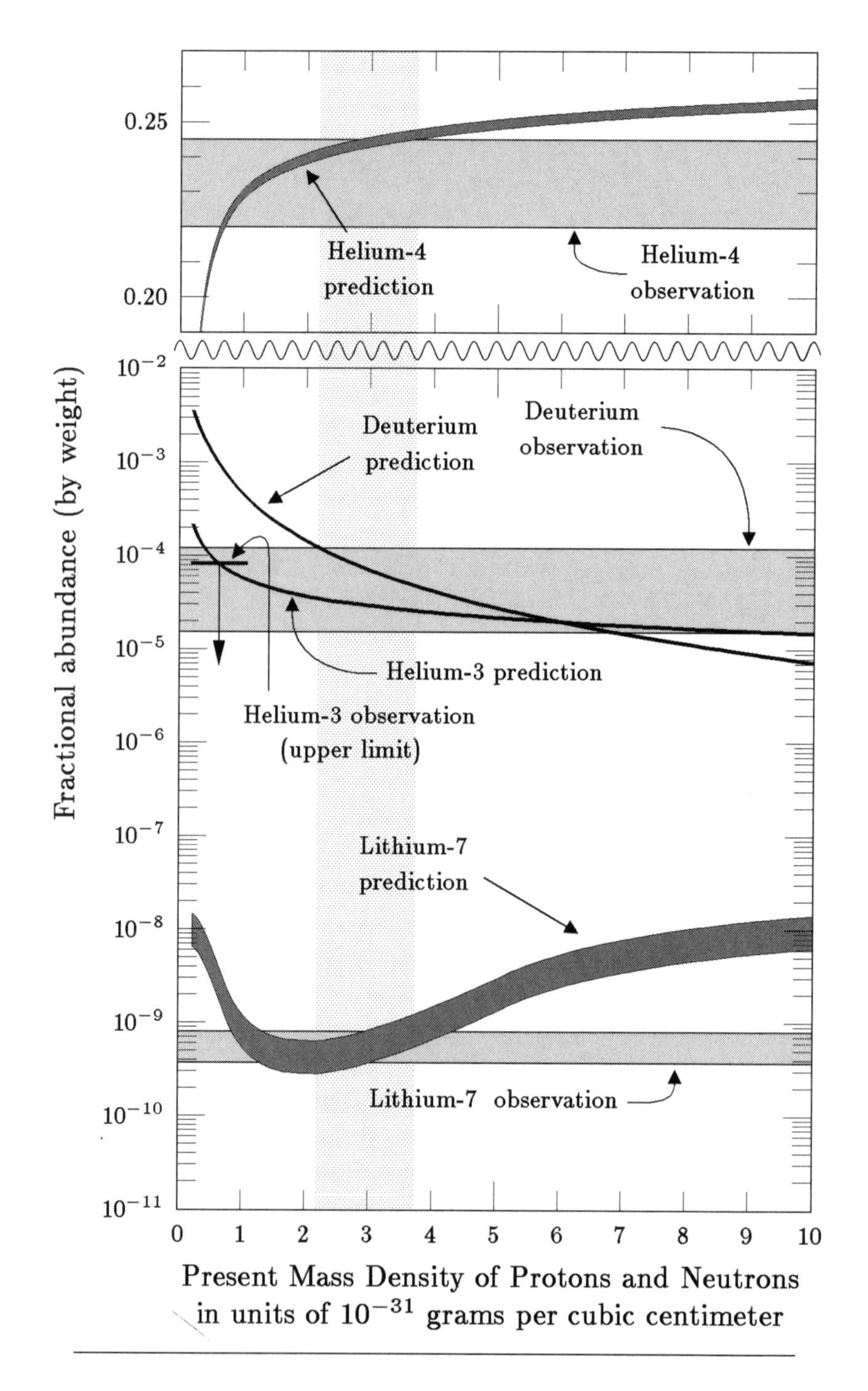

The status of modern nucleosynthesis calculations is shown in Figure 5.5 [19]. The graph shows how the predicted abundances depend on the density of protons and neutrons in the universe. The more protons and neu-

Figure 5.5 **(facing page) Big-bang nucleosynthesis.** The graph shows the predictions and observations for the quantities of the light chemical elements that were produced in the big bang. The predictions (but not the observations) depend on the present mass density of protons and neutrons in the universe. Included on the graph are those nuclei believed to have been produced primarily in the big bang: helium-4 (2 protons and 2 neutrons), deuterium (1 proton and 1 neutron), helium-3 (2 protons and 1 neutron), and lithium-7 (3 protons and 4 neutrons). The abundances are expressed as a fraction of the weight of all known matter in the universe. The fractions do not add to one, because hydrogen is not shown; the big bang production of elements heavier than lithium is believed to be totally negligible. The predictions for helium-4 and lithium-7 are shown as thick lines, indicating the uncertainty caused by the estimated error in the measured reactions rates that are used in the calculation. The uncertainties in the deuterium and helium-3 predictions are not shown, since they are negligible compared to the uncertainty in the observations. The observed abundances of helium-4, deuterium, and lithium-7 are shown as horizontal shaded bars, with a thickness that indicates the uncertainty in these observations. Observations of helium-3 produce only an upper limit on the abundance produced by the big bang, as indicated by the short horizontal line, with a downward arrow extending from it. For the range of densities indicated by the lightly shaded vertical bar, every prediction is in agreement with the corresponding observation.

---

trons there are, the more frequently they collide, and the more helium-4 is produced. The predicted abundance of deuterium and helium-3 goes down with increasing density, however, since these nuclei are formed by an incomplete sequence of reactions. If the processes shown in Figure 5.4 were given a chance to continue, then all of the deuterium and helium-3 would be converted into helium-4. Lithium is produced by several reactions, resulting in a more complicated curve on the graph.

It is very difficult to astronomically estimate the density of protons and neutrons in the universe, so the nucleosynthesis calculations must be carried out for a range of possible values. For the range shown by the vertical gray band on Figure 5.5, the agreement between the predictions and the observations is excellent. While none of these are high-precision measurements, the fact that the abundances span such a large range of values makes the agreement very impressive. If we had no theory of big-bang nucleosynthesis, it would seem perfectly plausible for 25% of the mass to be lithium, and for only one part in a billion to be helium. The theory, however, predicts that it has to be helium that is 25% and lithium that is one part in a billion, and the observations agree. Most cosmologists see the results of big-bang nucleosynthesis as a strong piece of evidence in favor of the big bang picture. Not only are the qualitative features confirmed, but the success of the predictions suggests that the details of the theory are accurate all the way back to one second after the big bang.

Besides helping to confirm the big bang theory, nucleosynthesis also gives the best estimate of an important property of the universe: the density of protons and neutrons. If the density of protons and neutrons were less than 2×10^{-31} grams per cubic centimeter, then the universe would have to contain more deuterium than it does. If the density of protons and neutrons were greater than 5×10^{-31} grams per cubic centimeter, then it would have to contain more helium-4 and lithium-7.

Cosmologists usually measure the mass density in comparison to the critical density, the density that would create a gravitational field just strong enough to eventually halt the expansion of the universe. The value of the critical density depends, however, on the Hubble constant, which is not very well known. As mentioned in Chapter 3, the Hubble constant is believed to lie somewhere in the range of 15–30 kilometers per second per million light-years. For this range of values, the critical density is somewhere between 4.5×10^{-30} and 1.8×10^{-29} grams per cubic centimeter. Thus, the big-bang nucleosynthesis calculations imply that the density of protons and neutrons is at most 11% of the critical density, and it could be as little as 1.1%.

As will be discussed later, the inflationary model predicts that the total density of matter should equal the critical density, so there is a potential conflict with the results of big-bang nucleosynthesis. There is also, however, astronomical evidence that strongly suggests (but does not prove) that the mass density is larger than 11% of the critical density. Unfortunately, it is very difficult to measure the mass density of the universe, because at least 90% of the mass is *dark;* i.e., it is invisible, and can be detected only by its gravitational pull on visible matter. In order to reconcile big-bang nucleosynthesis with inflation and the astronomical indications of a high mass density, we will have to assume that some or all of the dark matter is not composed of protons and neutrons. If inflation and big-bang nucleosynthesis are both correct (and I believe that they are), then most of the universe must be made from some substance that is not yet identified, and is perhaps not even known.

CHAPTER 6

MATTERS OF MATTER AND ANTIMATTER

During the fall of 1978, Henry Tye continued to try to interest me in the question of grand unified theories and magnetic monopole production in the early universe. I continued to put him off, partly because the project sounded ill-defined, and partly because I was busy with other projects. I still had no idea that Henry's project would lead to anything more than dubious conclusions founded on questionable hypotheses. The following spring, however, a visit to Cornell by a highly respected physicist would help to overcome my skepticism, sending my career in a new direction.

In hindsight my reluctance to work with Henry was foolish, but at the time there were other issues that seemed important. While grand unified theories were gaining momentum elsewhere, the attention of the theory group at Cornell was focused on another important development, *lattice gauge theory,* a method of studying the forces between quarks, that had been invented in 1974 by Cornell's Kenneth Wilson. Although we did not know that four years later Ken would win the Nobel Prize, it was easy to see that he was extremely creative and influential. And lattice gauge theory was clearly a promising assault on the vexing question of how quarks bind together to make protons, neutrons, and other particles. In the fall of 1978 I was actively working on a lattice gauge theory project that would eventually lead to two publications. I thought this work was important, and the citation history suggests that I was right—each of these papers has been referenced about a hundred times, far more than any paper I had previously written.

I was still concerned about jobs as well, since I could not afford to miss an opportunity to apply for an assistant professorship. I responded to virtually every job advertisement in my field. Since job applications require letters of recommendation, I went to my senior colleagues with the usual requests. I was crestfallen when one Cornell theorist, whom I thought I had favorably impressed, informed me that the strongest letter he could honestly write would be "kindly, but vague." And one Columbia physicist told me that he had already agreed to recommend two theorists currently at Columbia for many of these positions, and did not feel that he could recommend me as well. It was not all grief, however, for several of my senior colleagues were distinctly encouraging. The bottom line, however, was a disappointment. I still have all the applications, and all the form letters with politely worded rejections that arrived sporadically during the coming spring.

While I was preoccupied with lattice gauge theory and jobs, Henry continued to be interested in the monopole question, and continued to remind me about it. He also was working on other things, however, so he did not actively pursue the problem.

Late the following April (1979), my attitude was changed by a visit to Cornell by Steven Weinberg, a leading theoretical physicist whose style I had grown to admire during my graduate student days at MIT. He had since moved to Harvard, and later would move to the University of Texas at Aus-

Steve Weinberg

tin. Steve gave two lectures on how grand unified theories can perhaps explain why the universe contains an excess of matter over antimatter.

By matter, Weinberg meant material like what we see around us: material that is made from the basic building blocks of protons, neutrons, and electrons. Antimatter refers to material constructed from antiprotons, antineutrons, and positrons (the antiparticles of electrons). An antiproton has the same mass as a proton but the opposite electric charge; if a proton and antiproton come together they can annihilate into pure energy, usually given off in the form of high energy photons, called gamma rays. An electron and positron similarly have the same mass but opposite charge, and also can annihilate into gamma rays. A neutron and antineutron both have the same mass and zero electric charge, but can be differentiated by their interactions: a neutron and an antineutron can annihilate into gamma rays, while two neutrons cannot.

Protons and neutrons are collectively called *baryons*, while antiprotons and antineutrons are called *antibaryons*. (There are also other types of baryons and antibaryons, which will be discussed in the next chapter. Since these particles decay in a small fraction of a second, however, they will not be relevant to the present discussion.) Weinberg was attempting to explain the number of baryons and antibaryons in the observed universe—the part of the universe that we can actually see. Roughly speaking, the observed universe is a sphere centered on our planet, with a radius of about 10 billion light-years. (We do not know, at present, how large the entire universe is. It could extend far beyond the observed universe, and might even be infinite.) Weinberg noted that the observed universe shows a startling asymmetry, containing about 10^{78} baryons but only a negligible number of antibaryons.*

* It is assumed that the distant stars and galaxies are all made of matter, but how can one know? It is not easy to tell, since the appearance and even the spectrum of an antimatter star or galaxy would look exactly like a star or galaxy made of matter. If the universe does contain regions of antimatter, then interstellar spaceships of the future will have to be on guard to avoid the immediate obliteration that would result from straying into such a region. Nonetheless, it appears that whatever dangers might threaten our space-traveling descendants, the calamity of antimatter annihilation will not be one of them. Cosmic rays arriving at the Earth are essentially all matter and not antimatter, indicating that our galaxy and probably neighboring galaxies as well are all made from matter. The space between galaxies, furthermore, is not completely empty, but instead is permeated by a tenuous gas. If separate regions of matter and antimatter existed, then the gas at the interface of these regions would undergo annihilation, emitting an identifiable spectrum of gamma rays. The density of gas in nearby clusters of galaxies is high enough so that the annihilation radiation would be detectable, but no sign of such radiation has been found. Thus, at least within nearby clusters of galaxies, the material is either all matter or all antimatter. Observation does not exclude the possibility that some clusters are matter while others are antimatter, but theorists argue that this option is unworkable. Although many cosmologists have tried, no one has succeeded in finding a sensible mechanism to separate matter and antimatter into clumps of such enormous proportion. Most cosmologists are convinced, therefore, that all the material in the universe is matter, and not antimatter.

Weinberg went on to explain that this dramatic imbalance between matter and antimatter can be traced back to a very tiny imbalance in the early universe. At 10^{-6} seconds after the big bang the universe was so hot (10^{13}°K) that protons and neutrons could not exist, so we would find only the quarks that would later combine to form baryons. The number of quarks in the observed universe was much higher then, about 10^{87}, with the number of antiquarks almost exactly equal. For every 300,000,000 quarks, there were 299,999,999 antiquarks. The quarks and antiquarks started to annihilate as the universe cooled, and since the production of quark-antiquark pairs by particle collisions decreased rapidly as the temperature fell, the pairs were not replenished. Essentially all the pairs disappeared, leaving only a single unpaired quark for every 300,000,000 quarks that were initially present. These unpaired quarks rapidly formed into the baryons that we observe today. Thus, the small excess of quarks over antiquarks in the early universe was crucial to our very existence: if the hot soup of the early universe were a perfectly equal mix of quarks and antiquarks, then their rapid annihilation would have led to a universe devoid of stars and planets. If the fraction of unpaired quarks had differed appreciably from one in 300,000,000, then the nucleosynthesis processes discussed in Chapter 5 would have been affected, producing a universe with a radically different chemical composition.

From the time that antiprotons were first produced in the Bevatron accelerator at Berkeley in 1955, physicists have found that antibaryons are produced only in combination with baryons. To express this succinctly, physicists define a quantity called *baryon number*, which is the number of baryons in a system, minus the number of antibaryons. In observed reactions, the total baryon number is conserved. A high-energy collision of two protons, for example, with a baryon number of two, can result in a final state containing three protons and one antiproton, which also has baryon number two. But it cannot result in a final state of two protons and one neutron, which would have a baryon number of three.

Since we believe that the observed universe has a baryon number of 10^{78}, the conservation of baryon number would imply that it *always* had a baryon number of 10^{78}. If the conservation of baryon number can never be violated, there would be no hope of explaining this number. We would be forced to believe that the baryon number was established when the universe was created, and nothing in the present framework of physics gives us a clue about how to understand such a number. Why should nature have chosen 10^{78}, rather than 6 or 4000 or 10^{4000}?

Experimentally, the conservation of baryon number is as well demonstrated as any conservation law in physics. Although physicists have looked very hard, no one has ever observed a process in which the baryon number has changed. One process that has been studied with enormous effort is the possible decay of the proton. Since the proton is the lightest baryon, any conceivable decay would produce products with baryon number zero. If baryon number is conserved, it follows that the proton must be absolutely stable. No experiment can ever prove absolute stability, but by 1979 it was known that the proton is either stable or has a half-life longer than 10^{30} years. Since 1979 the limit on the half-life has been increased to more than 10^{32} years! Since the age of the universe is only about 10^{10} years, these numbers are fantastic. If the proton does not live forever, it is fair to say that it lives at least "almost" forever.*

Weinberg's lectures, however, emphasized that grand unified theories suggest a totally new perspective on the baryon number of the universe. In grand unified theories, baryon number is not *exactly* conserved. The theories are consistent with the observations, predicting that under normal conditions baryon-number changing processes are extremely rare. At extraordinarily high temperatures, however, in the vicinity of 10^{29}°K, processes that violate baryon number become very common. At this temperature the average thermal energy of a particle is about 10^{16} GeV, which is a typical energy scale for the new particles and new effects that are introduced by grand unified theories. According to standard cosmology, the universe had a temperature of this order at about 10^{-39} seconds after the big bang. Grand unified theories open the possibility, therefore, that the baryon number of the universe no longer has to be postulated. If baryon number is not conserved, then the universe might have started out with zero baryon number. The observed baryon number could have been created, by baryon-number nonconserving processes, during the first 10^{-39} seconds of the history of the universe! Such a process is called *baryogenesis*.

As Weinberg explained, however, to understand baryogenesis we need to know more than the mere fact that baryon number is not conserved. There are two additional conditions that must be met:

1. The underlying laws of physics must contain a fundamental asymmetry between matter and antimatter.

2. The particle reactions must be slow, compared to the expansion and cooling of the universe, so that equilibrium densities are not achieved.

* Since the limit on the half-life is 10^{22} times longer than the estimated age of the universe, obviously no one has observed a proton for that long. A limit in this range can be set, however, by observing a huge number of protons. The largest proton decay detector in the world, the Irvine-Michigan-Brookhaven detector at the bottom of the Morton-Thiokol salt mine near Fairport, Ohio, monitors about 2×10^{33} protons contained in a tank of 9,000 tons of purified water.

Without the first condition—a fundamental asymmetry between matter and antimatter—particle reactions would be just as likely to produce antibaryons as baryons. Chance alone would produce an imbalance in their numbers, but not nearly enough to explain the observed baryon excess. In 1958 Alpher and Herman [1] showed that the statistically expected imbalance would be less than the number of baryons in a single star. Their result was actually very conservative; modern cosmologists would use the same argument, with a more reasonable estimate for the size of the observed universe, to conclude that the random baryon excess would not even be enough for a large asteroid. To explain the observed baryon number, the laws of physics must provide that baryons are produced with a higher probability than antibaryons.

Since the properties of matter and antimatter are essentially identical, it seems strange to even suggest that there might be some small difference between the two. It would be as if nature had attempted to be completely symmetric, but had bungled some fine detail. Nonetheless, a small but undeniable difference between matter and antimatter was observed in 1964 at the Brookhaven National Laboratory by James H. Christenson, James W. Cronin, Val L. Fitch, and René Turlay [2]. They observed the effect in the decay of a particle called the neutral *K* meson, or kaon. The discovery was unexpected, extremely surprising, and rapidly verified by at least three other groups. The senior members of the team, Cronin and Fitch, were awarded the Nobel Prize in 1980, immediately after it became apparent that this result, combined with the theoretical edifice of grand unified theories, could provide the key to the matter-antimatter asymmetry of the universe. The asymmetry discovered in this experiment has since been incorporated into modern particle theories, including the grand unified theories. These theories can easily accommodate the asymmetry, yet they shed no real light on the puzzle of why nature has chosen to be ever so slightly asymmetric. In Cronin's Nobel lecture, he expressed the hope that "at some epoch, perhaps distant, this cryptic message from nature will be deciphered."

The second condition—the requirement that particle reactions be slow—is linked to the concept of equilibrium that was discussed in the previous chapter in the context of nucleosynthesis. If baryon number is not conserved and the first condition is met, then one can easily construct theories for which the production rate of quarks would exceed that of antiquarks. But, we have to ask how this process would terminate. A viable answer can be found if the reactions were slow compared to the cooling of the universe, because then the reactions could be stopped by the cooling after the desired baryon excess is attained. If the reactions were fast, on the other hand, then the quark excess would build up rapidly, leading to more frequent collisions of excess quarks with each other and with other particles.

Since baryon number is sometimes destroyed in these collisions, the rate of baryon number destruction will increase. A steady-state, or equilibrium, condition will be reached when the rate at which baryon number is destroyed is equal to the rate at which it is produced. The equilibrium value of the baryon number density is determined entirely by reaction rates, and is independent of the previous history.

Using underlying symmetry principles believed to hold in all particle theories consistent with both relativity and quantum theory, Weinberg showed that the equilibrium baryon number density is zero. A nonzero baryon number can be achieved only if the reactions are slow enough so that the cooling of the universe cuts them off prematurely, before they have time to reach equilibrium.*

Since the early universe was rapidly expanding and cooling, it is not hard to invent scenarios in which the equilibrium densities are not reached. Weinberg investigated the possibility that some of the superheavy particles predicted by the grand unified theories, with mass/energy in the vicinity of 10^{16} GeV, are unstable but long-lived. As the universe cools, these particles become less likely to be produced, so their equilibrium density drops precipitously. But if the half-life is long compared to the time required for the

* The argument was based on a symmetry called *CPT.* CPT symmetry is consistent with all experiments, and is predicted by relativistic quantum theory to be exact and unavoidable. The meaning of CPT symmetry can be understood most simply by imagining a motion picture of any physical process, such as a supernova explosion, the decay of an elementary particle, or the bouncing of a ball. Assume that the motion picture shows infinite detail, so that each elementary particle of the system can be seen and labeled. Now imagine showing the motion picture, but only after transforming it by the following three steps. First, change each particle label to the corresponding antiparticle, so "electron" labels are replaced by "positron," etc. Some particles such as photons are their own antiparticles, so these labels are to be left alone. Next, arrange the theater so that the movie screen will be viewed through a mirror, reversing left and right. Finally, show the movie backwards in time, so that the last frame is shown first, and the first frame is shown last. CPT symmetry is the statement that the viewer of the transformed filmstrip will see a sequence of events that are completely consistent with the laws of physics. (The name CPT is assembled from the physicist's names for the three steps of the transformation: C for *charge conjugation*, P for *parity*, and T for *time reversal.*) Thus, CPT symmetry implies that even if an antiproton is different from a proton, it will necessarily behave the same as a proton seen in a motion picture that is mirror-reflected and time-reversed.

To understand Weinberg's conclusion, suppose for the sake of argument that the equilibrium baryon number density were positive (i.e., more baryons than antibaryons). Imagine filming the equilibrium system, and transforming the film by the three-step procedure described above. The relabeling would imply that the transformed film shows a negative baryon number density. Since particle densities are time-independent once equilibrium is reached, the transformed (time-reversed) film will also show time-independent densities. The viewer of the transformed film would therefore conclude that she is watching an equilibrium system with negative baryon number density. However, since CPT symmetry guarantees that the transformed film shows valid physics, and since the equilibrium state is unique, this contradicts our supposition that the equilibrium baryon number density is positive. The only consistent assumption is that the system in equilibrium has zero baryon number density.

universe to cool, the particles will continue to exist in large numbers. When the particles finally decay, the system is far from equilibrium and a baryon excess can be established.

Finally, we come to the bottom line: can GUT baryogenesis explain the observed baryon number of the universe? That is, can GUT processes explain why the early universe contained one extra quark for every 3×10^8 quark-antiquark pairs? Answer: a definite maybe. The ambiguities begin with the fact that "grand unified theories" is a plural expression. There is no unique grand unified theory, but instead there is a large class of theories. Weinberg emphasized this point, although at the time many more presumptuous physicists were narrowing their attention to the earliest and simplest of the grand unified theories, the one proposed in 1974 by Howard Georgi and Sheldon L. Glashow [3] of Harvard. Even with the choice of a definite theory, however, there is still the problem of unknown parameters. Any theory contains some number of parameters that must be measured before the theory can be used to make predictions, and the grand unified theories contain several parameters that can be measured only by experiments at energies near 10^{16} GeV. Since baryogenesis depends on these parameters, nothing definite can be said. One finds, at least, that the correct baryon number prediction can be obtained for plausible values of the unknown parameters.

Weinberg's current work, in 1979, was by no means the first on this topic. In fact, Weinberg himself had forecast the possibility of baryogenesis in a remarkably prescient lecture given at the Brandeis Summer Institute in 1964 [4], when I was just finishing high school. In this lecture Weinberg pointed out that baryon number differs in two important ways from electric charge, which physicists firmly believe *is* conserved. First, electric charge creates a long-range electric field, while baryon number creates no long-range field whatever. The electric field is described classically by a set of equations first written by James Clerk Maxwell in 1864. Remarkably, these equations would become blatantly inconsistent if electric charge were not exactly conserved. Weinberg had pursued this issue from the point of view of modern relativistic quantum theory, finding deeper support for Maxwell's result: the existence of long-range fields must always be associated with an exactly conserved quantity. Thus, electric charge must be exactly conserved, while baryon number has no reason to be conserved. Second, Weinberg pointed out that "it is only necessary to look around us to see a clue that baryon . . . number may not be as exactly conserved as charge, for the visible part of the universe has a tremendous preponderance of baryons over antibaryons, . . . while its electric charge density appears to be extremely small." These arguments, according to Weinberg's 1964 lecture, "lead us to the strong suspicion that baryon . . . number [is] not precisely conserved. However, there seems no immediate possibility of testing this conjecture."

The earliest attempt to explain how the baryon excess could be generated by physical processes, based on a crude model of baryon-number nonconservation, was published in 1967 [5] by Andrei D. Sakharov, the well-known Soviet physicist and human rights advocate.* In 1978 Motohiko Yoshimura [6] of Japan became the first to apply grand unified theories to the calculation of the matter-antimatter asymmetry. Although his paper missed the necessity of a departure from equilibrium, it was this work that sparked the new wave of interest in grand unified theories and cosmology.

For me, Weinberg's visit had a tremendous impact. I was shown that a respectable (and even a highly respected) scientist could think about things as crazy as grand unified theories and the universe at 10^{-39} seconds. And I was shown that really exciting questions could be asked. Although the grand unified theories could not give a firm prediction for the baryon number of the universe, they at least provided a framework for calculation. Until this time the baryon number of the universe was taken as a given quantity, determined by initial conditions about which science had nothing to say. Now the number was in principle calculable. Some day, when we discover the correct grand unified theory and determine the values of its parameters, the baryon number of the universe will be calculated.

The day after Weinberg left, I went to Henry to talk about grand unified theories and magnetic monopoles. He gave me some papers to read on grand unified theories, and the two of us began to work in earnest on the production of magnetic monopoles in the early universe.

* My only meeting with Sakharov, when he visited MIT one Saturday afternoon in November, 1988, led to an embarrassing experience. Visits by Soviet scientists were still rare at that time, so the meeting was carefully orchestrated. Sakharov had sent word in advance that he wanted to talk physics, not politics. The MIT theoretical physics group and distinguished friends from neighboring institutions were there in full force. Sakharov arrived a few minutes early, and the seminar room instantly filled to capacity. The translation was handled by a Russian émigré who was visiting our group, Shimon Levit. Jeffrey Goldstone acted as chairman, and asked Sakharov what he wanted to do. Sakharov responded that he had nothing to say, but wanted to bring news of the exciting developments in physics back to Moscow. He mentioned several topics including the inflationary universe, so I was coaxed to the blackboard to give the first impromptu presentation. I made the mistake of beginning with some observational cosmology, planning to move immediately into theory. But Sakharov launched into questions, and I spent a full hour discussing issues I knew very little about. While Sakharov was telling his questions in Russian to Shimon, I was fretting about how I *should* have answered the previous question. At one point when Sakharov stopped speaking, I turned to Shimon for the translation. With a baffled expression on his face, Shimon stared at me and exclaimed, "But Alan, that *was* English!"

CHAPTER 7

THE PARTICLE PHYSICS REVOLUTION OF THE 1970s

In coming up to speed on grand unified theories, I had a great deal to learn. To appreciate the motivation for these theories, one had to understand them as the culmination of a revolution that had taken place in particle physics during the 1970s. After decades of confusion, physicists had finally developed a spectacularly successful theory known as the *standard model of particle physics.* At least in principle, this theory could account for the results of any particle physics experiment that was then feasible. The "standard model" pointed the way toward the unification of fundamental forces, and at the same time provided a firm foundation from which theorists could boldly extrapolate to ever higher energies. Before I could understand grand unified theories, I would need to understand the standard model, the springboard from which they were launched. In 1979 I was well aware that important developments had taken place, but I had not followed the details. Nor had I fully appreciated the magnitude of the progress that had been made.

When I was a graduate student, back in the medieval period before the revolution, particle physicists described the world in terms of four separate, fundamental interactions of nature. The word *interaction* is used here in the broadest possible sense, referring to literally anything that an elementary particle can do. For example, an unstable particle might decay into two or more lighter particles, or two particles might undergo a collision that deflects their motion. A collision might lead to the production of new particles, since special relativity implies that energy can be used to create mass. If a particle collides with its own antiparticle, the pair might annihilate, with all of the energy from the original particles channeled into the production of new particles. Elementary particle physicists use the word "interaction" to refer to any of these processes.

From the weakest to the strongest, the known interactions are gravitation, the weak interactions, electromagnetism, and the strong interactions:

Gravitation is well known to all of us, although it is not apparent from our everyday lives that gravity is the weakest of the known forces. Certainly no one who has recently tried to move a refrigerator would believe that gravity is weak. Gravity gives the illusion of strength because it is the only one of the known forces that is both long-range and always attractive. To lift a refrigerator we must oppose the attractive force between every atom of the refrigerator and every atom of the planet Earth. The force of gravity acting between two elementary particles, on the other hand, is so weak that it has *never* been detected. The gravitational force between a proton and an electron, for example, is 2×10^{39} times weaker than the electrical attraction.

The *weak interactions* were first observed in the radioactive decay of many kinds of nuclei. An example is beta decay, discussed in Chapter 1, in which a nucleus emits an electron and an antineutrino. Beta decays may not seem very important to us, but they are actually crucial to the sequence of energy-generating nuclear reactions in the sun. Neutrinos are subject only to the weak force and the much weaker force of gravity, so physicists can study the weak force by observing the scattering of a beam of neutrinos colliding with a target nucleus. The weak force is never noticed directly in our everyday experience, because it has a range that is only about 1/100 of the size of an atomic nucleus.

Electromagnetism, like gravity, plays an obvious role in our everyday lives. Electromagnetism includes both electricity and magnetism, which were shown during the 1800s to be inextricably linked: a changing magnetic field produces an electric field, and a changing electric field produces a magnetic field. Electrical forces bind the electrons of an atom to its nucleus, providing the basic structure of all the matter around us. The light with which you are reading this page is composed of a wave of electric and magnetic fields, as are microwaves, radio waves, and X rays.

Although the power of a large electric motor is impressive, the full strength of electrical forces is never seen in our everyday experience. The force between a positive and negative charge is attractive, but the force between like charges is repulsive. Since ordinary matter contains almost exactly equal amounts of the two charges, the attractive and repulsive forces almost exactly cancel. To understand the full power of the electrical force, imagine removing all the negative charge from two one-pound iron balls. At one foot apart, the two balls would repel each other with a force of 7×10^{18} tons!

The *strong interactions*, with a range roughly the size of an atomic nucleus (about 10^{-13} centimeters), bind the protons and neutrons inside a nucleus. The protons and neutrons are each believed to be composed of three quarks, also held together by the strong interactions. Strong interactions account for the tremendous energy release of a hydrogen bomb. In addition, many short-lived particles produced in particle accelerator experiments are observed to undergo strong interactions.

Before the 1970s, only one of these types of interactions—electromagnetism—was described by a completely successful theory. At the level of classical physics, the first full description of electromagnetism was published by the British physicist James Clerk Maxwell in 1864. Although Maxwell knew nothing of the theory of special relativity that Einstein would invent in 1905, the equations he proposed were completely consistent with special relativity. It was actually Maxwell's equations, and the description of electromagnetic waves implied by these equations, that motivated Einstein to develop special relativity.

Our view of electromagnetism was modified with the advent of quantum theory, starting with Einstein's suggestion in 1904 that the energy of an electromagnetic wave is concentrated in indivisible bundles called photons. Einstein did not, however, propose a full theory that would reconcile the existence of photons with the success of Maxwell's equations. For some time it was mysterious why light would sometimes behave as if it were made of particles, while under other circumstances it would behave like the continuous wave described by Maxwell's equations. This problem was not solved until the 1920s, when Erwin Schrödinger of Austria and Werner Heisenberg of Germany developed the full theory of quantum mechanics. Quantum theory generally predicts no unique outcome for an experiment, but instead predicts the probabilities for alternative outcomes. The behavior of any one photon is unpredictable, but the average behavior of a large number of photons closely mimics the continuous waves described by Maxwell's equations.

The full theory of relativistic electrons and positrons interacting with photons—called *quantum electrodynamics,* or QED—was formulated in the 1930s, but was soon found to lead to serious technical problems. Since the theory could not be solved exactly, physicists resorted to a method of successive approximations called a *perturbation expansion.* The first approximation was found to give answers that agree quite well with experiments—typically to 1% or better. The second approximation, however, was found to produce mathematical expressions that are infinite, and hence meaningless.

The full success of QED was not achieved until the 1940s, with the work of Richard Feynman, Shin'ichiro Tomonaga, and Julian Schwinger, who were awarded the Nobel Prize for their contributions in 1965. These three physicists, working independently, used a technique called *renormalization* to reformulate the theory in a way that avoids the infinities. QED is undoubtedly the most precise physical theory ever devised by science: some of its predictions can be checked experimentally to an accuracy of eleven decimal places, and they are correct!

In contrast to electromagnetism, the other three fundamental interactions were only partially understood. Gravitation had been the first interaction to be described theoretically, beginning with the work of Newton in 1687. Newton's theory was superseded in 1915 by Einstein's theory of general relativity. General relativity is believed to be highly accurate, although it cannot be tested at anywhere near the same level of precision as quantum electrodynamics. General relativity is inadequate as a fundamental description of nature, however, since it is a classical theory, inconsistent with the quantum view that is essential in understanding the atomic and subatomic world. As in the early years of QED, attempts to construct a quantum theory of gravity have been plagued by the appearance of infinite expressions. The renormalization technique that proved effective for QED does not work for general relativity.

The status of the weak interactions in 1970 was similar to that of gravity—there was a very successful theory, but it was not totally acceptable. The theory, dating from 1956, was similar in form to QED. The first approximation of the perturbation expansion gave results that agreed excellently with experiments, but the next term produced infinities. In contrast to QED, the infinities here could not be eliminated by renormalization. The experimental success certainly meant that the theory was on the right track, but the problem of infinities remained unsolved.*

For the strong interactions, the level of ignorance was even higher. Various patterns of behavior had been noticed, and the important idea that strongly interacting particles are composed of quarks had been suggested in 1964. Nonetheless, nobody understood how quarks interact or why we had

* Since theorists are free to invent theories, the reader might wonder why no one proposed that the first term in the perturbation calculation should be accepted as the final answer. This proposal would have been consistent with all experiments performed in the 1950s and 1960s, and the problem of infinities would be evaded by fiat. To understand the flaw of this approach, one must remember that we are discussing a quantum theory, a theory that predicts probabilities for alternative outcomes of an experiment. If the calculation were stopped after the first term in the perturbation calculation, the sum of the probabilities for all possible outcomes would not add exactly to one. Although the discrepancies were very small—typically less than one part in a billion for the kinds of experiments that were feasible at the time—logical consistency requires that the probabilities sum exactly to one. Furthermore, the discrepancies become larger when the theory is applied to processes at higher energy. For energies about 100 times higher than those available at the time (a dream that became real in the 1980s), the probability sum for some processes would go up to 2 and beyond.

never seen a quark, so the entire area of strong interactions was shrouded with confusion.

With the advent of the 1970s, however, particle physics went through a period of extraordinary progress, culminating in a theory that has come to be known as the standard model of particle physics. While this theory is not believed to be the ultimate theory of nature, for reasons that will be discussed later, it is nonetheless the most successful theory that physicists have ever devised. The weak, electromagnetic, and strong interactions are all described with sufficient accuracy to agree with every reliable experiment that has been performed to date. The model proposes that all of nature can be described by a short list of fundamental building blocks and a set of equations that describe how they interact. The fundamental particles of the standard model are listed in Table 7.1.

The new ideas concerning the weak interactions had their roots in a paper written in 1967 by Steven Weinberg [1], who was then at MIT. Weinberg proposed a unified theory of the weak and electromagnetic interactions, now often called the theory of *electroweak* interactions. The same theory was proposed independently in 1968 by Abdus Salam [2], a Pakistani-born physicist working at Imperial College in London. In this theory the weak interactions are transmitted by new particles, in a manner analogous to the way electromagnetic interactions are transmitted by photons. The transmitters of the weak interactions are the neutral Z^0 and a pair of oppositely charged particles, the W^+ and W^-. While the photon always moves at the fixed speed of light, the W^+, W^-, and Z^0 are heavy and usually move slowly.

Both Weinberg and Salam conjectured that the new theory might be *renormalizable*, which would mean that the problem of infinities could be cured by the same techniques used in QED. Neither physicist, however, was able to turn this hope into a reality. Frustrated by his failed attempts, Weinberg turned his attention to other topics. Over the next four years Weinberg's paper was cited in the scientific literature only five times. Salam's work was published only as part of a collection of lectures, but never in a standard physics journal.

In 1971, however, a Dutch graduate student named Gerard 't Hooft developed [3] a new method for renormalizing the infinities that arise in theories of the type proposed by Weinberg and Salam. While 't Hooft's original paper fell short of giving a complete proof that all the infinite expressions are eliminated, it went far enough to be extremely persuasive. In the words of Harvard University's Sidney Coleman [4], it "revealed Weinberg and

	1st Generation	2nd Generation	3rd Generation
Quarks:	Up Down	Charm Strange	Top Bottom
Leptons:	Electron Electron-Neutrino	Muon Muon-Neutrino	Tau Tau-Neutrino
Force Carriers:	Weak Interactions:		W^+, W^-, Z^0
	Electromagnetism:		Photon
	Strong Interactions:		8 Gluons
1 Higgs Particle			

Table 7.1 **The particle content of the standard model of particle physics.** The standard model proposes that everything in nature is constructed from a short list of fundamental building blocks. Each of the quarks listed above exists in three "colors," and for each quark and lepton there exists a corresponding antiparticle. The Z^0, photon, and Higgs particle are each their own antiparticle, the W^- is the antiparticle of the W^+, and the list of 8 gluons includes the antiparticle for each particle on the list. The Higgs particle will be discussed in the next chapter.

Salam's frog to be an enchanted prince." The renormalizability of the electroweak theory was established conclusively over the next several years [5].

During the same period, our understanding of the strong interactions was also transfigured. By the 1960s the number of known strongly interacting particles was growing so fast that physicists were running out of letters to name them. A table of the low-mass strongly interacting particles is shown as Table 7.2 (which the reader is not expected to memorize!). And the list is still growing. The particle data tables [6] show 153 established strongly interacting particles, plus more than a hundred other suspects of which the existence is not yet confirmed. With the exception of the proton and neutron, these particles are all short-lived. They are produced in particle accelerators, persist for lifetimes ranging from 10^{-24} to 10^{-7} seconds, and then disappear by decaying into lighter particles. Despite the brevity of the existence of these strongly interacting particles, physicists can measure their properties.

With a faith in the simplicity of nature, physicists suspected that they were looking not at a table of elementary particles, but rather at a chart more akin to the periodic table of chemical elements. The breakthrough occurred in 1964, when Murray Gell-Mann [7] of Caltech and George Zweig [8] of CERN independently suggested that the complexity of strongly interacting particles could be explained by assuming that they are constructed from three fundamental building blocks. Zweig called them *aces*, but Gell-Mann's term *quark** has survived. The most novel feature of the proposal was the peculiar assignment of electric charges: one type of quark had a charge $\frac{2}{3}$ that of a proton, and the two other types each had a charge $-\frac{1}{3}$ times that of a proton, as shown in Figure 7.1. In contrast, all known particles have charges that are integer multiples of the proton charge: an electron has a charge that is –1 times that of a proton, the particle called delta plus plus (Δ^{++}) has a charge that is 2 times that of a proton, etc. Known particles, with their integer charges, are described in the quark theory as either bound states of three quarks or as bound states of a quark and an antiquark. Since fractional charges had never been observed, Gell-Mann preferred to think of quarks as mathematical constructs—fictional entities which are nonetheless useful for describing the properties of strongly interacting particles. He also discussed, however, the possibility that they might be real particles. "A search for stable quarks . . . at the highest energy accelerators," he wrote in his original paper, "would help to reassure us of the non-existence of real quarks."

If quarks with fractional charge exist, then they should be easy to identify in accelerator experiments. Searches were carried out, and the results were indeed reassuring, in the sense of Gell-Mann: No quarks were found. The dominant particle physics textbook of the mid-1960s—*Elementary Particle Physics,* by Stephen Gasiorowicz (1966)—includes only one paragraph on quarks, concluding with:

> A search for new particles of this type has so far proved unsuccessful. In view of this failure and the difficulty of inventing a mechanism which would bind a quark and an antiquark, or three quarks, but not two quarks, for example, we will not discuss the quark model further.

The tone of the story gradually began to change in 1966, however, when the largest electron accelerator in the world—the Stanford Linear Accelerator

* Gell-Mann footnoted the name with a citation to James Joyce's *Finnegans Wake*, which contains the line "Three quarks for Muster Mark!" on p. 383.

Baryons

Particle	Mass	Charge	Spin	Half-life
p	0.938	+1	½	$>10^{32}$ yr
n	0.940	0	½	616
Λ	1.1156	0	½	1.82×10^{-10}
Σ^+	1.1894	+1	½	0.55×10^{-10}
Σ^0	1.1926	0	½	$7–8 \times 10^{-20}$
Σ^-	1.1974	–1	½	1.48×10^{-10}
$\Delta(1232)$	1.230–1.234	–1,0,1,2	3⁄2	$3.6–4.0 \times 10^{-24}$
Ξ^0	1.315	0	½	2.9×10^{-10}
Ξ^-	1.321	–1	½	1.64×10^{-10}
$\Sigma(1385)$	1.382–1.388	–1,0,+1	3⁄2	$11–15 \times 10^{-24}$
$\Lambda(1405)$	1.403–1.411	0	½	$9–10 \times 10^{-24}$
$N(1440)$	1.430–1.470	0,+1	½	$1–2 \times 10^{-24}$
$N(1520)$	1.515–1.530	0,+1	3⁄2	$3–4 \times 10^{-24}$
$\Lambda(1520)$	1.518–1.520	0	3⁄2	30×10^{-24}
$\Xi(1530)$	1.531–1.536	–1,0	3⁄2	$46–51 \times 10^{-24}$
$N(1535)$	1.520–1.555	0,+1	½	$2–5 \times 10^{-24}$
$\Delta(1600)$	1.550–1.700	–1,0,1,2	3⁄2	$1–2 \times 10^{-24}$
$\Lambda(1600)$	1.560–1.700	0	½	$2–9 \times 10^{-24}$
$\Delta(1620)$	1.615–1.675	–1,0,1,2	½	$3–4 \times 10^{-24}$
$N(1650)$	1.640–1.680	0,+1	½	$2–3 \times 10^{-24}$
$\Sigma(1660)$	1.630–1.690	–1,0,+1	½	$2–11 \times 10^{-24}$
$\Lambda(1670)$	1.660–1.680	0	½	$10–20 \times 10^{-24}$
$\Sigma(1670)$	1.665–1.685	–1,0,+1	3⁄2	$6–11 \times 10^{-24}$
$\Omega^-(1672)$	1.672	–1	3⁄2	0.82×10^{-10}
$N(1675)$	1.670–1.685	0,+1	5⁄2	$2.5–3.3 \times 10^{-24}$
		.		
		.		
		.		

Mesons

Particle	Mass	Charge	Spin	Half-life
π^0	0.134974	0	0	$7.8–9.0 \times 10^{-17}$
$\pi^\pm$	0.139568	–1,+1	0	2.603×10^{-8}
$K^\pm$	0.49365	–1,+1	0	1.24×10^{-8}
K_L^0	0.4977	0	0	$5.1–5.2 \times 10^{-8}$
K_S^0	0.4977	0	0	0.892×10^{-10}
η	0.547	0	0	$3.5–4.2 \times 10^{-19}$
$\rho(770)$	0.768	–1,0,+1	1	3.0×10^{-24}
$\omega(783)$	0.782	0	1	54×10^{-24}
$K^*(892)$	0.892–0.896	–1,0,+1	1	$8.9–9.3 \times 10^{-24}$
$\eta'(958)$	0.958	0	0	$2.1–2.5 \times 10^{-21}$
$f_0(975)$	0.972–0.977	0	0	$8–12 \times 10^{-24}$
$a_0(980)$	0.981–0.985	–1,0,+1	0	$7–10 \times 10^{-24}$
$\phi(1020)$	1.01941	0	1	103×10^{-24}
$h_1(1170)$	1.15–1.19	0	1	$1.1–1.4 \times 10^{-24}$
$b_1(1235)$	1.22–1.24	–1,0,+1	1	$2.8–3.1 \times 10^{-24}$
$a_1(1260)$	1.23–1.29	–1,0,+1	1	1×10^{-24}
$f_2(1270)$	1.27–1.28	0	2	$2–3 \times 10^{-24}$
$K_1(1270)$	1.26–1.28	–1,0,+1	1	$4–7 \times 10^{-24}$
$f_1(1285)$	1.28–1.29	0	1	$17–22 \times 10^{-24}$
$\eta(1295)$	1.291–1.299	0	0	$8–10 \times 10^{-24}$
$\pi(1300)$	1.2–1.4	–1,0,+1	0	$0.8–2.3 \times 10^{-24}$
$a_2(1320)$	1.317–1.319	–1,0,+1	2	4×10^{-24}
$\omega(1390)$	1.377–1.411	0	1	$1.7–2.4 \times 10^{-24}$
$f_0(1400)$	1.4	0	0	$1–3 \times 10^{-24}$
$K_1(1400)$	1.395–1.409	–1,0,+1	1	$2.4–2.8 \times 10^{-24}$
		.		
		.		
		.		

Table 7.2 **(facing page) Table of strongly interacting particles.** The particles fall into two classes, *baryons* and *mesons*, where the baryons are on average heavier. The first two baryons on the list, p and n, are the familiar proton and neutron. Following the habit of particle physicists, the masses are indicated by their equivalent energy, in units of GeV (billion electron volts). As can be seen, 1 GeV is about the mass of a proton. The charge is shown in units of the proton charge, and the half-life is shown in seconds. The spin is measured in the standard unit of quantum spin, defined in Appendix D. All mesons have integer spin, while all baryons have spins that are midway between two integers. For each baryon listed there is a corresponding antiparticle, while the list of mesons includes both the particles and antiparticles.

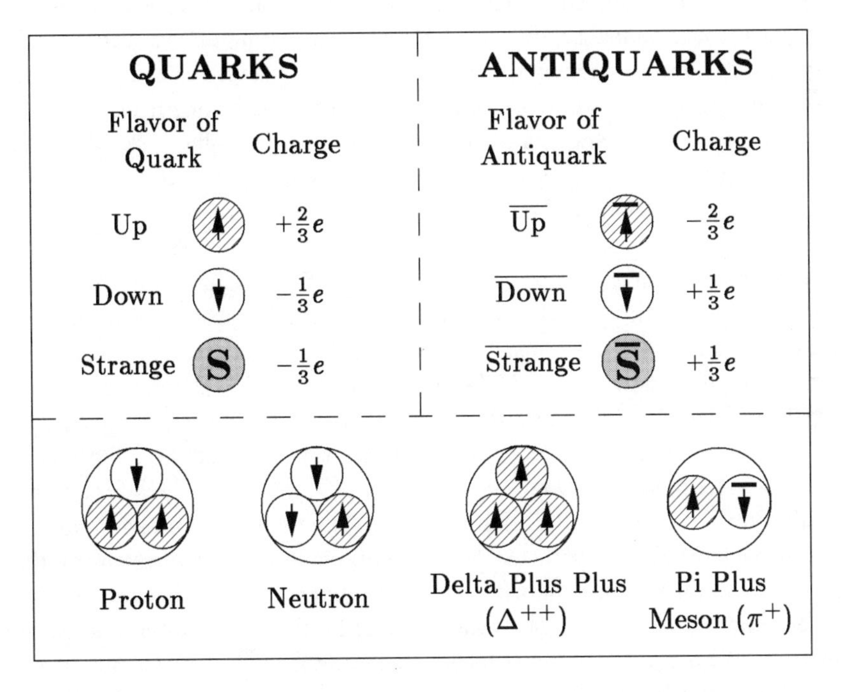

Figure 7.1 **The original quark model.** The quark model of Gell-Mann and Zweig proposed that all strongly interacting particles are composed of quarks and/or their antiparticles, called antiquarks. The quarks exist in three types, or *flavors,* which Gell-Mann named *up, down,* and *strange.* The top part of the diagram shows the six types of quarks and antiquarks. The electric charge of each is shown in units of e, the charge of the proton. The bottom half of the diagram shows how the proton and neutron are each composed of three quarks, as is the short-lived particle called the delta plus plus. The particles called mesons are each composed of one quark and one antiquark.

Center—began operation just outside the main Stanford campus. The two-mile long device could accelerate electrons to an energy of 21 GeV, about three times higher than any previous electron accelerator. The first major project to be carried out with the new accelerator came to be known as the SLAC-MIT experiment, headed by Jerome Friedman and Henry Kendall of MIT,* and Richard Taylor of SLAC. (The trio shared the 1990 Nobel Prize in physics for this work.) Using a bank of mammoth detectors in End Station A, the largest experimental hall, the experimenters measured the distribution of electrons that scattered from protons and neutrons in their targets. In a series of experiments conducted from 1967 to 1973, the group found that protons and neutrons appear to be composed of smaller, essentially point-like particles, and that these particles appear to have exactly the fractional charges predicted by the quark theory. Furthermore, they found a surprising effect: Although the quarks are bound so strongly that they were never observed to escape from the proton or neutron, the electrons of the SLAC beam were nonetheless scattered with a distribution of angles and energies that indicated collisions with quarks that are virtually free. This last property was initially baffling to theorists who were trying to understand the forces that hold the quarks together.

The eventual resolution of the quark interaction problem hinged on a type of theory that was first constructed in 1954 by Chen Ning Yang and Robert Mills, working at Brookhaven National Laboratory in Upton, New York. In an elegant paper just four and a half pages long, Yang and Mills had set in motion what would become a revolution in theoretical physics [9]. The *Yang-Mills theories*, also known as *gauge theories*, are patterned after the electromagnetic theory of Maxwell, but they describe a more complicated system with several types of charge that mutually interact in a highly symmetric way. A full description of Yang-Mills theories would take us too far from the main thrust of this chapter, so it will suffice to say that the construction of these theories was motivated mainly by their mathematical elegance. The original goal was to describe the strong interactions, but for many years no one had any idea how to adapt these theories into a realistic description of anything. Nonetheless, the rich mathematical structure was a fascination to theorists, many of whom were enamored with the theory from the start. This fascination eventually led to success, as the unified electroweak theory described earlier is an example of a Yang-Mills theory.

The relevance of Yang-Mills theory to the strong interactions was discovered in 1973, by David J. Gross and Frank Wilczek [10] at Princeton,

* Friedman and Kendall remain at MIT. In addition to continuing an active role in research, Kendall has spent a great deal of time working on environmental and disarmament issues with the Union of Concerned Scientists, which he founded and chaired for many years. When I met Henry at a lunch truck outside MIT one day last year, I asked him if he ever had time to think. "Oh, yes, . . . thinking." He paused and then continued, "I had two minutes to think just a few years ago, and I cherish the memory."

and independently by H. David Politzer [11] at Harvard. Gross was the old man of the trio at age 32, while Wilczek and Politzer were both still graduate students, working for Gross and Sidney Coleman, respectively. By performing some rather complicated calculations to extract the predictions of Yang-Mills theories for the collisions of very high energy particles, they concluded that Yang-Mills theories behave differently than all theories that had previously been examined. While other theories describe interactions that become stronger and stronger at high energies, the interactions of particles described by Yang-Mills theories become weaker and weaker. This property, which Gross and Wilczek called *asymptotic freedom*, became the key to a new theory of the strong interactions. The asymptotic freedom of the Yang-Mills theories is precisely what is needed to explain the SLAC-MIT experiments, which showed that the quarks inside a proton appear to be very weakly bound when they are struck by extremely high energy electrons. The new theory was dubbed *quantum chromodynamics*, or QCD, a name that was coined to mimic the name quantum electrodynamics and its abbreviation QED. Within a few years QCD became the generally accepted description of the strong interactions.

QCD proposed that each of the three flavors of quarks—up, down, and strange—exists in three variations, called *colors* [12]. The color of a quark of course has nothing to do with the way it appears to the human eye—quarks are invisible to the human eye. The word "color" was adopted only because it distinguishes three variations, in analogy with the three primary colors. The names for the three colors are usually taken as red, green, and blue, the primary hues that are blended to create the full-color images on a television screen. Thus, instead of just three types of quarks, the new theory assumed that there are nine, each specified by a color and flavor combination: red-up, green-up, blue-up, red-down, etc. The colors play the role of the charges for the Yang-Mills theory. Yang-Mills theories can describe a variety of possible interactions, but the strong interaction theory is built on one in particular, labeled by the mathematical symbol $SU(3)$[13]. The name *gluon* was given to the 8 particles that act as carriers of the Yang-Mills force, since these particles "glue" the quarks together inside a proton or neutron. The gluon plays the same role for the strong interactions that the photon plays for electromagnetism, and that the W^+, W^-, and Z^0 play for the weak interactions.

Asymptotic freedom means that quarks interact very weakly in high energy processes, including the collision of a high energy electron with a stationary proton, as in the SLAC-MIT experiment. But the proton itself is a low-energy system, so the quarks inside the proton are bound strongly to each other. These genuinely strong interactions are very difficult to calculate, but most physicists are convinced that the theory leads to *confinement*, meaning that quarks can literally not exist independently, but are forever confined within strongly interacting particles. Stated more generally, it is believed that the only systems that can exist in isolation are *colorless*. The precise definition of "colorless" is phrased in mathematical terms, but the bottom line is that colorless combinations can be achieved in two ways. In the particles called baryons, such as the proton or neutron, colorless states are achieved by a combination of three quarks, one of each color. In mesons, on the other hand, colorless quark-antiquark combinations are achieved by combining a red quark with its antiquark, a blue quark with its antiquark, and a green quark with its antiquark [14]. Thus, the property of confinement explains why we see precisely the quark combinations that we do see, without any possibility of seeing a free quark or any particle with a noninteger value of the charge. The possibility of confinement was suggested vaguely by Gross and Wilczek in their original paper, and more explicitly by Weinberg [15] in an article submitted one month later.

The next step in the particle physics revolution was the discovery of the fourth quark, known as the *charmed* quark. The idea of a fourth quark was first suggested in 1964 by James D. Bjorken of SLAC and Sheldon L. Glashow of Harvard [16], but the first persuasive argument was published in 1970 by Glashow, with John Iliopoulos and Luciano Maiani [17]. Although no direct sign of the charmed quark had ever been observed, the threesome reasoned that a particle of this sort might nonetheless contribute as an intermediate step in reactions that had been measured. Further, they argued that the observed rates for various reactions could not be understood except by assuming the existence of the charmed quark.

My thesis advisor, Francis Low, was an early enthusiast of the charmed quark idea, having recently worked with Gell-Mann and others to explain these same reaction rates. Curiously Gell-Mann and his coworkers chose not to consider quarks, but instead developed a baroque theory [18] requiring the invention of over thirty new particles. The relative simplicity of a single new quark was very attractive. My Ph.D. thesis might have involved further work on this topic, but neither Francis nor I had any particularly good ideas about how to extend it.

Since even the idea of quarks was not yet well-accepted in 1970, the physics community responded to the charmed quark with interest but not with enthusiasm. But everything changed dramatically on November 16, 1974, when Samuel C.C. Ting of MIT and Burton Richter of SLAC held a joint press conference to announce a spectacular discovery. A team at Brookhaven Laboratory led by Ting, and a team at SLAC led by Richter, had independently discovered a new and unexpected particle, with a mass/energy of 3.105 GeV (about 3 times the mass of a proton), a surprisingly long lifetime for a particle so heavy, and a very large rate of production in electron-positron collisions. According to the press announcement,

> The suddenness of the discovery coupled with the totally unexpected properties of the particle are what make it so exciting. . . .

> The theorists are working frantically to fit it into the framework of our present knowledge of the elementary particles. We experimenters hope to keep them busy for some time to come.

The particle was named the *J* by the Brookhaven group and the psi (ψ) by the SLAC group, so today it is known by the compromise symbol J/ψ. For a short time most physicists were confused by the discovery, although the group at Harvard decided immediately that the particle was a bound state of a charmed quark and a charmed antiquark: *charmonium*. Just two days after the press conference, the *Physical Review Letters* received a paper describing charmonium by Thomas Appelquist and David Politzer, both of Harvard. A week later another Harvard paper, by Glashow and Alvaro De Rujula, claimed that charm had been found. Lunchtime discussions in physics groups around the world were quickly elaborated into publications, and on January 6, 1975, the *Physical Review Letters* published a spate of nine papers offering at least five different explanations of the new particle.

Meanwhile, the group at SLAC had already found a second particle, at 3.695 GeV, a particle that was soon dubbed ψ' and identified as a higher energy state of the charmonium system. The beauty of charmonium lies in the heaviness of the charmed quarks. This causes them to move slowly inside the bound system, so physicists can calculate the expected energy states of charmonium with much better accuracy than any other quark system. Over the next two years a number of higher energy particles were found, and all of them matched well against the calculated pattern of charmonium states. In the end, the J/ψ did more than simply convince the physics community to accept a fourth quark. It was really the J/ψ that convinced the physics community to take seriously the idea of quarks in the first place. In hindsight it seems clear that the SLAC-MIT experiment provided strong evidence for quarks, but the J/ψ measurements were more straightforward to interpret, and much more persuasive at the time. As Glashow wrote in *The New York Times Magazine* in July 1976, "As a result of the discovery of the J/ψ and its kin, the quark model has become orthodox philosophy." In 1976, less than two years after the discovery, Ting and Richter shared the Nobel Prize.

With the addition of the charmed quark, the set of known fundamental particles fell into a simple pattern. Besides those already discussed, the set included the muon, which was discovered in 1937 but not properly identified until 1945. It appears to be identical to the electron in both its electromagnetic and weak interactions, and like the electron it does not participate in the strong interactions. Despite these similarities, the muon is 207 times more massive than the electron. There is also a neutrino associated with the muon, which was shown in 1962 to be distinct from the neutrino associated with the electron. With the inclusion of the charmed quark, the full theory had a balance between the quarks and the non-strongly interacting particles, called *leptons.* The four flavors of quark—up, down, charmed, and strange—could be matched against the electron and its neutrino, and the muon and its neutrino. In fact, it is now known that this matching between quarks and leptons is essential in avoiding the problem of infinities that was discussed earlier. The up and down quarks, and the electron and its neutrino are called the first generation of quarks and leptons. The charmed and strange quarks, and the muon and its neutrino, are called the second generation. The properties of the second generation seem to be identical to those of the first generation, except that the particles of the second generation are much heavier. (The neutrino of the second generation, however, has a mass which is either zero or too small to have so far been measured.)

To many it seemed time for the proliferation of particles to stop, but nature had other intentions. In 1975 a team led by Martin Perl at SLAC discovered an even heavier lepton which they called the tau (τ), with a mass of 1.784 GeV, about 3500 times as heavy as the electron. This finding upset the balance between leptons and quarks, but the balance was restored in 1977 with the discovery of the upsilon (Υ) by a team led by Leon Lederman at Fermilab (the Fermi National Accelerator Laboratory). This particle, with a mass of 9.46 GeV, is interpreted as the bound state of a fifth quark, called the *bottom*, and its own antiparticle. The remaining third generation quark, called the *top*, was discovered in 1995 by two independent groups (the "D-Zero" collaboration and the "Collider Detector at Fermilab" collaboration) working at Fermilab. Although the neutrino associated with the tau lepton has not been shown to be distinct from the other neutrinos, physicists now have enough confidence in the pattern of the three generations to feel certain that the tau-neutrino is an independent particle.*

* Since the third generation was found without any prediction of its existence, the reader might wonder if the list might go beyond three, maybe even continuing without limit. No one knows for sure, but there is observational evidence that the list stops with exactly three generations. Since particle experiments tell us nothing about the existence or nonexistence of extremely massive particles, which cannot be produced due to insufficient energy, the evidence for additional generations hinges on the neutrinos. The neutrinos of the three known generations all have masses that are either zero or too small to have been detected, so we assume that any additional generation would probably also contain a light neutrino. Cosmologically, extra types of light neutrinos would contribute to the mass density of the early universe and affect the rate with which gravity slows the expansion. The slowing, in turn, would affect the predictions of big bang nucleosynthesis. For most of the 1980s it appeared that one additional generation would be tolerable, but the possibility of more than one is excluded by the observed nuclear abundances. Starting in 1989, accelerator experiments have confirmed and strengthened this prediction from cosmology. By observing the decays of the Z^0 particle, physicists at CERN and SLAC have ruled out the possibility of even a single additional light neutrino. At about the same time improved measurements of the cosmic helium abundance and the neutron half-life allowed cosmologists to reach the same conclusion.

When the theory of quantum chromodynamics for the strong interactions is combined with the electroweak theory, the resulting combination is called the standard model of particle physics. The construction of the standard model is the spectacular achievement of the particle physics revolution of the 1970s.

It is difficult to overstate the triumphant feeling of accomplishment that the standard model brought to the particle physics community. There are at least three reasons why the impact was so great.

First of all, the pace had been staggering. In 1970, only one of the four fundamental forces—electromagnetism—was understood. (Although Weinberg and Salam had published their model of the weak interactions, no one, not even Weinberg and Salam, appreciated the significance.) By 1976, however, the standard model gave what appears to be a complete description of three out of the four known types of interactions.

Second, the power of the theory was unanticipated and unprecedented. The only known force that lay outside the scope of the theory is gravity, which is so weak on the subatomic scale that its effects are in any case imperceptible. On larger scales gravity is of course very important, but quantum theory can be ignored on such scales, and the classical theory of general relativity is completely adequate. Thus, the standard model comes close to being a complete description of the laws of nature, from which all phenomena in the universe can in principle be derived. At this time there is no reliable experiment that is inconsistent with the standard model.*

Third, at least for many theorists, the standard model was seen as a validation of our intuition. This concept may sound surprising, since we are often led to believe that intuition is important only in areas such as art, politics, or stock market investment. The cold hard logic of science, we are told, has no place for intuition. Nothing could be further from the truth! As Steven Weinberg writes in *Dreams of a Final Theory* [19],

> Time and again physicists have been guided by their sense of beauty not only in developing new theories but even in judging the validity of physical theories once they are developed. It seems that we are learning how to anticipate the beauty of nature at its most fundamental level. Nothing could be more encouraging than we are actually moving toward the discovery of nature's final laws.

* Lest I make things sound too simple, I should point out that even correct theories are frequently contradicted by experiment. In an essay on the 1979 Nobel Prize in physics (cited earlier), Sidney Coleman pointed out that the standard model was shown to be experimentally incorrect at least three times between 1973 and 1978. In each case, however, the experiments were later shown to be in error.

In the case of the standard model, physicists take pride in recognizing the mathematical beauty of the abstract theory constructed by Yang and Mills in 1954. Although it was 13 years before these equations were incorporated into a successful physical theory, the interest in Yang-Mills theories did not fade away. Similarly, many physicists took pride in recognizing the beauty and simplicity of the quark model. And finally, the prediction of the charmed quark was particularly striking. The paper of Glashow, Iliopoulos, and Maiani appeared in 1970, before the other aspects of the standard model were in place, so the argument depended on many assumptions. Yet the charmed quark was soon found, and its properties were precisely in accord with the 1970 proposal. Thus, even an argument that was indirect and uncertain, when guided by the right intuition, was found to correctly anticipate the behavior of nature.

Needless to say, not all particle physicists could realistically take personal pride in the development of the standard model. Some physicists were advocating alternative approaches, and many, like myself, were working on other topics altogether. Nonetheless, the development of the standard model was a community effort, in which the ideas of many physicists—far more than those mentioned in this brief summary—were melded together to create a single theory. This aspect of the enterprise was emphasized by Sheldon Glashow in his 1979 Nobel lecture, titled "Towards a unified theory: threads in a tapestry" [20]:

> Tapestries are made by many artisans working together. The contributions of separate workers cannot be discerned in the completed work, and the loose and false threads have been covered over. So it is in our picture of particle physics.

Pride in the standard model was contagious. I learned about the model after it was developed, but I could easily tell myself that *if* I had been working on these questions, I would have proceeded the same way. Thus, even to physicists who had not taken part in the discovery, the success of the standard model was viewed as a monument to our collective vision.

It was from this atmosphere of enormous success that grand unified theories were born.

CHAPTER 8

GRAND UNIFIED THEORIES

In an article for *Scientific American* in April, 1950, Einstein explained that there are two motivations for developing new physical theories. The first is straightforward: "New theories are first of all necessary when we encounter new facts which cannot be 'explained' by existing theories." However, Einstein went on, "there is another, more subtle motive of no less importance. This is the striving toward unification and simplification of the premises of the theory as a whole." Indeed, unification has been the inspiration for many of the greatest achievements in the history of science, from the atomic theory of Democritus to Newton's law of universal gravitation, Maxwell's theory of electromagnetism, and the electroweak theory of Glashow, Weinberg, and Salam. Many physicists, including the author, hope that this list of momentous accomplishments will someday include grand unified theories.

The basic goal of a grand unified theory, or GUT, is to describe the weak, electromagnetic, and strong interactions of the standard model of particle physics in terms of a single, fully unified interaction. Gravity is omitted, but as discussed in the previous chapter, the force of gravity between two elementary particles is so much weaker than the other forces that it has never even been detected. While the ultimate theory of nature must clearly include gravity, grand unified theories can be expected to be an excellent approximation.* The first theory of this kind was proposed early in 1974 by Sheldon L. Glashow and Howard Georgi [1], both of Harvard.

* Although gravity has proven far more difficult to understand at the quantum level than the other interactions of nature, in the last decade very significant progress has been made in unifying gravity with the other forces. While the force of gravity between elementary particles is feeble at ordinary energies, it grows rapidly at high energies and becomes comparable to the other forces at about 10^{19} GeV, a scale called the Planck energy. (Expressed in terms of fundamental

While the standard model of particle physics is very well established, the same cannot be said for grand unified theories. Even if the idea of grand unification is correct, we certainly do not know which of the many conceivable grand unified theories is likely to be the right one. Nonetheless, GUTs are considered highly attractive for a variety of reasons, and here I will explain two.

First, GUTs are the only known theories which predict that the charges of the electron and the proton are equal in magnitude. This statement may at first seem unimpressive, since we all learned in high school that those two charges are equal, and it never seemed very consequential. High school teachers are notorious, however, for neglecting to tell their students that prior to grand unified theories, nobody had even a fuzzy idea about *why* those two charges are equal. Experimentally, the charges are known to be equal to the incredible accuracy of twenty-one decimal places! In all theories developed prior to GUTs, however, the two charges could each have had any value whatever. The precise equality of these charges had to be attributed either to chance, or to some mysterious principle that we do not yet understand.† Grand unified theories, on the other hand, contain a fundamental symmetry that relates the behavior of electrons to the behavior of the quarks which make up the proton. This symmetry guarantees that the charges are equal. If this symmetry were violated by even the smallest amount, then the theory would no longer be mathematically well-defined. Thus, if the electron and proton charges were found to differ in the twenty-second decimal place, then grand unified theories would have to be abandoned. But if suc-

constants, the Planck energy is $\sqrt{hc^5/(2\pi G)} = 1.22 \times 10^{19}$ GeV, where h is Planck's constant, c is the speed of light, and G is Newton's gravitational constant.) It is possible, therefore, that gravity unifies with the other forces at the Planck scale. A theory that unifies gravity with the other forces of nature is more ambitious than a grand unified theory, and is sometimes called a *superunified* theory. It has not been easy to construct a superunified theory, but the difficulty is generally viewed as an encouraging sign. If there were many consistent theories they might be easy to find, but we would have no way of knowing which was right. But if there is only one consistent superunified theory, it might be very hard to find. If we found it, however, we might then expect to know the ultimate laws of nature, the "theory of everything." There is a possibility that the solution already exists, in the form of what is called *superstring theory*. As the name suggests, the fundamental entity in this theory is an ultramicroscopic string-like object, with a length of typically 10^{-33} centimeters and effectively zero thickness. Superstring theory offers an intriguing hope of learning the true fundamental laws of nature, but at present our understanding of string theory is very limited. The simplest predictions of superstring theory concern processes at the Planck energy, and so far very little is known about the consequences of string theory at lower energies. It is expected that at energies of about 10^{16} GeV and below, the predictions of superstring theory would closely approximate those of grand unified theories. Since superstring theory is only just beginning to play a role in cosmological theories, the discussion here will focus on grand unified theories.

† For completeness, I mention that if the standard model were reduced to a single generation of quarks and leptons, then the requirement of renormalizability would imply that the electron charge is the opposite of the proton charge. With two or more generations, however, there is more freedom. The electron charge could be larger, for example, provided that the muon charge were smaller.

cessively more accurate experiments continue to confirm the equality of the charges, then these results would be further evidence for grand unification, or at least for some theory beyond the standard model.

A second reason for the attractiveness of GUTs is a prediction that these theories make for the strengths of the known particle interactions. The standard model of particle physics describes the observed properties of the weak, electromagnetic, and strong interactions in terms of three fundamental interactions. One is the color interaction of quantum chromodynamics (QCD), describing the strong interactions, and labeled by the symbol $SU(3)$. The weak and electromagnetic interactions, although unified, are nonetheless described in the theory by two fundamental interactions, labeled $SU(2)$ and $U(1)$. (To a particle physicist these symbols have a mathematical meaning that describes the detailed form of the interactions, but here the symbols will be used only as labels for the different interactions.) The electroweak theory is unified in the sense that the $SU(2)$ and $U(1)$ interactions are twisted together to describe both the weak and electromagnetic interactions. Each of the three interactions—$SU(3)$, $SU(2)$, and $U(1)$—has a different strength. These strengths are free parameters, which means that they are not determined by the theory, but instead must be fixed by experiment. It is found that the $SU(3)$ interaction is much stronger than the $SU(2)$ interaction, which in turn is slightly stronger than the $U(1)$ interaction. The strengths vary with energy, but once the strengths of the three interactions are measured at one energy, the theory determines the strengths at any other energy.

The first calculation of the variation of interaction strengths with energy for the full standard model was performed several months after the original Georgi-Glashow paper, by Georgi, Helen R. Quinn (then at Harvard, now at SLAC), and Steven Weinberg (then at Harvard, now at the University of Texas) [2]. They found that although the strengths of the three interactions are very different at the energies of experimental particle physics, the strengths approach each other at high energies. The data available at that time was very crude, but it was consistent with the possibility that all three interactions have the same strength at some extraordinarily high energy, somewhere between 10^{11} and 10^{17} GeV.

Figure 8.1 shows a recent analysis by Paul Langacker and Nir Polonsky [3] of the University of Pennsylvania, using the best data currently available. All three lines appear to meet rather accurately at a single point, at an energy just above 10^{16} GeV. While two curves will typically cross at a point, there is no reason for all three curves to meet at a single point unless there is some underlying relation between them. Grand unified theories provide just such an underlying relation! Based on the idea that the three interactions of the standard model of particle physics arise from a single fundamental interaction, GUTs imply that all three curves must meet. That means, for example,

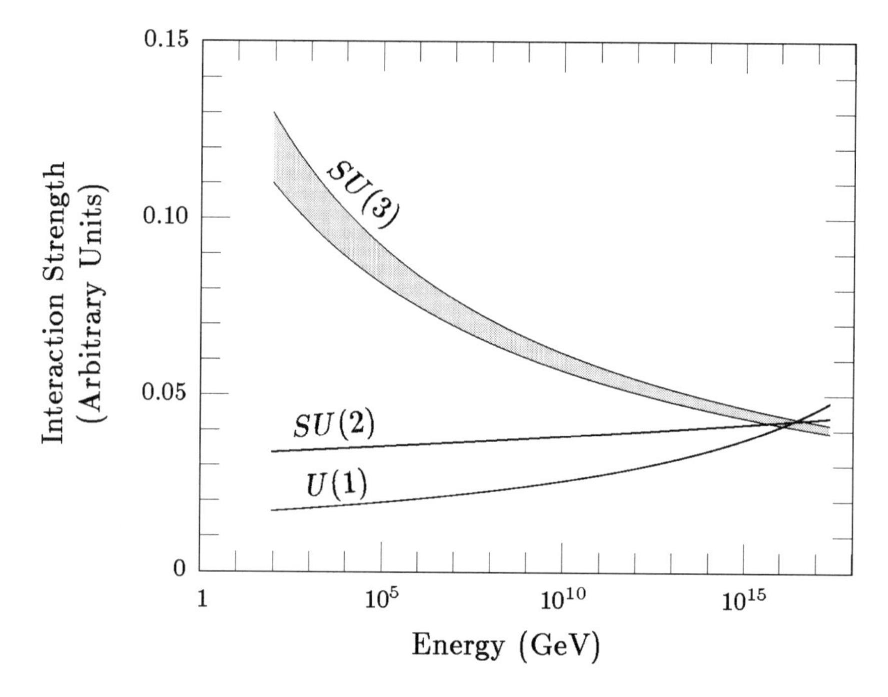

Figure 8.1 **Dependence of interaction strengths on energy.** The strengths of the three types of fundamental interactions of the standard model of particle physics—*SU*(3), *SU*(2), and *U*(1)—are measured at energies near 100 GeV. The curves show the calculated strengths of these interactions at higher energies. The thickness of the *SU*(3) curve indicates the experimental uncertainty, which is negligible for the *SU*(2) and *U*(1) curves. The meeting of all three curves at an energy of about 2×10^{16} GeV suggests that at this energy all three interactions meld into a single grand unified interaction.

---

that if any two of the interaction strengths are measured, then the third interaction strength can be predicted by insisting that its curve must pass through the point where the first two curves cross each other. This prediction of the grand unified theories is found to agree with the experimental result to an accuracy of about ¼%, while the estimated experimental uncertainty of the test is about 1½%.*

* The calculation of Figure 8.1 is actually based on a version of particle physics called the *minimal supersymmetric extension of the standard model.* An extrapolation based on the standard model without elaboration leads to a discrepancy between theory and experiment of about 10%. For this reason the "supersymmetric" model now looks like a better bet for grand unification. Nonetheless, since supersymmetry has little relevance to most of the issues discussed in this book, I will pursue it no further.

To present the full picture, I should mention that grand unified theories suffer from one important drawback, known as the *hierarchy problem*. The problem is that the enormous energy scale of unification—the 10^{16} GeV energy at which the interaction strengths meet—has to be "put in by hand." That is, we know no a priori reason why this energy scale is so many orders of magnitude larger than the other energy scales of relevance to particle physics. Recall, however, that there is a clear *experimental* reason for believing that the energy scale of unification is very high: Figure 8.1 shows that the energy scale of unification is determined by the measurements of the strengths of the three interactions at energies of about 100 GeV. The unification energy is high because the measured values of the three interaction strengths are very different from each other.

To understand how particle theorists view the hierarchy problem, one must realize that grand unified theories are not seen as the ultimate fundamental theory of nature. First, the ultimate theory must include gravity, which grand unified theories do not. Second, the grand unified theories are too ugly to be serious candidates for the ultimate theory. In particular, even the simplest GUT contains over twenty free parameters (i.e., numbers, such as the charge and mass of an electron, that must be measured experimentally before the theory can be used to make predictions). Most particle physicists believe that the ultimate theory will be much simpler, with few if any free parameters. Thus, we expect that someday the correct grand unified theory will be derived as an approximation to the ultimate theory, so the energy scale of grand unification will be calculable. When we say that the unification scale is "put in by hand," we mean that we presently have no understanding of why this hypothetical calculation of the future will give such a large number. So the advocates of grand unified theories are hoping that someday the reason will be found.

To complete the story of grand unified theories, we still need to understand how, at energies below 10^{16} GeV, the single "unified" interaction can simulate the existence of three very distinct types of interactions—the strong, weak, and electromagnetic. The secret behind this masquerade is a phenomenon called *spontaneous symmetry breaking*. The description of it, which will take us into the inner workings of the grand unified theories, is a bit more intricate than most of the subjects discussed in this book. It will, however, play an important role in the chapters that follow. While I was boning up to work with Henry Tye, I was well aware that the magnetic monopoles that we were planning to study owed their existence to spontaneous symmetry breaking. I was of course not yet aware that spontaneous symmetry breaking would also figure prominently in the development of a new cosmological theory called the inflationary universe. Some readers may wish to only skim the rest of this chapter, but those who follow it closely will find the information useful later on.

Spontaneous symmetry breaking is not a new idea associated with grand unified theories. The concept was used also in the construction of the electroweak theory, and is intimately related to the Higgs particle of the standard model that is mentioned but not explained in Table 7.1. Although the Higgs particle is yet to be found, the confidence that physicists have in the theory remains strong. The *W* and *Z* particles were not discovered until 1983,* sixteen years after the electroweak theory was proposed, but their masses and interactions agreed perfectly with the predictions. It is hoped that the Higgs particle of the standard model will be discovered when the Large Hadron Collider is completed at CERN, at about the turn of the twenty-first century.

According to the general definition, a spontaneously broken symmetry is a symmetry which is present in the underlying theory describing a system, but which is hidden when the system is in its equilibrium state. While spontaneous symmetry breaking occurs in esoteric theories such as the electroweak theory or grand unified theories, it also occurs in much more familiar situations, such as the formation of a crystal. I will use this analogy to help explain how spontaneous symmetry breaking works.

To make the analogy between spontaneous symmetry breaking in crystals and in grand unified theories as clear as possible, we will consider a particularly simple type of crystal, called *orthorhombic*, an example of which is the mineral topaz. These crystals have a rectangular structure, as illustrated in Figure 8.2, so all the angles are right angles. Unlike a cubic crystal, the three principal lengths of an orthorhombic crystal are all different, a feature that will make the analogy with grand unified theories a little closer.

An outline of the crystal/GUT analogy is shown as Table 8.1, which can be used as a checklist to help keep track of the following discussion. Starting at the top of the table, the first row indicates the symmetry that is involved. For the case of the crystal, the relevant symmetry is rotational invariance. The physical laws that describe the system are rotationally invariant, which means that they make no distinction between one direction of space and another. In the case of the grand unified theory, the symmetry is more abstract, having nothing to do with orientation in physical

* The original discovery of the *W* was a spectacular tour de force, involving a team of 135 physicists under the leadership of Carlo Rubbia. They analyzed data from about one billion collisions at the CERN accelerator to discover a total of six *W* particles. A subsequent effort of similar magnitude produced five particles. Using higher energy accelerators, however, more recent experiments have detected over 20,000 *W* particles and over 20 million *Z* particles.

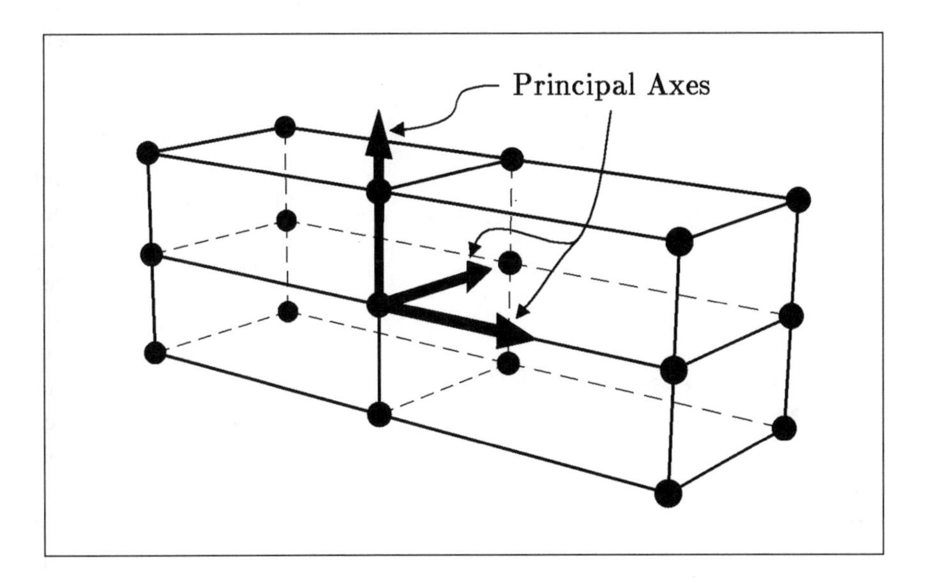

Figure 8.2 **The structure of an orthorhombic crystal.** This type of crystal has a rectangular structure, in which all three principal lengths are different. The formation of such a crystal is an example of spontaneous symmetry breaking, analogous to the spontaneous symmetry breaking in grand unified theories.

space. Instead the symmetry is what is sometimes called an *internal* symmetry, one that relates the behavior of one type of particle to the behavior of another. In this case, the underlying symmetry of the grand unified theory has two manifestations. First, the symmetry implies that the three interactions of the standard model—$SU(3)$, $SU(2)$, and $U(1)$—are really one interaction, and hence indistinguishable. Second, the GUT symmetry implies that the underlying laws of physics make no distinction between an electron, a neutrino, or a quark. Both manifestations of the GUT symmetry are analogous to the indistinguishability of the different directions of space for the crystal.

The second row of Table 8.1 describes the mechanism of spontaneous symmetry breaking—what is it that breaks the symmetry? In the case of the crystal, the atoms arrange themselves along crystal axes which are picked out randomly by the first few atoms as the crystal starts to grow. Thus, the three directions of the principal axes (see Figure 8.2) become distinguishable from each other and from all other directions. In the construction of GUTs, theorists include a set of fields specifically for the purpose of spontaneously breaking the symmetry. These fields are known as *Higgs fields* (after Peter

	CRYSTAL	GUTs
SYMMETRY	Rotational Invariance	Three interactions indistinguishable Electron, neutrino, and quark indistinguishable
SPONTANEOUS SYMMETRY BREAKING	Crystal axes pick out three distinguishable directions	Higgs fields pick out three distinguishable particles — electron, neutrino, and quark — and also three distinguishable interactions
LOW ENERGY PHYSICS	Three fundamental axes of space Three fundamental speeds of light	Three distinguishable particles Three distinguishable interactions
HIGH TEMPERATURE PHYSICS	Crystal melts— rotational invariance restored	Phase transition at $T \approx 10^{29}\,^{\circ}\mathrm{K}$ — Symmetry restored

Table 8.1 Spontaneous symmetry breaking: the crystal–GUT analogy.

W. Higgs of the University of Edinburgh), and the spontaneous symmetry breaking mechanism is called the *Higgs mechanism.*

While Higgs fields were novel when they were first introduced into the electroweak theory, the general concept of a field had for many decades been part of the particle physicist's toolkit. In the standard model of particle physics, for example, a field is introduced for each of the fundamental particles, as listed in Table 7.1 (page 120). The electromagnetic field associated with the photon is familiar, but it is perhaps not so well known that the photon is the prototype on which the rest of our theories are based. The standard model includes a field associated with the electron, a field associated with the green-bottom quark, and so on. All these fields, including the electromagnetic and Higgs fields, are treated on an equal footing: Each field is postulated to exist, and to evolve according to a specified set of equations. Quantum theory implies that the energy of the electromagnetic field is concentrated in bundles, called photons, which can be interpreted as particles of light. Similarly, quantum theory implies that the energy of the electron field is concentrated in bundles, and these bundles are interpreted as the particles that we call electrons. In present-day particle theories, every fundamental particle is described as a bundle of energy of some field. The energy of the Higgs fields is concentrated into particles that are, not surprisingly, called Higgs particles. The Higgs particles associated with the breaking of the grand unified symmetry are expected to have masses corresponding to energies in the vicinity of 10^{16} GeV, which means that they are far too massive to be produced experimentally in the foreseeable future. The Higgs particle of the standard model, on the other hand, is expected to have a mass in the vicinity of 10^3 GeV.

Although ideas from quantum theory were needed in the previous paragraph to explain how particles arise from fields, the mechanism of spontaneous symmetry breaking itself can be described classically. Physicists designed grand unified theories so that they will lead to spontaneous symmetry breaking, by formulating them so that the energy density of the Higgs fields behaves in a peculiar way. For most fields, such as the electric and magnetic fields, the energy density of the field has its lowest possible value—zero—when the field vanishes. For the Higgs fields, however, the theories are constructed so that the energy density is lowest when the Higgs fields have nonzero values. The Higgs fields in empty space—often called *the vacuum*—will therefore have nonzero values, since they will settle into the state of lowest possible energy density.

To illustrate this point, Figure 8.3 shows a sample energy density diagram for a set of two Higgs fields. (The simplest grand unified theory actually requires 24 Higgs fields, but a set of two Higgs fields is sufficient to describe how the mechanism works.) I will call the two Higgs fields A and B. Since the Higgs fields interact significantly with each other, the total energy density depends on both of the Higgs fields, and cannot be expressed as a sum of the energy density of Higgs Field A and the energy density of Higgs Field B. The three-dimensional diagram shows the energy density of the Higgs fields, for any specified pair of values for the two Higgs fields. For example, suppose that Higgs Field A has the value 3 and Higgs field B has the value 2. To find the energy density of the Higgs fields from the diagram, first find the point in the base plane corresponding to this pair of values, as shown. Then determine the energy density from the height of the surface above the point in the base plane.

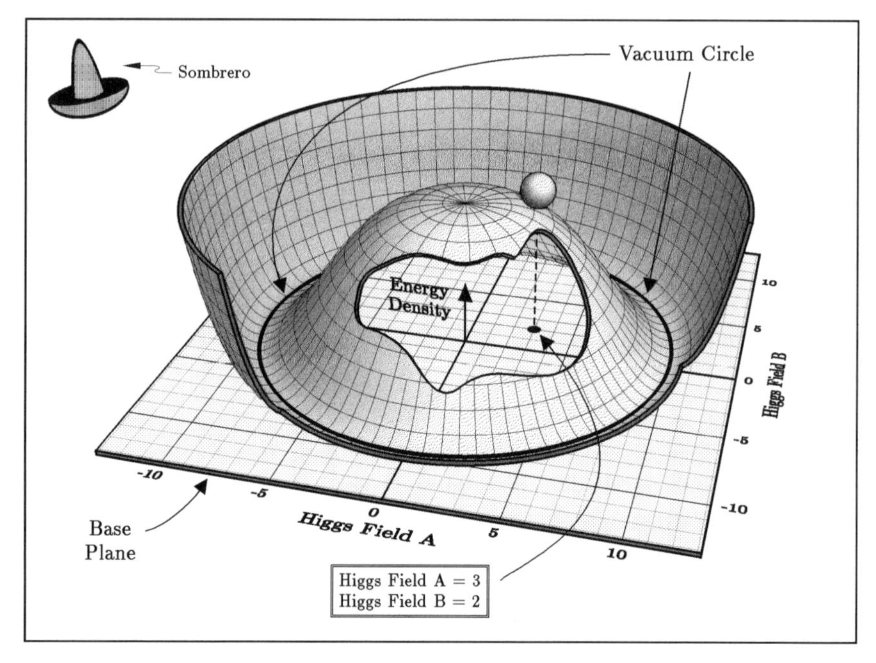

Figure 8.3 **Energy density of Higgs fields.** The graph shows the energy density for an illustrative theory that includes two Higgs fields. The motion of a ball rolling on this surface, as shown, is very similar to the evolution of the Higgs fields. The energy density has a large value if both of the Higgs fields are zero, but the energy density vanishes when the Higgs fields correspond to any point on the "vacuum circle." A random point on the vacuum circle is determined by random processes in the early history of the universe, spontaneously breaking the symmetry of the underlying theory.

To understand the implications of Figure 8.3, imagine a small ball located on the surface, as shown, directly above the point in the base plane corresponding to the values of the pair of Higgs fields. Since energy is needed to lift the imaginary ball, the gravitational energy of the ball increases with its height. The gravitational energy of the ball is therefore proportional to the energy density of the Higgs fields, which on the diagram is shown by the height of the surface. Although gravitational energy may not seem familiar, the *effects* of gravity are easy to visualize. The ball will be pulled downward by gravity, toward the curve labeled "vacuum circle." The evolution of the Higgs fields is in fact very similar to the motion of this ball.

The shape of the surface shown in Figure 8.3 is often described as a Mexican hat. Part of the Mexican hat has been cut away to improve the visibility, but the full diagram would include the entire hat, and even more: the brim would have no edge, but instead would extend upward to infinity. All units shown on the graph are arbitrary, since the goal is merely to show qualitatively how Higgs fields behave.

If both Higgs fields vanish, then the imaginary ball is sitting at the center of the Mexican hat, and the energy is relatively high. (It may sound strange that a large energy is required for the fields to vanish, but particle physicists have found nothing contradictory with this concept. Furthermore, the success of the electroweak theory provides strong evidence that, at least in this case, nature actually behaves in this peculiar way.) The state of lowest possible energy density is not unique: any of the points on the "vacuum circle" correspond to zero energy density. The imaginary rolling ball could eventually come to rest at any point on this circle. The values of the Higgs fields in the vacuum are therefore not determined by energy considerations. Just as the atoms in the crystal can align equally well along any of an infinite number of possible orientations, the set of Higgs fields in the vacuum can settle equally well at any point on the vacuum circle. Some particular point on the vacuum circle would be chosen randomly in the early history of the universe, just as the directions of the crystalline axes are chosen randomly as a crystal begins to form. This random choice of nonzero Higgs field values breaks the grand unified symmetry, just as the orientation of the crystal breaks the rotational symmetry. In both cases the underlying laws of physics remain exactly symmetric—the symmetry is broken "spontaneously," in the sense that it is only an accident of history that chooses the orientation of the crystal or the point on the vacuum circle for the Higgs fields.

The other particles in the theory interact with the Higgs fields, so they are affected by the random choice of Higgs field values. Since different particles interact with different Higgs fields, distinctions arise between particles that would otherwise be indistinguishable. For example, suppose the fields in the vacuum settle at the point for which Higgs Field A has the value 10 and Higgs Field B has the value 0. Then, as one might guess, the particles that interact with Higgs Field A will behave very differently from those particles that interact with Higgs Field B. Just as a finger pressing against a violin string can radically alter the pattern of its vibrations, the large value of Higgs Field A in this example can radically change the patterns of vibration of the other fields with which it interacts. Since particles are the bundles of energy of the vibrating fields, the properties of the particles are dramatically affected. In particular, the mass of a particle is determined by its interaction with the Higgs fields, so the masses of the particles that interact with Higgs Field A will become different from the masses of the particles that interact with Higgs Field B.*

In a full grand unified theory with a larger number of Higgs fields, some of the particles are caused to act like electrons, some like neutrinos, and others like quarks. Similarly some of the force-carrying particles will be caused to act like the gluons that carry the strong interactions; others will be caused to act like the W^+, W^-, and Z^0 of the weak interactions, and one will be caused to act like the photon of the electromagnetic interactions. Thus, the distinction between the strong, weak, and electromagnetic interactions is caused entirely by the ways that the force-carrying particles interact with the different Higgs fields. Since the masses of the force-carrying particles are determined by the interactions with the Higgs fields, the large masses of the *W* and *Z* particles are attributed to the Higgs mechanism. The *W* particle weighs 85 times as much as a proton, while the *Z* weighs 97 times as much. According to quantum theory, the range of a force is inversely proportional to the mass of the force-carrying particle, so the short range of the weak interactions is understood as a consequence of the Higgs mechanism.

Returning to Table 8.1 on page 138, we see that the third row describes the behavior of low energy physics in the two systems. Here the analogy can be made more illustrative by whimsically imagining a world of intelligent creatures living inside an orthorhombic crystal—we will call them Orthorhombons. Let us assume that the Orthorhombons can somehow move about and carry on the task of scientific investigation, but that they cannot muster enough energy to melt or even significantly perturb the crystal in which they live. Since the crystal would be immutable, the Orthorhombons would not recognize it as an object. Instead, the crystalline structure would

* There is a subtle point involved here, which for the benefit of footnote-readers I will try to clarify. Without looking at the detailed mathematics, one would most likely conclude that if Higgs Field A and Higgs Field B had equal values, then the symmetry would remain unbroken. The truth, however, is that the physical properties of the system are always the same, regardless of which point on the vacuum circle is chosen. To understand this claim, consider the crystal analogy. Before the crystal forms, all directions in space are equivalent, because the laws of physics make no distinction between one direction in space and another. Consequently, when the crystal forms, its internal properties do not depend on how the crystal is oriented in space. Sound waves, for example, might propagate faster along the shortest principle axis than along the longest, but the truth of this statement will not depend on how the crystal is oriented. Similarly, before the nonzero field values are established for the Higgs fields, all directions in the base plane of Figure 8.3 are equivalent. The directions labeled Higgs Field A and Higgs Field B were in fact chosen arbitrarily. As with the crystal, the internal physical properties of the system will therefore not depend on the orientation chosen by the Higgs field pair in this diagram. If Higgs Field A is equal to Higgs Field B, for example, then the theory is sufficiently complex to contain a field (or perhaps several) that interacts with the sum of Higgs Fields A and B, and also a field that interacts with the difference. After the Higgs fields acquire their nonzero values, the field interacting with $A + B$ will behave differently from the one interacting with $A - B$. When the rules of quantum theory are included, the two fields describe two types of particles, with different masses.

be taken as a fixed property of space. An Orthorhombon physics book would make no mention of rotational symmetry, but would instead contain a chapter discussing the properties of space and its three primary axes. The orthorhombic structure would affect the propagation of light through the crystal, so an Orthorhombon table of physical constants would list three speeds of light, one for each primary direction.

If grand unification is correct, then our universe is similar to this crystal world. The Orthorhombons live inside a crystal, the effects of which they mistakenly view as fixed attributes of space. We live in a region permeated by Higgs fields, the effects of which we mistakenly view as fixed attributes of the laws of physics. Our tabulation of the different properties of the strong, weak, and electromagnetic interactions is analogous to the Orthorhombon tabulation of the three different speeds of light. Similarly, the distinct properties that we observe for electrons, neutrinos, and quarks are not fundamental—they represent the different ways that particles can interact with the fixed Higgs fields that exist throughout the visible universe.

Finally, the last row of Table 8.1 describes the high temperature behavior of the two systems. If a crystal is heated sufficiently, it will melt and become a liquid. The melting of a crystal is an example of what is called a *phase transition*—a sudden change in the behavior of a material as the temperature is varied. The distribution of molecules in the liquid phase is rotationally symmetric, looking the same no matter how the liquid is turned. At high temperatures, therefore, the rotational symmetry is restored.

In grand unified theories, an analogous symmetry-restoring phase transition happens at extraordinarily high temperatures. To visualize this transition, recall that at zero temperature the two Higgs fields in our sample theory assume a pair of values on the vacuum circle of Figure 8.3. As before, we can represent the values of the two Higgs fields as the position in the horizontal plane of a small ball, which at zero temperature lies motionless at some point on the vacuum circle. As the temperature is raised, the Higgs fields acquire thermal energy and begin to oscillate. For visualization, we can imagine that the surface in Figure 8.3 is connected to a vibrator, and the vibrations cause the rolling ball to jiggle along the surface. When the temperature is low, the ball will undergo only small oscillations, which will continue to be centered on some point on the vacuum circle. Since the average values of the Higgs fields described by the oscillating ball are nonzero, the symmetry remains spontaneously broken. Once the temperature exceeds a certain value, however, the ball begins to thrash about wildly, sometimes crossing over the central peak. Its average position becomes the center of the Mexican hat, and all evidence of the initial zero-temperature values of the Higgs fields is lost. Each Higgs field then has an average value of zero, so the grand unified symmetry is restored. The $SU(3)$, $SU(2)$, and $U(1)$ interactions all merge into a single interaction, and there is no distinction whatever between electrons, neutrinos, and quarks.

For a typical GUT, this transition occurs at a temperature in the vicinity of 10^{29}°K. (This temperature corresponds to an average thermal energy of—you guessed it—10^{16} GeV.) If we lived at such extraordinary temperatures, our observations would directly exhibit the full grand unified symmetry.

A temperature of 10^{29}°K is of course outrageously high, even by the standards of astrophysics. The center of a hot star, for example, is only about 10^{7}°K. The application of grand unified theories, however, forces us to consider such outrageous temperatures. In the *Scientific American* article quoted at the beginning of this chapter, Einstein pointed out that there is an important advantage to theories for which the basic concepts lie close to our experience. With such theories there is much less danger in going completely astray, since they are relatively easy to test. "Yet more and more," Einstein continued, "we must give up this advantage in our quest for logical simplicity and uniformity in the foundations of physical theory." With grand unified theories, the "advantage" of staying close to our experience has been completely abandoned.

Returning now to the interaction strength diagram of Figure 8.1, we can now understand a bit more about how the diagram relates to the structure of a grand unified theory. The unification scale, where the three lines meet, is identified with the typical energies associated with the Higgs fields. More precisely, the important issue is the effect of the Higgs fields on the force-carrying particles. For interactions of particles at energies higher than the unification scale, the nonzero values of the Higgs fields are irrelevant. Such high energy particles are unaffected by the Higgs fields, just as a high speed bullet passes undeflected through a sheet of paper. Since the Higgs fields are unimportant at these energies, the force-carrying particles all behave the same way and the observed interactions reflect the full GUT symmetry. In particular, at these energies all of the force-carrying particles interact with the same strength. For the interaction of particles at energies below the unification scale, however, the effect of the Higgs fields is substantial. Some of the force-carrying particles behave like *SU*(3) force carriers, while others carry the *SU*(2) or *SU*(1) forces, and thus the standard model of particle physics is reproduced. The strengths of the three interactions are all related, however, because they must all meld into the single grand unified interaction at the unification scale.

In his 1950 *Scientific American* article, Einstein summarized his conclusions as follows:

> The skeptic will say: "It may well be true that this system of equations is reasonable from a logical standpoint. But this does not prove that it corresponds to nature." You are right, dear skeptic. Experience alone can decide on truth.

Einstein was writing of course about his own attempts to construct a unified field theory, attempts which never proved satisfactory. We can hope that grand unified theories will be more successful.

As we will see in the next chapters, the cosmological consequences of grand unified theories seem very attractive. While the ideas remain speculative, it is very encouraging to see that a consistent picture emerges. Grand unified theories not only provide a plausible description of fundamental particle interactions, but through the inflationary universe theory they can also help to explain the origin and structure of the universe around us.

CHAPTER 9

COMBATTING THE MAGNETIC MONOPOLE MENACE

At the beginning of May 1979, as I was still learning about the grand unified theories discussed in the previous chapter, I finally began to work with Henry Tye on the project that he had been touting since the previous November. We sought to estimate, under the assumptions of grand unified theories, how many magnetic monopoles would have been produced in the big bang. As is usually the case in starting a new problem, we had no idea where the results might take us. Magnetic monopoles had never been seen, so if the calculation predicted a large number of monopoles, it could spell trouble for GUTs. If the calculation led to a moderate number of monopoles, it would stir experimenters to search harder for these unseen relics of the big bang. Alternatively, we might find that only a negligible number of monopoles would have been produced, in which case the main value would be to save other theorists the time of repeating the calculations.

As we began to think about the problem, I found it fraught with difficulties. Whenever I start a new problem, I worry that I might miss the most obvious and important effects. In a territory as unfamiliar as the early universe, my natural feelings of insecurity were quadrupled. These uncertainties were further compounded by the fact that I was still shaky in my knowledge of grand unified theories.

At least one fact, however, was unambiguous: Magnetic charge is conserved, like electric charge. That is, the total magnetic charge of any system of particles cannot change with time. Since the magnetic charge of a monopole cannot disappear, a monopole cannot decay into ordinary matter. Nonetheless, magnetic monopoles can disappear. A south monopole, which is the antiparticle of a north monopole, has a negative magnetic charge that exactly cancels the magnetic charge of a north monopole. Since the total magnetic charge of a north and south monopole is precisely zero, the two particles can collide and annihilate. All the energy is then converted into ordinary particles—particles that are not magnetic monopoles.

To estimate the number of magnetic monopoles that are expected to survive from the early universe, there are two processes that need to be understood. The first is the creation of monopoles, and the second is their annihilation. Henry and I understood neither.

Nonetheless, within a week we made our first crude estimate of the monopole density from the early universe. We finessed the production problem by assuming that monopoles had an initial abundance approximately equal to that of photons. While we had no grounds for confidence in this assumption, we knew that it was not totally crazy. According to the standard big bang theory, at very early times all the known fundamental particles had the same abundance as photons to within about a factor of 2. Since we did not understand how monopoles would be produced, we assumed that they would fall into the same pattern as everything else. For the annihilation, we used a rough estimate based on the formula for the annihilation of electrons and positrons. Our result: The monopoles and antimonopoles that have not yet annihilated should be about 10,000 times more abundant today than protons or neutrons.

Since monopoles had never been seen, it was obvious that the universe could not possibly contain as many monopoles as this calculation indicated. There were three possibilities: Grand unified theories might be wrong, standard cosmology might be wrong, or our estimate might be wildly off target. Unfortunately, at this point the third possibility seemed at least as likely as either of the first two.

The uncertainties were enormous. For example, just two days after Henry and I formulated this estimate, Ken Wilson told us at lunch that he was not even convinced that monopoles are a valid prediction of grand unified theories. All the calculations leading to monopoles had been based on a classical description, and Ken suspected that the effects of quantum theory would change things completely. Several weeks later Ken agreed that his original arguments were probably wrong, but his doubts added weight to my own uncertainties.

At this time Henry and I did not even know if monopoles might possess fractional electric charges, in addition to their magnetic charges. If so, they could be confined in a manner similar to quarks, and the monopole overproduction problem would disappear. (Later we learned that monopoles have integer electric charges, like electrons or protons.) We discovered that in addition to the expected magnetic field, GUT monopoles would produce another field that is similar to the magnetic field, but which arises from quantum chromodynamics, the theory of the strong interactions. We had to worry whether this field might lead to confinement similar to quarks, but

Henry discovered a reason to believe that it would not. At the end of June we discovered a paper by Gerard 't Hooft (the co-inventor of the monopole) that reached the same conclusions.

The probability of annihilation was also an important uncertainty. We knew that the attractive force between monopole and antimonopole would be very effective if the monopoles were moving slowly, and since the monopoles are so massive they would move *very* slowly. We were not sure, however, how to compute the annihilation of such slowly moving particles. Two weeks after our first estimate Henry worked out an alternative, based on a textbook calculation of how free electrons and protons can interact to emit a photon and form a hydrogen atom. This time he found that monopoles should be 10^{12} times more *rare* than protons or neutrons! Meanwhile I considered a treatment of the problem that completely ignored the effects of quantum theory, which seemed a reasonable approach for such slowly moving, massive particles. My new estimate was very close to our first, but we did not understand why we were obtaining such different answers. We were also uneasy about the fact that these annihilations were not happening in empty space, as our calculations presupposed, but instead were occurring in a hot, dense gas of charged particles—a plasma. There is a whole branch of physics associated with the complicated behavior of plasmas, and Henry and I knew none of it.

Finally, there was the uncertainty in the production rate of monopoles. They would have been produced during the grand unified theory phase transition at about 10^{29}°K, which was described in the previous chapter. Our initial guess—that monopoles would be about as abundant as photons—could be grossly in error, since monopoles are really very different from all the other particles in the theory.

In the middle of May we learned from Paul Ginsparg, a Cornell graduate student, that a friend of his, John Preskill, was working on the same problem. Preskill was a graduate student at Harvard, working under the supervision of Steve Weinberg. We were cocky enough not to feel threatened by a mere graduate student, but the involvement of Weinberg in the problem made us worry about the competition. (We soon learned, however, that the major competition on this issue would come from Preskill himself.)

At the end of May, Henry discovered an important paper on magnetic monopoles in the early universe, published the previous fall by the prominent Soviet astrophysicist Yakov B. Zeldovich and a younger colleague, Maxim Yu. Khlopov [1]. The paper gave a very thorough treatment of the annihilation of monopoles, which had been one of our major uncertainties. Since Zeldovich had experience in plasma physics, we expected that his treatment could be trusted. The paper, however, did not attack quite the right problem, for it considered only the possibility of much lighter monopoles, with a mass/energy of about 10^4 GeV, roughly a trillion times lighter than a GUT monopole. The conclusions of the paper depended crucially on the monopole mass, so they were not directly applicable to the monopoles of grand unified theories. Interestingly, however, Zeldovich and Khlopov concluded that the predicted monopole abundance was so large that they would easily have been seen in monopole search experiments. To avoid the discrepancy, Zeldovich and Khlopov proposed that "a mechanism similar to quark confinement should forbid free monopole existence."* By this time Henry and I were pretty confident that no such mechanism exists.

At this point I was very interested in the monopole question, but the majority of my time was still committed to other projects. The work on lattice gauge theory, which was mentioned in Chapter 6, was continuing at a fast pace, as the level of my work schedule accelerated by several notches. My first lattice gauge theory paper had been completed just a few days before Steve Weinberg's visit to Cornell. Much of my time in June and July was spent drafting a second lattice gauge theory paper, which was completed just two weeks before I left for California.

In the middle of June, Henry and I received a preliminary draft of a paper by Preskill, and by this time we were also in telephone contact. A week and a half later we received the final version [2]. Preskill had adapted the methods of Zeldovich and Khlopov to GUT mass monopoles, and had also developed a convincing estimate of magnetic monopole production, at least for some of the theories of how the GUT phase transition in the early universe might have taken place. His conclusion: Under the standard assumptions of grand unified theories and cosmology, the number of monopoles expected from the big bang is about equal to the number of protons and neutrons. Without even considering the status of experimental searches for magnetic monopoles, Preskill argued that the universe could not possibly contain such a huge abundance of monopoles. Since the monopoles would each have a mass about 10^{16} times larger than a proton, their enormous gravitational attraction would have caused the expansion rate of the universe to plummet. This implies that in the past the universe expanded much faster than its present rate, allowing the galaxies to reach their present separation in a very short time. If there were as many monopoles as protons and neutrons, the universe could not be more than 1200 years old! To be consistent with realistic estimates of the age of the universe, between 10 and 20

* Unfortunately I never had a chance to meet Zeldovich, the fiery, self-educated man who for many years dominated Soviet cosmology, and who is remembered with great affection by his younger colleagues. Maxim Khlopov told me that Zeldovich could never stop correcting his own manuscripts, so at some point he would ask his coauthors not to show him the text. When Zeldovich was preparing a collection of his articles to be published by the Soviet Academy of Sciences in commemoration of his 70th birthday, he called Maxim and said, "Your work on magnetic monopoles is, certainly, the pearl of my collection of works, but I don't understand the conclusion of this work." Maxim reminded him that he had written the conclusion himself, and tried to explain how the conclusions were justified. Zeldovich was not satisfied. "Okay, I'll correct it," he snapped, ending the conversation.

billion years, Preskill concluded that the monopole abundance would have to be about a hundred trillion (10^{14}) times lower. Thus, Preskill's paper announced to the world what became known as the *magnetic monopole problem*—standard grand unified theories lead to a fantastic overproduction of magnetic monopoles in the early universe.

Preskill had clearly been very careful in his work, and Henry and I were impressed. We invited John to visit Cornell, and he came for three days in the middle of July. He gave a seminar on his work, and we discussed related issues. Shortly before John's visit Henry had discovered that the original grand unified theory of Georgi and Glashow probably leads to two early universe phase transitions, rather than just one. We showed this work to John, and discussed a number of questions that arose from it.

John Preskill in his office at Caltech.

During his visit, John also explained to us a new method he had invented for estimating the production of magnetic monopoles in the early universe. The new method was aimed at treating all types of phase transitions, while the results in his paper were restricted to the most straightforward class.

To understand Preskill's new method for estimating monopole production, we will need to have a picture of the internal workings of a grand unified theory magnetic monopole. Fortunately, the important features of the monopole can be shown pictorially, without any mathematics.

In a full grand unified theory the picture of a monopole becomes rather intricate, mainly because of the large number of fields. To avoid this sort of irrelevant complication, theoretical physicists often employ what are called *toy theories*—mathematical theories that are too simple to describe reality, but which are nonetheless useful to study because they incorporate important features of reality. (Rigorously speaking, even a theory such as general relativity could be considered a toy theory, since we believe that the ultimate description of gravity must include quantum effects that are not contained in general relativity. In practice, however, the phrase "toy theory" is reserved for theories that are much more blatantly unrealistic.) The modern picture of a magnetic monopole was actually discovered and described by 't Hooft and Polyakov in the context of a toy theory, and I will use the same toy theory to describe the monopole here.

The magnetic monopole is constructed from Higgs fields, like those described in the previous chapter. (For the sake of readers who skip around, however, I will repeat here all the properties of Higgs fields that are necessary to understand the monopole construction.) The simplest theory that gives rise to magnetic monopoles includes three Higgs fields, which I will unimaginatively call Field A, Field B, and Field C. Unlike an electric or magnetic field which points in some particular direction in space, a Higgs field has no directional properties. While an electric field is specified by its value and its direction, each Higgs field is completely specified by a single number—its value. At each point in space each of the three Higgs fields must have a value; the value can be zero, but in any case it is well defined.*

To proceed, we will benefit from having a graphical way of describing the value of the three Higgs fields. Since there are three dimensions to space,

* In quantum theory one is forced to deal with probabilities rather than certainties, so we would have to allow for a probability distribution of values for the Higgs field. The monopole, however, is rather large by the standards of elementary particle physics, so a description that ignores quantum theory is reasonably accurate. The original papers of 't Hooft and Polyakov gave only a classical description, although 't Hooft commented that quantum corrections should be calculable, but difficult.

we can conveniently represent the values of the three Higgs fields by drawing a three-dimensional graph, as in Figure 9.1. Using this graph, the values of all three Higgs fields at any *one* point of space can be represented by an arrow in three dimensions. To completely describe the Higgs field, however, we need to specify the values of all three Higgs fields at *every* point in space. In principle we must draw a three-dimensional arrow at each point in space, but in practice we settle for an arrow drawn at a representative sample of points.

There is an energy density associated with the Higgs fields, and, as discussed in the previous chapter, the energy for the three fields is not the sum of an energy density for each of the individual fields. Instead, the energy density depends on the length of the Higgs field arrow. The energy density is high when the arrow has zero length, and has a minimum value when the arrow has a specific nonzero length—called the vacuum Higgs value—as shown in Figure 9.2. (This is completely analogous to the energy density

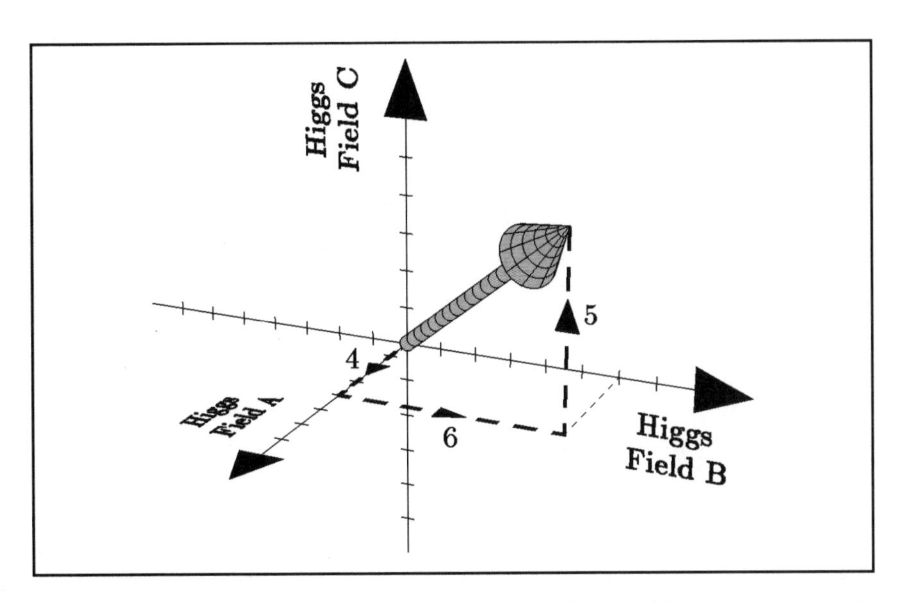

***Figure 9.1* Graphical representation of three Higgs fields at one point in space.** At any one point in space, the values of each of the three Higgs fields in the toy theory can be represented by a single arrow. Shown explicitly is the arrow corresponding to the case in which Field A has value 4, Field B has value 6, and Field C has value 5. The base of the arrow lies at the origin, the center point of the diagram. The tip of the arrow is located by starting from the origin, traveling a distance of 4 units along the Field A direction, a distance 6 along the Field B direction, and a distance 5 along the Field C direction. To represent the Higgs field everywhere, at each point in space an arrow would have to be drawn to show the values of the Higgs fields at that point.

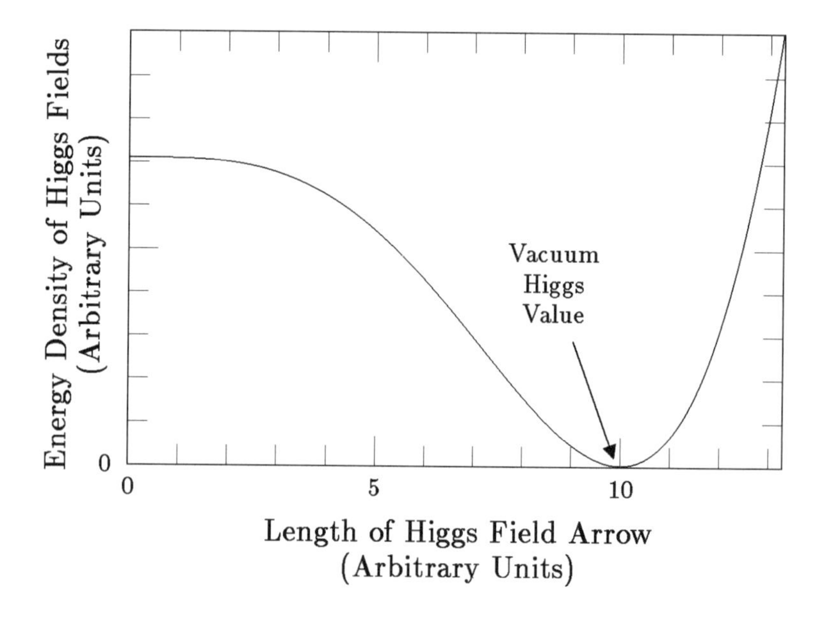

Figure 9.2 **Energy density of Higgs fields.** The graph shows the energy density of the three Higgs fields in the toy theory. The energy density depends on the length of the Higgs field arrow, as shown in Figure 9.1. The value of the length for which the energy density vanishes is called the *vacuum Higgs value*. All the units are arbitrary, since the toy theory is used only for illustration, and does not describe any real physical system.

described by the Mexican hat surface of Figure 8.3, which depends only on the distance between the center of the base plane and the point representing the Higgs field pair. The only difference is that now we are discussing three Higgs fields, rather than two.)

In the vacuum, the energy density is by definition at its lowest possible value. The length of the Higgs field arrow, therefore, must be the vacuum Higgs value. The direction of the arrow, however, is arbitrary, since the Higgs fields would have the same energy density for any direction. The direction of the Higgs field arrow is therefore not fixed by energy considerations, but is determined instead by random processes in the early universe. In addition to the energy density of the Higgs fields shown in Figure 9.2, the theory provides that any variation of the Higgs fields from one point in space to another also contributes to the energy. Since the vacuum has the least possible energy, the Higgs fields in the vacuum cannot vary from place to place. Figure 9.3(a) shows one possible picture of the Higgs fields in the vacuum. In this picture the Higgs arrows all point up, which according to Figure 9.1 indicates that only Higgs Field C is nonzero. The picture is not

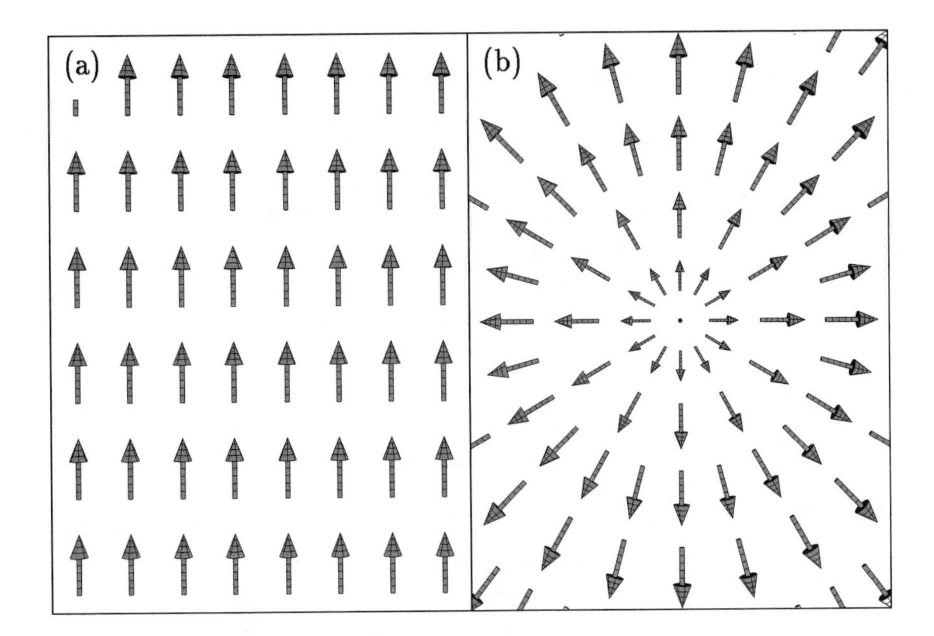

Figure 9.3 **Higgs fields for the vacuum and for a magnetic monopole.** At the left is a picture of the Higgs fields in the vacuum. Each arrow is a graphical representation for the values of the three Higgs fields, as in Figure 9.1. The arrows describing the vacuum could just as well point in any direction, but the length must be the vacuum Higgs value, and all the arrows must point in the same direction. On the right is a picture of the Higgs fields of a magnetic monopole. The fields are shown on a plane going through the center of the monopole. All three Higgs fields vanish at the center, and the Higgs arrows everywhere else point directly away from the center.

unique, however, because the upward direction for the Higgs arrows was chosen arbitrarily. Any other direction would correspond to an equally possible Higgs field for the vacuum.

The Higgs fields of the monopole are a little more difficult to visualize, since the monopole is a three-dimensional object. Figure 9.3(b) shows the Higgs fields on a two-dimensional cross section of the monopole—a plane that cuts through the center point of the monopole. All three Higgs fields vanish at the center, so the Higgs arrow has zero length, and is not shown. This creates a large energy density, which contributes a large part of the enormous mass of the magnetic monopole. Everywhere else the Higgs arrow points directly outward from the center. The length becomes larger as the distance from the center increases, but it does not become larger indefinitely. Instead, as the distance from the center is increased, the length becomes closer and closer to the vacuum Higgs value, the length for which the energy density vanishes. In a three-dimensional picture, the Higgs field arrows would similarly all point directly away from the center. Polyakov called the configuration a hedgehog, alluding to the way this porcupine-like creature curls itself into a ball for protection, with its sharp quills pointing outward.

Figure 9.3(b) shows the Higgs fields of the monopole, but it is the magnetic field (which has a similar appearance, shown in Figure 2.5(c), page 29) which justifies the name "magnetic monopole." The magnetic field arises from the complicated way that the Higgs fields interact with the force-carrying fields of the grand unified theory. Recall that these interactions determine which of the force-carrying fields behaves as the electromagnetic field, which behave as the gluons of the strong interactions, and so on. Because the Higgs fields in Figure 9.3(b) vary from point to point, the determination of which force-carrying field behaves as the electromagnetic field is made differently at each point. The detailed mathematics shows that the Higgs field configuration of Figure 9.3(b) leads to a magnetic monopole field, with a magnetic strength exactly equal to what Dirac found in 1931 to be the minimal unit of magnetic strength that is consistent with quantum theory. This magnetic strength is 68.52 times stronger than the electric charge of an electron. Since the force between two charged particles is proportional to the product of the two charges, the force between two monopoles is $(68.52)^2$, or 4695 times stronger than the force between two electrons at the same separation.

With this understanding of the structure of a magnetic monopole, we can now understand the logic of John Preskill's new method for estimating the production of monopoles in the early universe. Consider the conditions in the very early universe, when the temperature was above 10^{29}°K. As discussed in the previous chapter, the Higgs fields were fluctuating wildly in the intense heat, just as molecules oscillate wildly at high temperatures. However, the average value of the Higgs field at any point, averaged over a brief interval of time, was equal to zero. As the universe cooled below about 10^{29}°K, the Higgs fields underwent a phase transition, in which the oscillations diminished and a nonzero average value was established.

The cooling was taking place throughout the universe, but the Higgs fields at a particular location could interact only with the fields in nearby regions. There were no forces that could transmit the effects of the Higgs fields over long distances. Thus, the random determination of the direction in which the Higgs arrow began to point was made independently in many different small regions of space.

Thus, the Higgs field shortly after the phase transition could not have looked anything like the highly organized pattern that describes the vacuum, as in Figure 9.3(a), but instead must have been a jumble in which the Higgs field arrow varied randomly from one small region to another. This initial jumble, however, would not survive for long. The variation of the Higgs field from place to place requires energy, and the energy density was decreasing as the universe expanded and cooled. The Higgs fields therefore smoothed out as the universe evolved, with the Higgs arrow in any region tending to align with the Higgs arrow in neighboring regions.

This tendency of the Higgs field arrows to align, however, could proceed only in regions for which the neighboring regions showed some degree of consistency. For some regions, however, this consistency would be lacking. For example, consider the point at the center of the monopole in Figure 9.3(b), at which the Higgs fields are all zero. Surrounding this central point the Higgs arrows point in every possible direction. If the Higgs fields at the central point became nonzero, then the Higgs arrow would have to point in some direction. If the arrow began to align with the fields in the region above the central point, the arrow would become further from alignment with the Higgs arrow below the central point. The net effect, when the mathematics is carried through, would be to increase the energy. In other words, once a Higgs field pattern such as in Figure 9.3(b) is encountered, the alignment process can continue no further.

The monopoles, therefore, are the surviving remnants of the chaos in the Higgs fields immediately after the phase transition. This basic idea was set forth very clearly by T.W.B. (Tom) Kibble in 1976 [3], but his paper did not specifically discuss grand unified theories.

To estimate the amount of monopole production, we must understand the degree of chaos in the Higgs fields that would result from the phase transition. Phase transitions, however, are always complicated things, and in this case we are trying to learn as much as possible without even knowing the details of the underlying particle theory. (Recall that grand unified theories are a class of theories, and that even if we chose the correct theory, we would still require access to a 10^{16} GeV accelerator to measure the necessary parameters.)

John, however, had an idea for at least estimating the amount of chaos in the Higgs fields. He pointed out that nothing travels faster than light, and the phase transition happened so early in the history of the universe that even signals traveling at the speed of light could not have gotten very far. Specifically, John invoked what cosmologists call the *horizon distance*, which is the total distance that a light pulse could have traveled from the instant of the big bang until the time under consideration. Suppose we imagine two points, which I will call *A* and *B*, immediately after the phase transition, as shown in Figure 9.4. If these two points are separated by more than the horizon distance, then there is no way that events at one point could have any influence on events at the other point. Furthermore, if the separation is more than two horizon distances, then it is not even possible for an event at some third point *C* to have a common influence on both points *A* and *B*. If the separation is

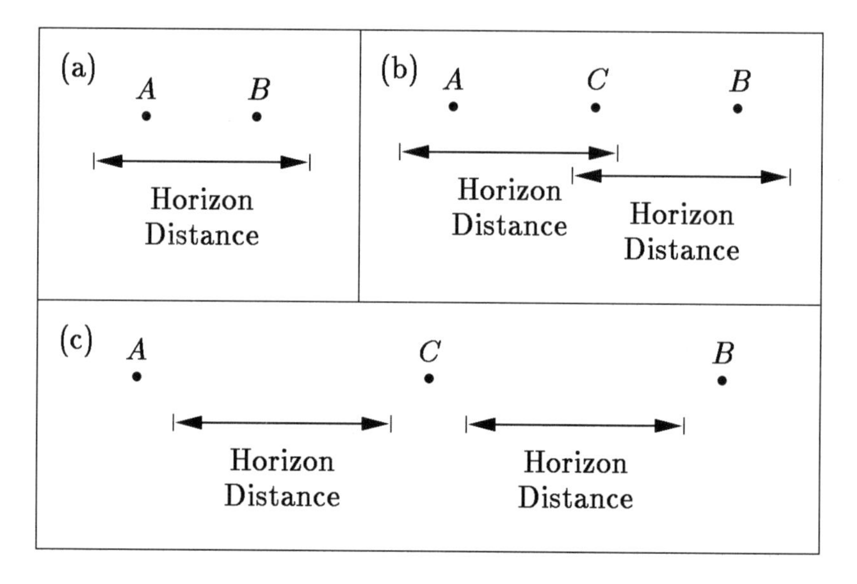

Figure 9.4 **The horizon distance and Higgs arrow alignment.** In part (a), two points *A* and *B*, immediately after the phase transition, are separated by less than the horizon distance, the total distance that a light pulse could have traveled since the instant of the big bang. Events that have taken place at one point can therefore affect the Higgs arrow at the other point, causing a tendency for the two Higgs arrows to align. In part (b), the two points are separated by more than the horizon distance, but less than twice the horizon distance. In this case, a tendency for the two arrows to align can be caused by events that have taken place at points such as C, which can influence the Higgs arrow at both points *A* and *B*. However, if the separation between points *A* and *B* is more than twice the horizon distance (as in part (c)), then there can be no tendency for the two Higgs arrows to align.

more than two horizon distances, the Higgs arrow at point *A* can have no tendency whatever to align with the Higgs arrow at point *B*. The statement that there is no alignment of Higgs arrows at distances more than twice the horizon distance provides a measure of the chaos in Higgs fields. The Higgs fields might have been more chaotic than this statement implies, since they might not have aligned over the full two horizon distances. They could not, however, have been less chaotic, since less chaos would require more alignment, which is forbidden by the restriction of the speed of light. Since monopoles are the remnants of the chaos, this statement about the degree of chaos can be converted into a statement about the number of monopoles produced. The estimate is a lower bound on the true number, which could have been larger than this estimate, but could not have been smaller.

Unlike the lecture by Dicke discussed in Chapter 2, which left an indelible mark on my brain, for some reason this particular conversation with John Preskill sank quickly into the lower levels of my subconscious. I doubt that I questioned the validity of the argument, but I probably thought it sounded weak. That is, I suspected that the true number of monopoles produced would be much larger than this bound. I made a brief note of the discussion in my diary, but then put it out of mind. Later Henry and I would use a version of this argument in our paper, and shortly afterward I would receive a friendly letter from John. John made some suggestions on the physics, and commented that he was surprised to see his argument used without acknowledgment. I then tried to reconstruct what had happened, and decided that John was right. I apologized to John (or at least I think I did), and in any case my next paper included an acknowledgment to John on this point.

Now that Preskill's paper was out, Henry was anxious for us to publish, too, so that we could claim at least some credit for what appeared to be an important result. I tended to be more cautious, reluctant to publish until I was confident that I understood a topic well. In any case, we agreed that we needed to shift the emphasis of our work. Since Preskill had already established the existence of the magnetic monopole problem, there would be no value in merely repeating his conclusions. At this point, the important question was whether the monopole problem could be avoided. Could we find some set of circumstances under which grand unified theories would be compatible with hot big bang cosmology? In the first week of July, Henry wrote a working draft of our results, but it was quite sketchy and seemed to me to be a long distance from a publishable paper.

By this time, however, I was becoming increasingly pressured by the practical chores of arranging the move to Stanford. Housing prices in California were astoundingly high, especially by Ithacan standards, so finding an affordable place to stay was not easy. SLAC maintained a housing office to assist, but each of the few possibilities that they offered was unacceptable for one reason or another. By the end of July there were still no promising leads, and I became nervous. At this point I learned that there are benefits to belonging to a nomadic tribe—the tribe of physics postdocs. I had three good friends who had moved to Stanford University or its immediate vicinity, so I asked them for help. I soon had a small brigade poring through newspaper ads, visiting campus housing offices, and checking out apartments and houses. It was not until August 13 that one of my friends found a promising lead: a rental listing of a small furnished house in nearby Menlo Park. The following day he visited the house, and reported to me that evening.* The house was a little more expensive than we could really afford, but it sounded perfect—it had an enclosed yard for my infant son to play in, it was within biking distance of SLAC, and there was a small extra room that I could use as a study. I called the owners, but they were hoping to find one tenant for the entire two-year period that they expected to be away. They had just recently made these plans, however, and they were beginning to appreciate that September was only two weeks away! By the next day they agreed to rent the house on a one-year lease.

After much family discussion, my wife Susan and I decided that I would drive the car to California, with our belongings stuffed into a rented trailer. Susan would meanwhile stay with our son Larry at her mother's, and they would later fly to California to join me.

When I made the trip, during the last week in August, it was the first time that I had been separated from Larry since he was born. He was only one and a half years old, so our telephone conversations consisted of my telling him how much I loved and missed him, and his making "horsey" noises and saying "Da-da, Da-da, Da-da!" I had one car breakdown along the way, but was lucky in finding a service station to replace my voltage regulator in Geneseo, Illinois. It was wonderful when the expedition was over, and I met Susan and Larry at the San Francisco Airport to bring them to our new home.

As I began to work at SLAC, it instantly became clear that I had come to the right place. Sidney Coleman was visiting SLAC for the entire year, on sabbatical from Harvard. Stanford had recently hired Leonard (Lenny) Susskind, one of the pioneers of lattice gauge theory and a number of other topics in theoretical physics. Paul Langacker, visiting for several months from the University of Pennsylvania, was writing what was to become the major review article on grand unified theories and proton decay [4]. So-Young Pi, with whom I would later collaborate (see Chapter 13), was then a SLAC postdoc with strong interests in grand unified theories and cosmology. Including regular SLAC and Stanford people and long-term and short-term visitors, there were maybe two dozen particle theorists actively at work. The atmosphere at SLAC, furthermore, was very conducive to interactions, so information and ideas were literally bouncing off the walls at all times.

About a week after I arrived, John Ellis, a leading GUT researcher from CERN, gave a seminar on "Grand Unification and Beyond." At the seminar Lenny Susskind asked about magnetic monopoles, insinuating that he

* Stuart Raby: In case I have forgotten to thank you—Thanks!

viewed the magnetic monopole problem as a serious hurdle for grand unified theories. At this point I chimed in, announcing that Henry and I were also looking at that question, and that we thought there were possibilities for evading Preskill's arguments. Afterward I spoke to Lenny for some time, and according to my diary notes, "I think I convinced him that grand unified models are not ruled out." Apparently Lenny brought up the limit imposed by horizons—the same argument Preskill had explained during his visit to Cornell in July. We did a rough calculation, however, confirming my belief that this argument leads to no strong conclusions. The next day, however, Lenny called to tell me that we had botched the calculation. In fact, the horizon argument leads to an estimate that is about 100 times smaller than the prediction in Preskill's paper, which means it is still about a trillion times too large!

The day after the telephone call from Lenny, I had a long telephone conversation with Henry. He was very eager to get us into print, so we agreed to write what we knew into a paper. It was still unclear, however, exactly what we knew. I explained Lenny's horizon bound to Henry, and after about a week he became convinced that it was a valid argument that we needed to address. (But neither of us remembered that Preskill had shown us essentially the same argument.)

Although the horizon bound on monopole production depends on none of the details of the phase transition, it can still be evaded under the right circumstances. Recall that the monopole estimate is large because the horizon distance is short, implying a large amount of chaos in the Higgs fields. Suppose, however, that the phase transition were somehow delayed. Then the horizon distance—the total distance that a light pulse could have traveled from the instant of the big bang until the completion of the phase transition—would have time to grow. The Higgs fields would have time to align over longer stretches of distance, so the degree of chaos would decrease. The horizon bound on the monopole production would similarly decrease. If Henry and I could find a situation in which the phase transition was significantly delayed, the monopole problem would melt away.

So Henry and I started sending drafts back and forth, and gradually our understanding crystallized. As was typical for us, Henry wanted to plow ahead, while I was inclined to dawdle and cogitate. There was a deadline, however, for Henry would be leaving just before Christmas for a short visit to Boston, followed by a six-week trip to the People's Republic of China. He and his wife would visit their families in Hong Kong, and then Henry had been invited to speak in Guangzhou at a conference on theoretical particle physics for Chinese-speaking physicists from around the world. Further complications were caused by Henry's situation at Cornell, where one senior theorist had told him that the work on monopoles was too "esoteric" to further his promotion case. He was applying for a newly created position as a staff theorist intended to facilitate interactions with the experimental particle physics group, so he was also working hard on a number of papers more closely related to experiments.

At the beginning of October, I received a strongly encouraging letter from Kurt Gottfried, one of the senior theorists at Cornell who had always been very helpful. He told me that David Schramm, who with Michael Turner had recently written a major article on "Cosmology and Elementary Particle Physics" for *Physics Today*, had visited Cornell and spoken to Henry about our work. I had not yet met Schramm, but I had learned enough about cosmology to know that he was one of the leaders in the field. Schramm was very interested in what Henry told him, and according to Kurt he was "appalled that it was not written up." Kurt urged me to waste no more time, and emphasized that we need not be embarrassed to say that there are important issues that are not yet resolved. Kurt was even tactful enough to add that he was giving the same advice to Henry, but I was sensible enough to recognize that it was targeted at me.

During October, Henry and I worked through a set of horrendous calculations to understand in detail the early universe phase transitions predicted by the original Georgi-Glashow grand unified theory. We verified what Henry had found earlier from a more approximate calculation: In this theory there would be two early universe phase transitions, not just one. The monopoles would be produced at the second phase transition, the one that occurs at the lower temperature. The exact temperatures of the phase transitions could not be determined, since they depend on parameters of the theory that could be measured only by experiments with a 10^{16} GeV accelerator. However, we found that *if* the values of unknown parameters were within a certain range, then the temperature of the second phase transition would be very low. A low temperature means a late phase transition, which is exactly what was needed to suppress the production of monopoles.

In the middle of November, Henry called, excitedly telling me that he had learned that Marty Einhorn (of the University of Michigan) was working on the same questions we were. Since we had already allowed Preskill to publish well ahead of us, we could certainly not afford to be outflanked again. From this point on we would be working at full throttle.

Two days later I showed our idea of a low-temperature second phase transition to Lenny Susskind and to Paul Langacker, and both thought that it looked like a plausible way to avoid the overproduction of magnetic monopoles. I began to write a draft based on this theme.

By Thanksgiving weekend, the draft was half-written and the pressure was mounting. Henry's trip was only one month away, and the following weekend I would be attending a conference at the University of California at

Irvine. I was not lecturing at the conference, but I would have contact with many people, and it would be very effective if I could tell them our results. I spent Thanksgiving day with my family, but worked at SLAC on the Friday, Saturday, and Sunday of that weekend. The draft was nearing completion, but as I set the details to paper, one flaw became apparent. The method that we were using to calculate the phase transitions was based on a high-temperature approximation, yet we were applying this method to a phase transition that occurs at a temperature which is very low, at least by the standards of GUTs. On Saturday I figured out a way to revise the approximations for low temperatures, and on Sunday afternoon I reworked the calculation with this method.

I then asked a question that we had not previously considered: How large is the range of values for the unknown parameters that would lead to the low-temperature second phase transition which we were hoping could explain the dearth of monopoles? The range proved to be incredibly narrow! It was so narrow that the parameters would have to be specified to twelve decimal places for the monopole suppression to work. While it is possible that the true values of the parameters lie in this range, we knew of no reason why they should. The low-temperature second phase transition began to look like a very implausible solution to the monopole problem!

There was an alternative possibility, however, that Henry and I had discussed earlier but never pursued in detail. The natural temperature of the second phase transition might be high, but the phase transition might occur only after a large amount of *supercooling*. Supercooling is a situation in which a phase transition is delayed, so the temperature falls well below the normal temperature of the phase transition before the transition takes place. Water, for example, can be supercooled by more than 20°C below its freezing point without it turning to ice. If the grand unified theory phase transition were postponed by supercooling, then we would expect monopole production to be strongly suppressed.

A phase transition with a large amount of supercooling is generally of a type that physicists call *first order*, precisely the type that was not considered in Preskill's paper. First order phase transitions are actually quite familiar, since the boiling of water is an excellent example. When water boils, the temperature of the water rises slightly above the boiling point. Bubbles of steam then form randomly in the hot liquid. As each bubble grows, it absorbs energy from its surroundings, preventing the temperature from rising much above the boiling point until the water boils away. The heat energy absorbed by the growing bubbles is used to convert water to steam, since steam is the higher energy phase.

The description of a first order phase transition in the early universe is very similar, except that the temperature is falling rather than rising. Thus, the temperature of the early universe would fall below the normal temperature of the phase transition. To suppress monopoles, it must fall very far below the normal temperature. Bubbles of the new phase would then begin to form randomly and to grow, just like the bubbles of steam. One obvious difference, however, is that bubbles of steam in a pot will rise to the surface, since gravity pulls downward more strongly on the water. In the early universe, however, the bubbles would just grow spherically, expanding at nearly the speed of light until they collide with other bubbles. Eventually these bubbles would merge to fill all space, completing the phase transition—or at least so we thought at the time. (As I will discuss in Chapter 11, the question of filling space becomes very complicated when inflation occurs.)

For GUT phase transitions, the phases can be described in terms of the behavior of the Higgs fields. In the high temperature phase before the transition, the Higgs fields were undergoing large oscillations, with an average value of zero. In the new lower temperature phase, inside the growing bubbles, the decrease in available energy causes the Higgs fields to settle into a low energy state. The Higgs field arrow continues to oscillate due to the thermal energy, but now the oscillations are smaller than in the high temperature phase, and are centered about an average value for which the length of the arrow is near the vacuum Higgs value. When each bubble forms, however, the direction of the Higgs field arrow inside that bubble is determined randomly. The Higgs field arrow is nearly constant within any one bubble, but the arrow of one bubble would have no tendency to align with the arrow of any other bubble. So, for a first order phase transition, the degree of chaos in the Higgs fields is determined by the distribution of bubbles. Each bubble expands at nearly the speed of light after it materializes, but the rate at which bubbles materialize depends sensitively on the details of the underlying particle theory. Rapid bubble materialization would lead to a dense foam of small bubbles with a high degree of chaos. If the bubbles materialized at a slow rate, however, then a small number of bubbles would have time to fill space, minimizing the degree of chaos. Since monopoles are the remnants of chaos, slow bubble formation would suppress their production.

On that Sunday afternoon I made a rough estimate of the rate of bubble formation, and decided that, for the original grand unified theory of Georgi and Glashow, there would probably be a wide range of parameters for which the rate would be sufficiently low. (This estimate was later confirmed by more detailed calculations.) I explained the new strategy to Paul Langacker, who was also working at SLAC that Sunday afternoon, and I was pleased that he thought it looked like a good idea. On Monday I spoke to Henry, and he quickly agreed that we needed to change the focus of our draft, concentrating on supercooling rather than a low-temperature second phase transition.

The problem of bubbles forming randomly and expanding in an expanding universe sounded very complicated, but Henry and I quickly found that it is simpler than it seems. With some reasonable approximations, we were able to calculate how randomly forming bubbles would gradually fill the universe. If the rate of bubble formation is sufficiently low (as was expected for a wide range of parameters), then the magnetic monopole problem would be solved.

On Tuesday afternoon, I began to rewrite our draft, continuing on Wednesday and working through the night, finishing at 7:30 A.M. On Thursday, John Ellis was again passing through SLAC, so I had a chance to show him our work. He was very interested, and even invited me for an extended visit at CERN.

The conference at Irvine was Friday and Saturday, and on Saturday I had lunch with Marty Einhorn, the physicist whom Henry had learned was working on the same problem. We had a very open and friendly discussion, and it became clear that Marty and his collaborators—Daniel Stein from Princeton and Doug Toussaint from Santa Barbara—were considering many of the same issues that Henry and I were. In particular, they had also realized that the Higgs arrow within each bubble of the phase transition would point in an independent random direction, so the bubble distribution can be used to estimate the monopole production. Henry and I could waste no time in turning our draft into a publication.

Early the next week I spoke to Henry, who had received the draft and was working on revisions. I find no record in my diary, but I remember that Henry made a crucially important remark. He suggested that an implicit assumption in our calculation—that the expansion rate of the universe would be unaffected by the supercooling—would need to be checked.

CHAPTER 10

THE INFLATIONARY UNIVERSE

Late on the following Thursday night (December 6, 1979), I sat down at my desk to pursue Henry's suggestion. By this time we were convinced that the supercooling of a delayed phase transition could ward off the glut of magnetic monopole production that would otherwise result from the combination of grand unified theories and hot big bang cosmology. The new problem was to understand the gravitational field produced by the supercooled matter, to see if it might affect the expansion of the universe.

Since the grand unified theory phase transition is controlled by the Higgs fields, the supercooling must be understood in those terms. If the energy density of the Higgs fields closely resembles the Mexican hat shape of Figure 8.3 (page 140), then a large amount of supercooling is not expected. Recall, however, that Figure 8.3 was presented only as an illustration. The actual shape of the Higgs field energy graph depends on the details of the underlying grand unified theory, which are not known. Since our goal was to learn whether there exists *any* version of a grand unified theory that avoids the monopole problem, we had much flexibility in choosing a Higgs field energy graph to consider.

For a theory to lead to a large amount of supercooling, the Higgs energy density should resemble the example shown in Figure 10.1, a shape that might be called a dented Mexican hat. As in the previous example, the state of lowest possible energy is achieved when the Higgs fields have values on the vacuum circle, for which the energy density is zero. The new feature, however, is the indentation in the middle of the figure.

To understand how the energy diagram of Figure 10.1 can lead to supercooling, we can visualize the values of the two Higgs fields as a ball rolling on the surface, as in Chapter 8. The effects of high temperature can

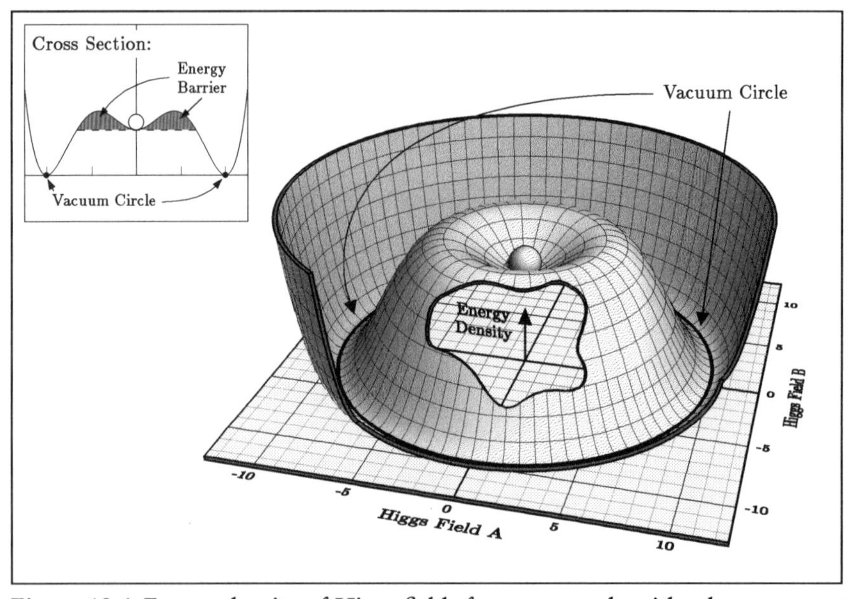

Figure 10.1 **Energy density of Higgs fields for an example with a large amount of supercooling.** The evolution of the Higgs fields is very similar to the motion of a ball rolling on this surface. Each point in the base plane signifies a specific pair of values for Higgs Fields A and B, and the height of the surface above that point indicates the energy density produced by fields with these values. The energy density vanishes when the Higgs fields correspond to any point on the vacuum circle—any such point can describe the Higgs fields in the vacuum, the state of lowest possible energy density. A ball is shown at the center of the dented Mexican hat, corresponding to the state called the *false vacuum*, in which both Higgs fields have the value zero. This state has a high energy density, but it is classically stable, since there is no energy available to allow the ball to jump over the energy barrier that surrounds the central point. For clarity, an insert at the upper left corner shows a two-dimensional slice through the center of the three-dimensional diagram.

---

be simulated by imagining that a vibrator is attached to the surface. If the temperature is high enough, then the ball is thrashed about so violently by the random thermal agitation that the central peak becomes insignificant, and the *average* position of the ball is at the center of the Mexican hat. At zero temperature, on the other hand, there would be no vibrations; if the ball were placed anywhere outside the dent at the center, it would minimize its energy by settling into a randomly determined point on the vacuum circle. At low temperatures, the average position of the ball would lie on or near the vacuum circle, and the ball would undergo small oscillations about this average position. As the system cools from high temperatures, the phase transition is marked by the average position of the ball moving away from the center of the Mexican hat.

If the ball starts at high temperature and cools, however, there is a chance that it can become stuck in the dent, as illustrated in the diagram. It is this situation that corresponds to extreme supercooling, since it is difficult for the ball to jump out of the dent. If the temperature fell all the way to absolute zero while the ball was in the dent, then all vibrations would cease, and the ball would have no energy to jump over the energy barrier that separates it from the vacuum circle. By the rules of classical physics, the ball would be stuck in the dent forever. Since the ball is used to visualize the values of the Higgs fields, this description implies that a region of space can supercool if the Higgs fields in the region have values near zero, corresponding to the ball near the center of the dent. If the temperature reaches zero, this pattern of fields would be stable under the rules of classical physics, because there would be no energy available to allow the Higgs fields to cross the energy barrier to reach the vacuum circle. This state, the product of extreme supercooling, is called the *false vacuum*.

To understand why the false vacuum is given such an odd name, we need to look a bit more carefully at its properties. Even though classical physics would imply that the state is absolutely stable, the deeper truth of quantum theory implies that it can decay. Although the fields do not have enough energy to jump the barrier classically, they can traverse it anyway by a quantum process called *tunneling*. The barrier does not disappear, but the exotic rules of quantum evolution make it possible, on rare occasions, for an object to appear on the far side of a barrier which classically would have been impenetrable.*

The "vacuum" of "false vacuum" must be interpreted in the language of the particle physicist, who uses the word to mean the state of lowest possible energy density.† Of course the false vacuum is not the state of lowest

* Although tunneling sounds bizarre, the reality of this phenomenon was recognized from the early days of quantum theory. In 1928 George Gamow, R.W. Gurney, and E.U. Condon used the concept of tunneling to explain an otherwise mysterious type of radioactivity called alpha decay. In this process an *alpha particle* (composed of two protons and two neutrons) escapes from inside a nucleus, despite the existence of an energy barrier similar to the one diagrammed in Figure 10.1. Today tunneling has many applications in electronics, including the scanning tunneling electron microscope, developed in the early 1980s by Gerd Binnig and Henrich Rohrer of the IBM Zurich Research Laboratory. Using this device to measure the rate at which electrons tunnel from the surface of a material to a needle-like probe, scientists can literally see the bumps and ridges of individual atoms on the surface.

† The typical dictionary definition, "a space entirely devoid of matter," is too imprecise for a physicist, since the word "matter" is ambiguous. The dictionary does not specify, for example, whether a nonzero Higgs field is considered "matter" or not.

possible energy, since the energy would be lower if the Higgs fields lay on the vacuum circle. However, the false vacuum can lower its energy only by quantum tunneling, which is usually a slow process. In a time interval too short to allow tunneling, the energy density of the false vacuum cannot be lowered. Thus, the false vacuum acts *temporarily* as a vacuum, so the word "false" in this context is used to mean "temporary."

The energy density of the false vacuum would be very high, since it is characteristic of the enormous energies associated with grand unified theories. Converted to an equivalent mass density, a typical number might be 10^{80} grams per cubic centimeter. This density is so enormous—about 65 orders of magnitude larger than the density of an atomic nucleus—that the false vacuum will clearly not be observed in the foreseeable future. Nonetheless, from a theoretical point of view the false vacuum seems to be well understood. The essential properties of the false vacuum depend only on the general features of the underlying particle theory, and not on any of the details. Even if grand unified theories turn out to be wrong, it is still quite likely that our theoretical understanding of the false vacuum would remain valid.

The false vacuum has a peculiar property that makes it very different from any ordinary material. For ordinary materials, whether they are gases, liquids, solids, or plasmas, the energy density is dominated by the mass of the particles, which according to special relativity is equivalent to an energy ($E = mc^2$). If the volume of an ordinary material is increased, then the density of particles decreases, and so does the energy density. The energy density of the false vacuum, however, is attributed not to particles, but rather to the Higgs fields.* Even as the universe expands, the energy density of the false vacuum remains at a constant value, provided that we do not wait long enough for the false vacuum to decay.

The idea of a material that can expand at constant energy density is very much at odds with our intuition, so the consequences must be carefully deduced. The first question that comes to mind is "where does the energy come from?" Let us consider, for example, the situation illustrated in Figure 10.2, which shows a piston chamber filled with false vacuum, surrounded by a region of ordinary vacuum. To keep the discussion as

* The astute reader is probably wondering how to reconcile this statement with the earlier statement that all fundamental particles are described in terms of fields: The particles are just quantum bundles of field energy. The distinction is that particles are concentrations of energy in the *oscillations* of a field. A light wave, for example, is composed of oscillating electric and magnetic fields, and the bundles of energy in the light wave are the particles called photons. If a reflecting box filled with light waves is enlarged, the oscillations of the electric and magnetic fields diminish, corresponding to a lower density of photons. For the false vacuum, on the other hand, the energy is stored in the *value* of the Higgs fields, not in the oscillations. If a box filled with false vacuum is enlarged, the energy density cannot become lower.

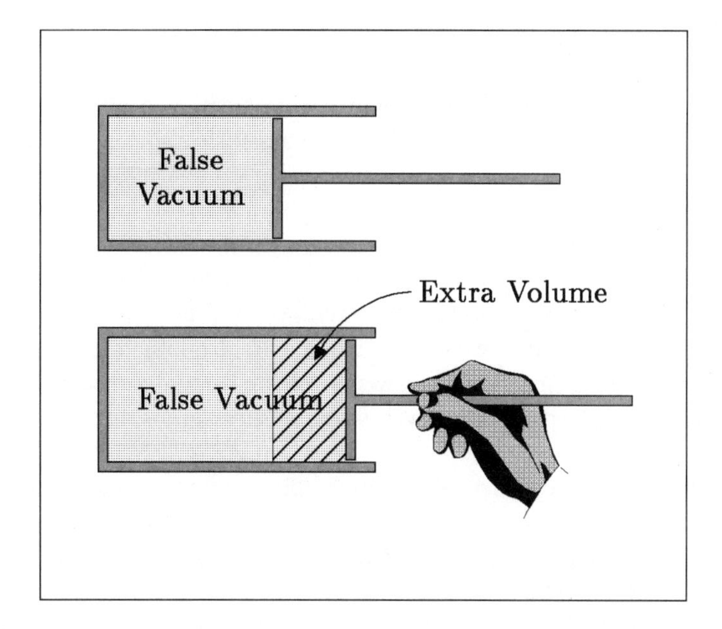

Figure 10.2 **A thought experiment to find the pressure of the false vacuum.** The false vacuum has the peculiar property that its energy density remains constant as it expands. If a piston chamber filled with false vacuum is enlarged, then the energy inside increases. The energy must be supplied by the hand that moves the piston, so the hand must be pulling against a force. The pressure of the false vacuum must therefore be negative, creating a suction that opposes the outward motion of the piston.

---

simple as possible, we will assume that the experiment is done quickly enough so that the false vacuum does not have time to decay. We will also assume that the apparatus is extremely small, so that the gravitational field created by the enormous mass density of the false vacuum can be ignored.* Consider what would happen if the piston were moved, so the volume of false vacuum in the chamber is enlarged, as in the lower part of the figure. Since the energy density of the false vacuum cannot decrease, the total energy inside the piston must increase. But where does the energy come from? There is only one possible source, and that is the hypothetical hand that moves the piston outward. This hand, evidently, must supply a

* Since the mass density of the false vacuum is expected to be in the vicinity of 10^{80} grams per cubic centimeter, the size of the piston would have to be smaller than 10^{-27} centimeters for the gravitational fields to be unimportant. Although such high mass densities and small sizes are far beyond the scope of present or foreseeable technology, theorists use "thought experiments" of this type to understand how the laws of physics work.

huge amount of energy, which indicates that it is pulling against a very large force. The only other force acting on the piston, however, is the pressure of the false vacuum. If this pressure were positive, as all normal pressures are, then it would push outward and assist the motion of the piston. To resist the motion, the pressure of the false vacuum must be large and *negative*!

So, the unusual notion of a material with a constant energy density has led to the bizarre notion of a negative pressure. The false vacuum actually creates a suction. In using the word "suction," however, we should remember that the suction created in drinking through a soda straw is really just a pressure below that of the surrounding air, while the false vacuum creates a suction even when no pressure is applied from the outside.*

Applying these ideas to a supercooled phase transition in the early universe, one might naively guess that the suction of the false vacuum would dramatically slow the expansion of the universe, maybe even reversing it. The truth, however, is exactly the opposite!

The pressure does not slow the expansion of the universe, because a pressure results in a force only if the pressure is nonuniform. For example, if a glass bottle is evacuated and sealed, then the air pressure on the outside will cause the bottle to implode if the walls are not thick enough. However, if the bottle is unsealed and air is allowed to enter, then the air pressure on the inside will very quickly match the pressure on the outside, and the walls will feel no force at all. The false vacuum in a supercooled universe would fill space uniformly, so the forces created by the negative pressure would cancel, like the air pressure inside and outside the open bottle.

Nevertheless, the negative pressure of the false vacuum leads to very peculiar *gravitational* effects. According to Newtonian physics, gravitational fields are produced only by masses. In general relativity, masses are described by their equivalent energy, and the theory implies that any form of energy creates a gravitational field. In addition, general relativity implies that a pressure can create a gravitational field. Under normal circumstances, this contribution is negligibly small: For air at room temperature, the gravitational field caused by the pressure is less than a hundred billionth of the gravitational field caused by the mass density (which is itself very small). In the early universe, however, the pressures were so high that the resulting gravitational fields were important. According to general relativity, a posi-

* The thought experiment discussed in the previous paragraph can be extended into a quantitative argument, demonstrating that the pressure of the false vacuum is equal to the negative of the energy density (provided, of course, that they are measured in the same units).

tive pressure creates an attractive gravitational field, as one might guess. A negative pressure, however, creates a repulsive gravitational field. For the false vacuum, the repulsive component to the gravitational field is three times as strong as the attractive component. *The false vacuum actually leads to a strong gravitational repulsion.*

Curiously, the gravitational effect of the false vacuum is identical to the effect of Einstein's cosmological constant, discussed in Chapter 3. Recall that Einstein introduced this term so that the repulsive force could prevent his static model of the universe from collapsing under the normal attractive force of gravity. There is, however, an important difference between the cosmological constant and the false vacuum: while the cosmological constant is a permanent term in the universal equations of gravity, the false vacuum is an ephemeral state that exerts its influence for only a brief moment in the early history of the universe.

A short calculation shows that the gravitational repulsion causes the universe to expand exponentially. That is, the expansion is described by a *doubling time*, which for typical grand unified theory numbers is about 10^{-37} seconds. In this brief interval of time, all distances in the universe are stretched to double their original size. In two doubling times, the universe would double again, bringing it to four times its original size. After three doubling times it would be eight times its original size, and so on. As has been known since ancient times, such an exponential progression leads rapidly to stupendous numbers.

The exponential sequence was the focus, for example, of the Indian legend of King Shirham, eloquently recounted by George Gamow in *One, Two, Three . . . Infinity.* The king wanted to reward his grand vizier Sissa Ben Dahir for inventing the game of chess, so he asked the vizier to suggest an appropriate gift. Sissa Ben responded with a surprising proposition. He asked for one grain of wheat for the first square of his chessboard, two grains for the second square, four grains for the third square, and so on, until all sixty-four squares would be covered.

The king, it appears, had never studied exponentials, so he happily agreed to this seemingly modest proposal. King Shirham's mathematical education was rapidly advanced, however, as his servants brought in bags and then cartloads of wheat in an attempt to comply with Sissa Ben's request. It soon became obvious, however, that all the wheat in India would not fulfill the king's promise to his vizier, as seen in Figure 10.3. How was the predicament resolved? According to Gamow:

> Thus King Shirham found himself deep in debt to his vizier and had either to face the incessant flow of the latter's demands, or to cut his head off. We suspect that he chose the latter alternative.

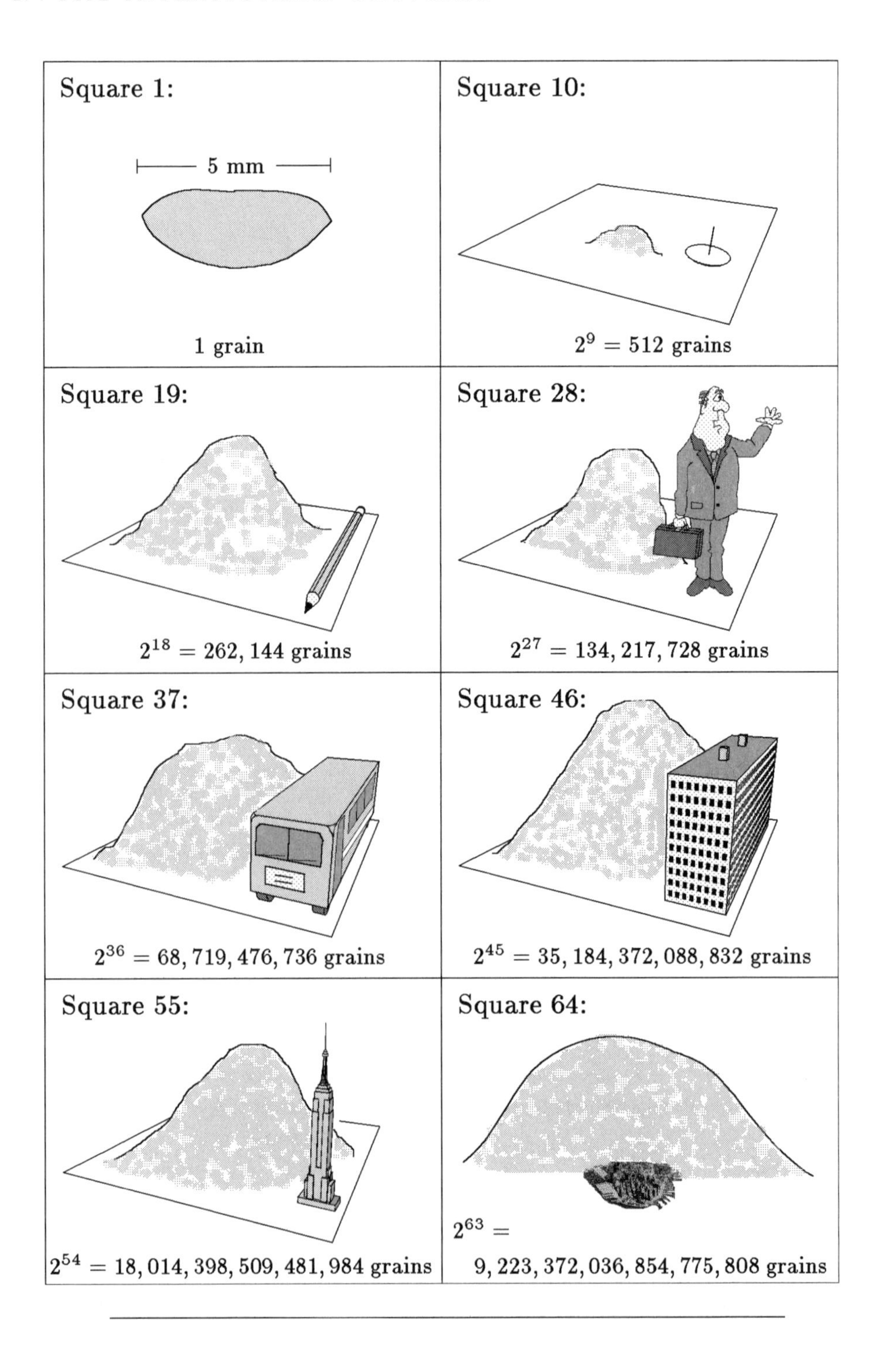
Square 1:
5 mm
1 grain
Square 10:
$2^9 = 512$ grains
Square 19:
$2^{18} = 262,144$ grains
Square 28:
$2^{27} = 134,217,728$ grains
Square 37:
$2^{36} = 68,719,476,736$ grains
Square 46:
$2^{45} = 35,184,372,088,832$ grains
Square 55:
$2^{54} = 18,014,398,509,481,984$ grains
Square 64:
$2^{63} =$
$9,223,372,036,854,775,808$ grains

Cosmological theories tend to go a bit further than Sissa Ben Dahir, typically invoking 100 or more doublings, rather than 64. After 100 doubling times—which is only about 10^{-35} seconds—the universe would be 10^{30} times its original size! For comparison, in standard cosmology the universe would grow during this time interval by only a paltry factor of 10.

Although I did not know it at the time, the exponentially expanding space that I discovered was hardly new—it was in fact one of the earliest known solutions to the equations of general relativity. I had rediscovered the equations of de Sitter's cosmology of 1917, written in a form that was introduced by Georges Lemaître in 1925 as part of his Ph.D. thesis at MIT [1]. I had never even heard of the de Sitter solution, but I would learn of the connection a few weeks later from Sidney Coleman. Even then, I did not appreciate the historical role of the de Sitter theory as a rival to Einstein's static universe. Nonetheless, the literature on the de Sitter spacetime would later become an invaluable resource.

Since the supercooled false vacuum state is not stable, the exponential expansion would not continue forever. Eventually the false vacuum would decay by quantum tunneling. Tunneling by a field is mathematically more complicated than tunneling by a single particle, although the physical principles are the same. The process by which the Higgs fields of the false vacuum can tunnel through the barrier in Figure 10.1 was described in Sidney Coleman's classic 1977 paper, "The Fate of the False Vacuum" [2]. The field does not tunnel everywhere in space at once, but instead follows the pattern of first order phase transitions, like the way water boils: bubbles of the new phase materialize randomly in space, just as bubbles of steam form randomly in water heated on the stove. Each bubble begins small, but the bubbles in false vacuum decay grow at nearly the speed of light until the bubbles merge to fill the space. Inside each bubble is essentially the ordinary vacuum, in which the Higgs fields have values at or very near a point on the vacuum circle. The rate at which the bubbles materialize depends

---

***Figure 10.3* (facing page) Sissa Ben Dahir's exponentially increasing wheat.** The figure illustrates the amount of wheat that would be needed for various squares of Sissa Ben Dahir's chessboard. For comparison, the piles of wheat are shown next to an assortment of common objects: a thumbtack, a pencil, a person, a bus, a 10-story building, and the Empire State Building. The wheat needed for the final square, more than 9 quintillion grains, is shown burying the island of Manhattan (New York City), with the tip of the island protruding from under the mountain of wheat.

very sensitively on the details of the theory, so the *decay of the false vacuum* can be very fast, or very slow, or something in between.

The energy that had been stored in the Higgs fields would produce high energy particles which would collide and create other particles. The product would be a hot soup of particles at high temperature, exactly the assumed starting point for the standard hot big bang cosmology. The excess of matter over antimatter could be established immediately after the decay of the false vacuum, by the grand unified theory processes that were discussed by Steven Weinberg in his lectures at Cornell (see Chapter 6).

I do not remember ever trying to invent a name for this extraordinary phenomenon of exponential expansion, but my diary shows that by the end of December I had begun to call it *inflation.*

So, after a few of the most productive hours I had ever spent at my desk, I had learned something remarkable. Would the supercooled phase transition affect the expansion rate of the universe? By 1:00 A.M., I knew the answer: Yes, more than I could have ever imagined.

The effect of the exponential expansion was clearly spectacular, but I had to figure out if it was good or bad. Would this enormous change in cosmological evolution lead to some catastrophe, implying that the universe could never have undergone such supercooling?

By this time I had developed a fairly good understanding of the basic equations of cosmology, so I was able to see on the same night that inflation would solve one other important problem. Inflation would not only wipe out magnetic monopoles, but it would also solve the flatness problem that I had learned from Dicke's lecture the year before.

The flatness problem, as you will recall from Chapter 2, concerns the quantity that astronomers call omega, the ratio between the actual mass density of the universe and the critical density. (The critical density, calculated from the expansion rate, is the density that would put the universe just on the borderline between eternal expansion and eventual collapse.) The problem is caused by the instability of the situation in which omega equals one, which is like a pencil balanced on its point. If omega is exactly equal to one, it will remain exactly one forever. But if omega differed from one by a small amount in the early universe, then the deviation would grow with time, and today omega would be very far from one. Today omega is known to lie between 0.1 and 2, implying that at one second after the big bang omega must have been between 0.999999999999999 and 1.000000000000001. Yet the standard big bang theory offers no explanation of why omega began close to one.

With inflation, however, the flatness problem disappears. The effect of gravity is reversed during the period of inflation, so all the equations describing the evolution of the universe are changed. Instead of omega being driven away from one, as it is during the rest of the history of the universe, during the period of inflation omega is driven toward one. In fact, it is driven toward one with incredible swiftness. In 100 doubling times, the difference between omega and 1 decreases by a factor of 10^{60}. With inflation, it is no longer necessary to postulate that the universe began with a value of omega incredibly close to one. Before inflation, omega could have been 1,000 or 1,000,000, or 0.001 or 0.000001, or even some number further from one. As long as the exponential expansion continues for long enough, the value of omega will be driven to one with exquisite accuracy.

To understand why inflation drives omega toward one, we can begin by recalling why this is called the flatness problem. According to general relativity, the mass density of the universe not only slows the cosmic expansion, but it also causes the universe to curve. If we assume that Einstein's cosmological constant is zero, then any mass density higher than the critical density causes the space to curve back on itself, forming a spatially closed universe, as described in Chapter 3. In such a universe, the sum of the angles in a triangle is more than 180°. If the mass density is less than the critical density, then the space is curved in the opposite sense: the sum of the angles in a triangle would be less than 180°. On the borderline between these two cases, when the mass density is equal to the critical density, the space is not curved at all. In this case the space is flat, meaning that ordinary Euclidean geometry is valid, and the sum of the angles in a triangle is exactly 180°. Thus, if we accept the relation between omega and geometry implied by general relativity, then we need only understand why inflation drives the universe toward a state of geometric flatness.

Once the question has been restated in terms of geometric flatness, the answer is as obvious as blowing up a balloon. The more we inflate the balloon, the flatter the surface becomes, as is illustrated by the sequence of drawings in Figure 10.4. To say it another way, inflation makes the universe look flat for the same reason that the surface of the earth appears flat, even though we know that the earth is really round. Since the earth is very large and we view only a small part of it at any one time, the curvature is completely imperceptible.

The standard cosmological evolution would resume at the end of inflation, so any deviation from flatness would begin to grow. The universe, however, would be so nearly flat at the end of inflation that it would remain essentially flat until the present day. Thus, the inflationary theory

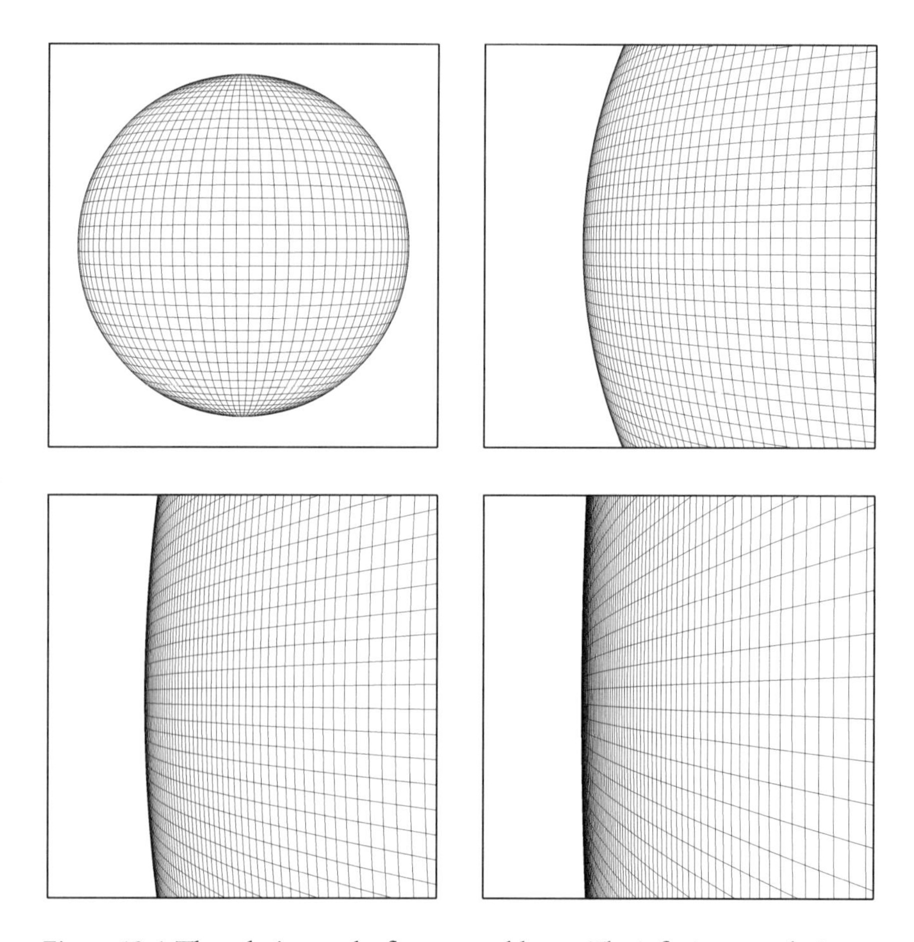

Figure 10.4 **The solution to the flatness problem.** The inflationary solution to the flatness problem is illustrated by this sequence of perspective drawings of an inflating sphere. In each successive frame the sphere is inflated by a factor of three, while the number of grid lines is increased by the same factor. By the fourth frame, it is difficult to distinguish the image from that of a plane. In cosmology, a flat geometry corresponds to a universe with omega equal to one. Therefore, as inflation drives the geometry of the universe toward flatness, the value of omega is driven toward one.

---

leads to an important prediction that is in principle testable: The present value of omega should be very precisely equal to one.*

* We still do not know the value of Einstein's cosmological constant, so it may be nonzero. If it is, then the description above becomes a little more complicated, but the conclusions remain the same. If the cosmological constant is nonzero, then the mass density that puts the universe on the borderline between eternal expansion and eventual collapse is no longer the same as the mass density that leads to geometric flatness. In that case, the effect of inflation is still to drive the universe to a state of geometric flatness. Figure 10.4 and the argument that goes with it are

The next morning I bicycled hurriedly to SLAC to start work, breaking my personal speed record with a time of 9 minutes and 32 seconds. The previous night I had understood the essential effects of exponential expansion, but I had not worked out the details.

Sitting down at my desk, I wrote at the top of the page:

SPECTACULAR REALIZATION:
This kind of supercooling can explain why the universe today is so incredibly flat — and therefore resolve the fine-tuning paradox pointed out by Bob Dicke.

I put the words in a double box—the only such embellishment that I can find in my notebooks. My instincts were telling me that I might be on to something big.

I then tried to reproduce the calculations behind Dicke's argument, and was pleased that I found the same numbers that Dicke had quoted: At one second after the big bang, omega must have equaled one to an accuracy of fifteen decimal places. With about 100 doubling times of exponential expansion, I calculated, the flatness problem would go away. My rough calculations from the night before were exactly on target.

I went to a lunchtime seminar, and after lunch I showed the new idea to Lenny Susskind. "You know," he replied after listening to the explanation, "the most *amazing* thing is that they pay us for this." Lenny's comment was undoubtedly true, but I had hoped that my new brainchild would evoke a more enthusiastic response. Most likely I was completely incoherent that afternoon, however, so in hindsight it is not surprising that my excitement failed to be contagious.

I was still nervous that the exponential expansion might lead to some disaster, and on the following Monday my fears were realized, at least temporarily. In an erroneous calculation, I found that if exponential expansion occurred, the phase transition would be completed in only a small fraction of the universe. "Don't know what to make of it," I wrote in my diary. Late the following afternoon, however, I discovered my mistake, and with great relief I added the words "All is okay!" to the diary entry.

still valid. The flatness problem is solved in this case, also, since again the value of omega before inflation can be almost anything. Regardless of the initial value of omega, inflation will drive the universe to a state of nearly perfect flatness. Although the deviation from flatness will begin to grow once inflation ends, it will remain imperceptible to the present day.

I called Henry to tell him about my ideas on extreme supercooling and exponential expansion, but he was no more enthusiastic than Lenny had been. It was now only ten days before Henry would be leaving for China, so it was not a good time to be considering major revisions to our paper. The important problem, at the time, was that our draft was 20% longer than the maximum length for *Physical Review Letters*, the rapid publication journal to which we intended to submit it. Since time was short, we agreed to publish the paper with no mention of exponential expansion. There were many phone calls that week, as we exchanged ideas for shortening the paper without obscuring the physics.

The following Monday I called Marty Einhorn, with whom I had not spoken since the conference at Irvine. The paper he was writing with Stein and Toussaint was also nearing completion, so we agreed to ask the journal to publish the two papers back-to-back. To my disappointment, I learned that Marty was also considering the possibility that the universe supercooled into a false vacuum state. My diary indicates clearly my reaction: "Ugh."

Meanwhile, Henry and I worked hard on finalizing the text of our paper, which we finished by Wednesday (December 19), just three days before Henry was to leave home. I rushed the paper to the SLAC secretarial pool, and was amazed by their efficiency. I was accustomed to waiting several weeks to get a paper typed, but the SLAC team had the paper finished and a diagram drafted by the next day. I proofread the paper that night, and on Friday the paper was mailed to the journal [3].

I called Marty Einhorn, who told me that their paper would not be typed until after New Year's Day. In the end their paper was published a few months after ours in the *Physical Review* [4], a journal somewhat slower than *Physical Review Letters*.

In my last phone conversation with Henry before his trip, I asked if it would be okay if I tried to publish a paper on extreme supercooling and exponential expansion while he was away. Since he recognized that there was serious competition, he said that it would be fine for me to proceed. I also mentioned that I was toying with the idea of making a trip to the East Coast to give seminars, hoping to drum up enough interest in our work to stimulate some job offers. Henry thought that such a trip could be very effective, and strongly encouraged me to do it.

December 1979 was my lucky month—a few weeks after the invention of inflation, I stumbled upon another key piece of evidence to support it. One day at the SLAC cafeteria, I had lunch with a group of physicists who were discussing a recent paper by Anthony (Tony) Zee, a particle theorist then at the University of Pennsylvania. The paper concerned a proposed solution to the *horizon problem*, a long-standing cosmological problem of which I was

Henry Tye and his grandmother in Shanghai, January 1980. This photograph was taken during the trip that caused Henry and me to hurriedly finish our paper on magnetic monopole production in the early universe.

totally unaware. Marvin Weinstein, a SLAC particle theorist, explained to me what it was.

The existence of horizons in cosmological theories was already familiar to me, since, as I described earlier, horizons were used as a method of estimating the production of magnetic monopoles. Since the universe in the big bang theory has a finite age, there is a maximum distance that light could have traveled since the beginning of time. This distance is called the *horizon distance*, where the word "horizon" is used in the sense of a limitation on our knowledge. As far as we know, nothing can travel faster than light, so the horizon distance is a rigid bound on the transfer of matter or information in the universe. Although I was aware of horizons, I did not understand that their existence could be viewed as a problem.

Probably the most persuasive statement of the horizon problem focuses on the release of the cosmic background radiation. Recall from Chapter 5 that until about 300,000 years after the big bang, the photons of the cosmic background radiation were constantly being scattered by collisions with the electrons in the hot plasma that filled the universe. At about 300,000 years, however, the universe cooled enough so that the electrons combined with atomic nuclei to form electrically neutral atoms. Such a gas is highly transparent to photons, so most of the photons of the cosmic background radiation have been traveling in a straight line since this time. The photons therefore provide a picture of the universe at an age of 300,000 years, just as the photons traveling from the book to your eyes provide an image of the letters on the page. The cosmic background radiation shows us, among other things, that the universe at 300,000 years was incredibly uniform, since the temperature of the radiation is found to be the same in all directions to an accuracy of about one part in 100,000.* It is natural to ask, therefore, whether we can understand how this extreme uniformity was established.

The general tendency of objects to come to a uniform temperature is well understood, and is often called by physicists the "zeroth law of thermodynamics." If a hot cup of coffee is placed on a table, it will gradually cool to room temperature. The speed with which heat energy can be moved from one place to another, however, certainly cannot exceed the speed of light, and so the transfer of heat in the early universe is limited by the horizon distance. At 300,000 years after the big bang, the horizon distance was about 900,000 light-years. (If the universe were static, the horizon distance would have been 300,000 light-years. In an expanding universe, however, photons can make extra progress during the early period when the universe was small, so the horizon distance is larger than one might expect.)

If we consider two photons arriving today from opposite directions in the sky, then we can use the mathematics of the big bang theory to trace back the trajectories to 300,000 years after the big bang. The calculation, which takes into account the expansion of the universe, shows that the photons were emitted from two points about 90 million light-years apart, as illustrated in Figure 10.5. In the diagram, *A* and *B* label the points at which these two photons were emitted. The uniformity of the cosmic background radiation temperature implies that the temperature was the same at points *A* and *B* (to an accuracy of one part in 100,000), yet they were separated from each other by about 100 times the horizon distance. Since nothing travels faster than light, in the context of the standard big bang theory there is no physical process that can bring these two points to the same temperature by

* The raw observations show a nonuniformity in temperature of about one part in 1,000, but the angular pattern of this nonuniformity indicates that it is caused by our motion through the background radiation. One infers that the entire Milky Way galaxy is moving through the background radiation at a speed of about 600 kilometers per second, or 0.2% of the speed of light, toward the constellation Hydra. After the effect of our motion is subtracted, then the remaining nonuniformities—attributed to variations in the cosmic background radiation itself—have a magnitude of only one part in 100,000.

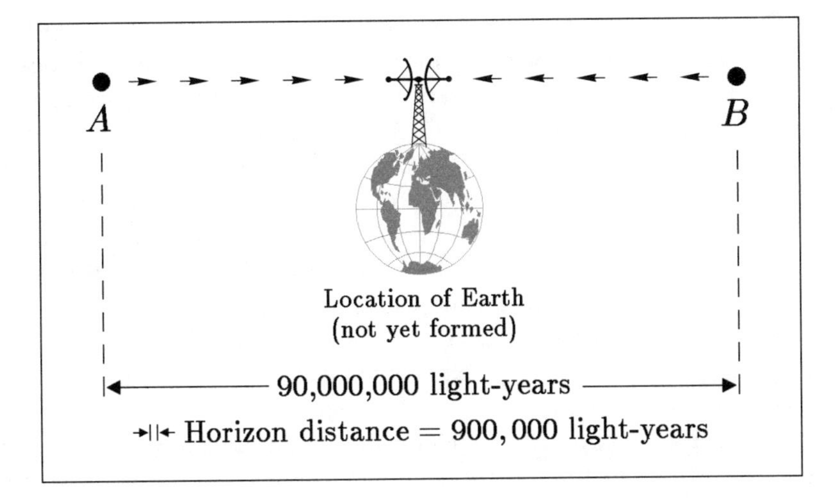

***Figure 10.5* The horizon problem of the standard big bang theory.** The diagram shows a picture of the universe at 300,000 years after the big bang, when the cosmic background radiation was released. At the center is the matter that will eventually become the Earth, and the points *A* and *B* are the sources of two photons that will arrive at Earth in the 20th century. The points *A* and *B* are separated from each other by about 90 million light-years, according to the standard big bang theory, while the horizon distance at this time is only 900,000 light-years. Since the points *A* and *B* are separated from each other by about 100 times the distance that light could have traveled since the big bang, there is no physical process that can explain why the two points are at nearly identical temperatures.

300,000 years after the big bang. We can imagine, if we wish, that the universe is populated by little purple creatures, each equipped with a furnace and a refrigerator, and each dedicated to the cause of trying to establish a uniform temperature. Even with the help of these creatures, the observed uniformity of the cosmic background radiation could not be established unless the creatures could communicate at 100 times the speed of light.*

* According to the big bang theory, the separation between points *A* and *B* in Figure 10.5 grew from zero to 90 million light-years during a time interval of only 300,000 years. The rate of separation, therefore, was much larger than the speed of light. For a subtle reason, however, this does not violate the stricture on faster-than-light travel. The complication stems from the fact that, in general relativity, space itself is plastic, capable of bending and stretching. In the big bang theory, the space stretches as the universe expands. The restriction on velocities remains valid, in the sense that no particle can ever win a race with a light beam. Nonetheless, the distance between two particles can increase due to the stretching of the space between them, and general relativity places no restriction on how fast the stretching can occur. As an analogy, one can imagine bugs crawling on a rubber sheet. No bug can crawl across the surface of the rubber at a speed faster than that of light, but the stretching of the sheet can carry bugs apart at unlimited speeds.

The horizon problem is not a failure of the standard big bang theory in the strict sense, since it is neither an internal contradiction nor an inconsistency between observation and theory. The uniformity of the observed universe is built into the theory by postulating that the universe began in a state of uniformity. As long as the uniformity is present at the start, the evolution of the universe will preserve it. The problem, instead, is one of predictive power. One of the most salient features of the observed universe—its large scale uniformity—cannot be explained by the standard big bang theory; instead it must be assumed as an initial condition.

Having learned about the horizon problem at lunch, I went home and thought about it. Eureka! The exponential expansion of inflation would obliterate this problem, too.

To understand how inflation eliminates the horizon problem, the first step is to recognize that the size of the presently observed universe is what it is, independent of one's theories of how the early universe evolved. Whether we believe in the standard big bang theory or inflation, or even the steady state theory, the most distant objects that we can detect are about 20–30 billion light-years away. (There are presumably other objects even more distant than this, but since we cannot see this far, these objects are outside the observed universe.) To trace the history of the presently observed universe, however, we need to adopt a theory of how the universe evolved. Figure 10.6 shows the history of the universe in both the standard hot big bang theory and in the inflationary theory. The two theories describe identical evolution for all times after the end of the inflationary period, so the two curves describing the radius of the observed universe coincide for all times later than about 10^{-35} seconds. During the brief period of inflation, however, the inflationary theory describes an enormous burst of expansion that is not predicted by the other theory. Thus, if we consider times earlier than the period of inflation, the size of this region in the inflationary theory is much smaller than in the standard theory.

In the inflationary theory, therefore, the universe started out incredibly small. Before inflation, the radius of the observed universe is shown in Figure 10.6 as only 10^{-52} meters. These numbers are not at all well-determined, however, so they should be taken only as an illustration of how inflation might work, and not as a prediction of the theory. Nevertheless, it is clear that, before inflation, the observed universe was incredibly small. The horizon problem therefore evaporates, since the speed of light imposes no barrier for such a small region. There was plenty of time for such a small region to come to a uniform temperature, by the same mundane processes by which a hot cup of coffee cools to room temperature. Then, once the uniformity was established in this very small region, the process of inflation stretched it to become large enough to encompass the entire observed uni-

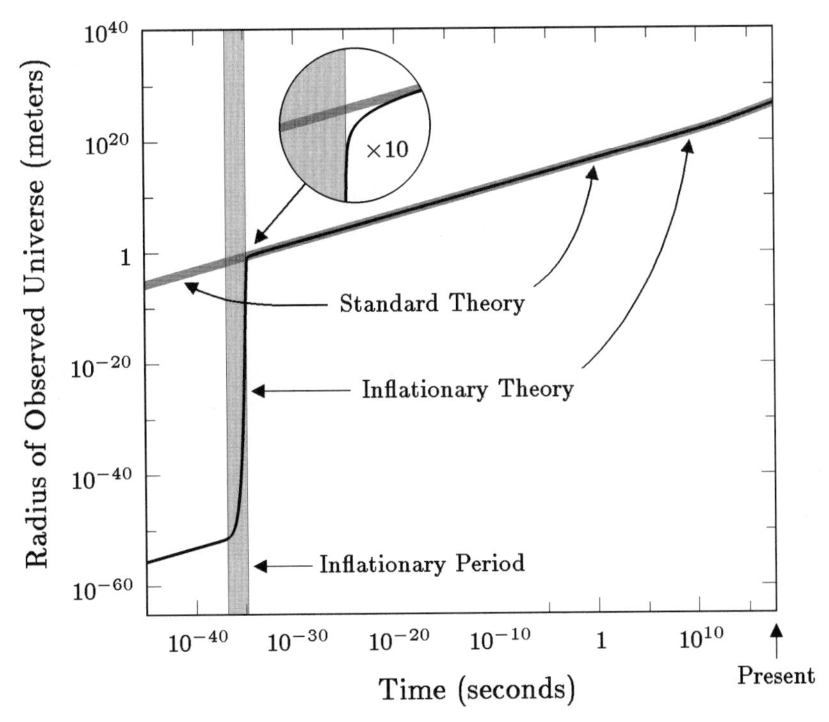

Figure 10.6 **Solution to the horizon problem: the size of the observed universe in the standard and inflationary theories.** The vertical axis shows the radius of the region that evolves to become the presently observed universe, and the horizontal axis shows the time. In the inflationary theory, the universe is filled with a false vacuum during the inflationary period, marked on the graph as a vertical gray band. The gray line describing the standard hot big bang theory is drawn slightly thicker than the black line for the inflationary theory, so that one can see clearly that the two lines coincide once inflation has ended. Before inflation, however, the size of the universe in the inflationary theory is far smaller than in the standard theory, allowing the observed universe to come to a uniform temperature in the time available. The inset, magnified by a factor of 10, shows that the rapid expansion of inflation slows quickly but smoothly once the false vacuum has decayed. (The numerical values shown for the inflationary theory are not reliable, but are intended only to illustrate how inflation can work. The precise numbers are highly uncertain, since they depend on the unknown details of grand unified theories.)

verse. Thus, the uniformity in temperature throughout the observed universe is a natural consequence of inflation.

Implicit in Figure 10.6 is a remarkable prediction of the inflationary theory. Due to the enormous expansion during the inflationary period, the size of the *observed* universe before inflation was absurdly small. There is no reason, however, to suppose that the size of the *entire* universe was this small. While the inflationary theory allows a wide variety of assumptions concerning the state of the universe before inflation, it seems very plausible that the size of the universe was about equal to the speed of light times its age, or perhaps even larger. If the universe were smaller than this, then it almost certainly would have already collapsed into a crunch. Applying this reasoning to the sample numbers shown on Figure 10.6, we find that the entire universe is expected to be at least 10^{23} times larger than the observed universe!*

These numbers are highly uncertain, since they depend sensitively on the duration of the period of inflation, which in turn depends on the decay rate of the false vacuum. Without knowing the correct grand unified theory and the values of all of its parameters, the decay rate of the false vacuum cannot even be approximated. Nonetheless, the qualitative behavior shown in Figure 10.6 seems to be typical of all inflationary universe calculations. If the inflationary theory is correct, then the observed universe is only a minute speck in a universe that is many orders of magnitude larger.

I discussed inflation with a number of people at SLAC over the following weeks, but I view the official debut of inflation as the seminar that I gave at SLAC on January 23, 1980. The title, "10^{-35} Seconds After the Big Bang," referred to the beginning of the inflationary period.† I had been somewhat nervous in preparing for the seminar, as I was still worried that some consequence of the theory might turn out to be spectacularly wrong. There was also the fear that I would reveal my status as a greenhorn cosmologist. To shore up my general background in cosmology, I had crammed from Steven Weinberg's excellent popular-level book, *The First Three Minutes*.

I began the seminar by recounting the standard big bang theory and the magnetic monopole problem, reviewing Preskill's arguments that indicated a large excess of monopole production. Then I described the work by Henry and me, showing how the monopole problem can be evaded by extreme

* To see how this number was estimated, note first that Figure 10.6 shows the beginning of inflation at about 10^{-37} seconds after the big bang. The speed of light times the age would then be 3×10^{-29} meters, so we suppose that the entire universe before inflation was at least this large. But Figure 10.6 shows that the observed universe was only 10^{-52} meters in radius at this time, which is 3×10^{23} times smaller. This ratio would persist, so today the entire universe would still be at least 3×10^{23} times larger than the observed universe. In the text I dropped the factor of 3, since the calculation is highly approximate.

† Figure 10.6 shows a slightly earlier time, based on more modern data from particle physics. In 1980 the energy scale of grand unified theories was believed to be 10^{14} or 10^{15} GeV, while today it is believed to be 10^{16} GeV.

supercooling. Finally I presented the inflationary theory, showing how it would solve not only the monopole problem, but also the flatness and horizon problems as well. The seminar went smoothly, although it ran one and a half hours, a half hour longer than scheduled.

The turnout for the seminar was large, and the reactions afterward were very encouraging. Sidney Coleman told me that it was the best seminar of the year. I asked him for suggestions on what to cut to bring the seminar down to the right length, to which he replied: "Nothing; every word was pure gold." Coming from a senior physicist whom I respected as much as Sidney, these words rang out a triumph that exceeded my wildest expectations.

My diary entry for the next day begins with a headline in capital letters: "HELL OF A GOOD DAY." To underscore the drama of the day, the California terrain provided a minor earthquake at 11:00 A.M., causing the filing cabinets in my office to sway back and forth. By the time of the earthquake, however, I had already received two significant phone calls. One was from a group of cosmologists on the Stanford campus, who had heard of my work and wanted me to explain it to them. The second was from Roman Jackiw, an MIT professor spending the year at the University of California at Santa Barbara, who invited me to give a seminar there. Afterward, a professor from the University of California at San Diego, visiting at SLAC, invited me for a seminar in San Diego later in the year. And all this was before lunch. At lunch, I told one of the permanent SLAC theorists that I would like to apply to stay at SLAC for another year. I had been planning this for some time, but decided to wait until after my seminar in the hope that my chances would be improved. It turned out that I had cut things very close, since the SLAC group was meeting that afternoon to discuss postdoc offers. Immediately after I returned from lunch, the head of the theory group, Sidney (Sid) Drell, rushed into my office. He had apparently not heard of my request, and came to ask if I might be interested in staying at SLAC. "Yes," I replied, "yes." In the middle of the afternoon, Jack Gunion, a regular SLAC visitor from the University of California at Davis, called me into his office at SLAC to tell me that I was at or near the top of the U.C. Davis list for a faculty appointment. In a phone call with a senior colleague at Columbia, I learned that the particle theory group at the University of Pennsylvania had become very interested in me for their faculty opening, after hearing about my recent work from Sidney Coleman. As I was leaving for the day, Sid Drell told me that the SLAC group had decided to offer me a three-year position, although it would not become official until approved by a faculty committee. When I went to bed that night, I knew that my career in physics had entered a new world.

CHAPTER 11

THE AFTERMATH OF DISCOVERY

On Friday, January 25, 1980, two days after I described the inflationary universe to an enthusiastic audience of SLAC particle theorists, I had my first encounter with the world of astrophysicists. As arranged by telephone the day before, I visited the main Stanford University campus to explain the new ideas to a group of three active cosmologists: Robert Wagoner of Stanford, Gary Steigman, who was visiting from the Bartol Institute in Delaware, and Gordon Lasher, who was visiting from IBM at Yorktown Heights, New York. I had not met any of them before, and I was a little nervous about the possibility that the professionals would find some crucial flaw in the theory.

I arrived about three o'clock, and we quickly went to a blackboard and started talking. I had fun explaining grand unified theories and spontaneous symmetry breaking and Higgs fields and magnetic monopoles, all of which they were eager to hear about. The discussion bogged down, however, when I began to describe the cosmological consequences of exponential expansion. When I tried to explain how inflation solves the horizon problem, communication seemed to break down completely. Although I am sure that we all understood the same basic framework for cosmology, they were accustomed to formulating it in a lexicon that was completely foreign to me. I gave them what to me was a perfectly sensible and self-contained explanation of how inflation would avoid the horizon problem, but they rephrased the argument in their own language and decided that it made no sense. Exponential expansion, they concluded, only makes the problem of horizons worse. I understood very little of what they were saying, so I had no idea why we were disagreeing. I kept repeating my own explanation, but they continued to rethink it in their own jargon and remained unpersuaded. They were nonetheless friendly and patient, and we talked until we were all exhausted; at about 6 P.M. I went home, totally drained, and took a nap. Although I was frustrated with the communication problem, I was somewhat relieved that they had not discovered any dreadful oversights in the theory. On the point of disagreement, I was 99% certain that I was right. The 1% uncertainty was not enough, under the circumstances, to keep me awake.

My faith in scientific communication was restored the following Tuesday, when Gordon Lasher called to say that they had understood my reasoning. He thanked me for explaining "how the early universe could be made to work."

On the Monday after my seminar at SLAC, I received the first of what became a series of job offers. Gino Segrè, of the University of Pennsylvania, called to tell me that the particle theory group wanted to offer me an assistant professorship. He invited me for a visit, which we arranged for the end of February.

Meanwhile my plans for a trip east to publicize the new theory and to seek job offers turned from speculation to reality. News of the inflationary universe had spread quickly through the particle physics community, I think largely due to the influence of Sidney Coleman. My itinerary grew to accommodate a steadily growing number of invitations. During the five-week tour that began in the third week of February, I lectured at the University of Minnesota, Northwestern, Fermilab, the University of Pennsylvania, Rutgers, Harvard, Princeton, Columbia, Cornell, and the University of Maryland.

The most memorable event of the trip was the "Theoretical Physics Get-Together" at Rutgers University. It was a one-day conference on Saturday, March 1, attracting physicists from most of the major East Coast institutions. The speaker list was short but highly prestigious: Lenny Susskind of Stanford, Murray Gell-Mann of Caltech, and Sheldon (Shelley) Glashow of Harvard. Both Gell-Mann and Glashow were Nobel laureates. I sat next to Lenny during Gell-Mann's opening talk, a summary of grand unified theories. Lenny asked me if I was speaking, to which of course the answer was "no." "But your work is the second most important thing in physics," he asserted. "What's first?" I asked, seeing no alternative to this obvious question. "My work, of course," he replied, with a boyish grin that dispelled any suggestion of vanity. During the question period following Gell-Mann's talk, Lenny brought up the subject of monopole production in the early universe, and then suggested to the organizers that they give me fifteen minutes to explain my proposal for solving this problem. They agreed to ten minutes, and I went to the front of the lecture hall to extemporize. I was somewhat disorganized, being taken by surprise, but was apparently coherent enough to get the main points across. When I came to the solution of the horizon problem, Gell-Mann understood the argument immediately and exclaimed, "You've solved the most important problem in cosmology!"

After the talks I chatted with Shelley Glashow, and happily learned that Harvard was intending to offer me a position. Shelley had learned about my work by telephone from Sidney Coleman, who was a Harvard faculty member on leave at SLAC. Shelley told me that he had decided inflation must be a good idea when he explained it to Steve Weinberg, who became "furious." "Did Steve have any objections to it?" I asked. "No," replied Shelley, who enjoyed poking fun at his colleague. "He just didn't think of it himself."

The trip was successful beyond my wildest expectations. By the time I returned I had official or unofficial job offers from the University of Pennsylvania, the University of Minnesota, Rutgers, Harvard, Princeton, and the University of Maryland. Meanwhile the Davis and Santa Barbara branches of the University of California had also offered me positions.

During the trip I visited my alma mater MIT, and I felt very comfortable and happy having lunch with so many old acquaintances in the theoretical physics seminar room. But MIT had not advertised any positions that year; I had not applied to MIT for a job, and the subject was never mentioned during my visit. Nonetheless, I felt that I would rather be at MIT than any of the places from which I had offers.

I spent the last three days of the trip at the University of Maryland, and on the final night I was taken out for a relaxed Chinese dinner with a small group of physicists. At the end of the meal we were served the fortune cookies that are traditional for Chinese restaurants, at least in the United States. The message in my cookie seemed surprisingly pertinent: "An exciting opportunity lies just ahead if you are not too timid."

After the long trip, it was a real joy to fly home the next morning to spend a relaxed weekend with Susan and Larry, who was now two years old. Susan made a filet mignon dinner to celebrate my homecoming, and Larry and I had loads of fun playing with blocks, Soma cubes, balloons, and bubbles, as well as building castles in the backyard sandbox.

On Monday, however, it was time to return to business. I thought about my job offers, and about MIT. Should I call and tell them that I would be interested in a job if they would offer one? It was not my style to be so pushy. If they wanted to offer me a job, they could have done it. On the other hand, the worst that could happen, if I called, is that they could say no. Emotionally I was hesitant to call, but the logic of the Chinese fortune cookie was compelling: If I called MIT, I would have nothing to lose and much to gain. There was no point in being timid. With my heart beating quickly, I dialed the number of Jeffrey Goldstone, one of the senior members of the particle theory group at MIT. "Well, . . . , um, . . . , uh," I explained, "there weren't any jobs at MIT so I didn't apply, but I wanted to tell you that I would be very interested in MIT, if you could consider me for a job." Jeffrey was very friendly, and my heart slowed down. He told me that they were hoping to find a replacement for my thesis advisor, Francis Low, who had been chosen to become the new Provost of MIT. They had been planning to look for a more senior person, but they had not yet made any decisions. He said that he was about to leave on a trip, but that he would tell the others in the group about my call, and somebody would get back to me.

When I returned from lunch the next day, there was a phone message from Arthur Kerman, the director of the MIT Center for Theoretical Physics. I nervously called back, and he told me that MIT would like to offer me an associate professorship! It was late in the spring and the bureaucracy was slow, so for the first year I would most likely be a "visiting associate professor" (visiting from nowhere, of course). In any case, they intended for it to become a regular faculty appointment. Fantastic! When I told my friends at SLAC that I was offered an associate professorship at MIT within twenty-four hours from a single phone call, they were flabbergasted. So was I. Was this all a dream, I wondered, or was it really happening?

According to my diary, that evening Larry learned to sing the letters of his name: "El, Ay, Ar Ar Why." Great kid.

My final choice focused on Pennsylvania, MIT, and Harvard. Pennsylvania offered the most in job security: They had promised to consider me for tenure within a year, and told me that I could count on a positive decision. They also had a strong group, including a number of people with whom I would enjoy working. MIT, on the other hand, would go no further than to say that my chances of tenure were probably better than the departmental average, which was 50%. The particle theory group at MIT, however, was larger and generally stronger than the group at Penn. The chances of tenure at Harvard were presumably the lowest of all, but one could not lightly dismiss the chance to work in a group that included Sidney Coleman, Shelley Glashow, and Steve Weinberg. Each of these three, however, was at that time considering offers at other institutions. (Steve Weinberg soon left, accepting a position at the University of Texas at Austin.) Susan encouraged me not to worry about security, but to choose the place I thought was best. I chose MIT. A few days later, at a Chinese lunch, my fortune cookie advised me that "You should not act on the impulse of the moment." But what would a Chinese fortune cookie know, anyhow?

I had not yet found time to publish anything about inflation, so I was taking a bit of a risk by lecturing instead of writing. If someone else had published similar ideas while I was meandering around the country, I could have lost credit for the discovery. At the time, however, I was not concerned about this possibility. Seminars are the fastest way to publicize new results, and I had decided in January that I could not waste time if I wanted to attract job offers for the coming year. I had adopted the somewhat risky strategy of "oral publication," assuming that if I gave seminars at enough places, my claims of authorship would be secure. There were a number of short jaunts in addition to my east coast expedition, so by the end of the spring semester I had spoken at no less than 18 major institutions.

I had one scare, however, during my East Coast travels. When I explained my work to a postdoc at MIT, he told me that someone else, whose identity he did not know, had visited recently and talked about work that sounded indistinguishable from mine. Only somewhat later did I learn that the mysterious competitor had been only an echo of myself, and that the unidentified visitor had been Sidney Coleman, speaking about my work.

Although inflation seemed a panacea for all of cosmology's ills, there was still an important unresolved issue: how exactly does inflation end?

The distribution of matter in the early history of the real universe is known to have been extremely uniform, as we can tell from the remarkable uniformity of the cosmic background radiation. One of the most attractive features of inflation was its ability to explain this uniformity. As inflation proceeds, the matter that was present at the beginning would be diluted to irrelevance, while space becomes filled with the exquisitely uniform mass density of the false vacuum.

The complication, however, is that inflation must end. The energy of the false vacuum must be released to produce the "ordinary" matter that populates the universe today. Would the uniformity produced by inflation survive the ending of inflation?

The false vacuum is an unstable state, which decays in a manner very similar to the way water boils. Small bubbles of normal matter—matter that is not false vacuum—would form in the midst of the false vacuum, just as bubbles of steam form in the midst of water heated to its boiling point. Once a bubble of normal matter materializes, it immediately starts to grow. The bubble wall moves outward at a speed that rapidly approaches that of light. If these bubbles could smoothly coalesce, the uniformity of the false vacuum could be preserved.

As the bubbles form and grow, the large energy density of the false vacuum (the energy of the Higgs fields) is released. The released energy, however, is not distributed uniformly through space. It is concentrated, instead, in the bubble walls, which acquire more and more energy as the bubbles expand. It is only when bubble walls collide that the energy can spread uniformly through space. The collisions convert the energy to spurts of particles, ejected in all directions, which in turn collide with each other. Through their random motion the particles can perhaps spread to fill space uniformly, much as molecules of air spread to fill the volume of a room. The important goal, therefore, is to understand the details of the bubble wall collisions.

If the space were static or expanding moderately, then bubble collisions would be frequent. Without the exponential expansion of inflation, all the bubble walls would soon collide with other walls or with the particles produced by wall collisions. The energy from the bubble walls would be converted to normal matter, and the decay of the false vacuum would be rapidly completed. There would be plenty of time for the particles to spread evenly through space, producing the uniform hot soup of matter that had been assumed as the starting point for the big bang theory.

When bubbles form during the exponential expansion of inflation, however, there is an important complication: As bubbles materialize randomly in space, the space between the bubbles continues to rapidly expand. Since the walls of a bubble move outward at essentially the speed of light, and nothing can ever move faster, the effects of a bubble cannot extend beyond its own wall. Even when the bubbles start to collide, the part of space not yet reached by any bubbles remains in the false vacuum state. The space outside the bubbles continues to exponentially expand, as if no bubbles had ever formed. As an example, the collision of two bubbles is illustrated in Figure 11.1.

If two bubbles materialize very near each other, the bubble walls would soon collide, and the energy in these colliding walls can be spread in all directions. But if two bubbles are not near each other when they materialize, then the space between them would expand so fast that the bubble walls, even moving at the speed of light, would never meet each other. The exponential expansion, therefore, drastically suppresses the collisions of bubbles. It was not clear, therefore, whether these collisions would be frequent enough to spread the matter uniformly through the universe.

By the beginning of January 1980 I had begun to worry about the behavior of the region outside the bubbles—the region that remains in the false vacuum. The disappearance of the false vacuum is exponential, which means that it can be described by a half-life, like the decay of a radioactive nucleus. After one half-life, the probability that a radioactive nucleus has not yet decayed is equal to $\frac{1}{2}$; during each succeeding time interval of equal length, the probability that the nucleus has not decayed is again cut in half. Similarly, if one follows a fictitious observer living in the false vacuum, after one half-life the probability that she remains in the false vacuum, rather than finding

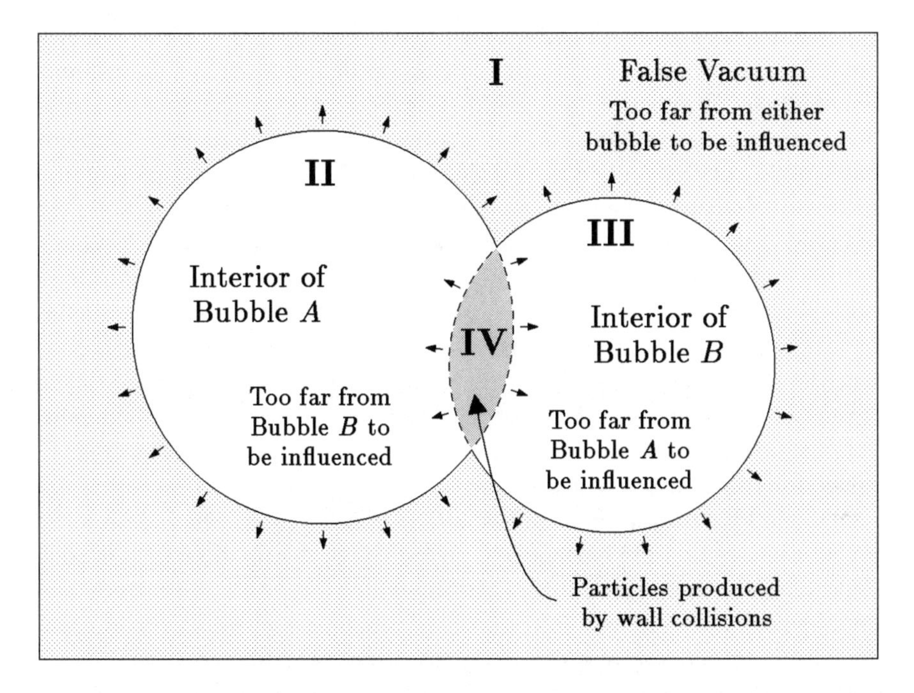

Figure 11.1 **The collision of two phase transition bubbles.** Since the bubble wall moves at the speed of light, the influence of the bubble cannot get beyond it. The collision is described by dividing space into four zones. Zone I (lightly shaded) is the outer region, too far from the center of either bubble to feel their influence. It is still exponentially expanding in the false vacuum state, oblivious to the formation of the bubbles. Zones II and III (white) are each close enough to one of the bubbles to be influenced by it, but beyond the range of influence of the other bubble. These zones contain an almost perfect vacuum—not a false vacuum, but an ordinary vacuum—since virtually all the energy released by the decay of the false vacuum is deposited in the bubble wall. The complicated spray of particles produced by the collision of bubble walls is restricted to Zone IV (darkly shaded), which is close enough to the centers of both bubbles to be influenced by them. The boundaries of this region are shown as dashed lines, tracing the motion that the bubble walls would have followed if they had not been broken up by the collision.

---

herself inside a bubble of normal matter, is $\frac{1}{2}$. During each time interval of equal length, the probability that she remains in the false vacuum is halved again. So, after many half-lives, the probability that any particular observer remains in the false vacuum becomes extraordinarily small. Nonetheless, while the false vacuum is decaying exponentially, those parts of it that have not yet decayed are continuing to expand exponentially. Furthermore, the exponential expansion is always much faster than the exponential decay. If we followed a region of false vacuum for one half-life, we would find that only half would remain false vacuum, but the volume of this half would be much larger than the volume of the whole region at the start. Even while the false vacuum is decaying, its volume increases!

The behavior of the false vacuum region seemed paradoxical. Since the fraction of the original false vacuum region that has not decayed becomes smaller and smaller with time, one would think that the phase transition to normal matter could soon be considered complete. Any remaining regions of false vacuum would presumably be so insignificant that they could be ignored. What difference could it make if, say, one trillionth of the original region of false vacuum has not yet decayed? On the other hand, since the volume of false vacuum *increases* with time, it might be a serious mistake to pretend that it completely disappears.

When I gave talks about inflation during this period, I had to admit to my audiences that I did not understand the ending of inflation. Absent such understanding, I proposed that when something like 90% of the false vacuum had disappeared, we could ignore the remaining false vacuum and proceed as if it had converted entirely to a uniform hot soup of normal particles. Under this assumption, the inflationary theory could resolve the monopole problem, and also the flatness and horizon problems.

In December 1979 an old friend and collaborator from Columbia University, Erick Weinberg (no relation to Steven), arrived to spend a semester at SLAC. The earlier work that I had done on magnetic monopoles, the work that had become my entrée into cosmology, had been done in collaboration with Erick and several others. We were both looking forward to the chance to work together again, and we agreed on several problems to pursue. The most important of these was the investigation of how inflation would end.

We quickly set our sight on a well-defined mathematical problem related to the way in which the growing bubbles merge into clusters. As a starting point, we had learned that if one placed equal-sized spheres randomly in space, allowing them to overlap, then a very curious phenomenon occurs as the density of spheres is increased. At very low densities, just a few spheres overlap with other spheres, while the vast majority are isolated. As the density is increased, clusters of two or perhaps three overlapping spheres become common (see the left side of Figure 11.2). If the density is further increased, the clusters continue to grow, at a rate that becomes precipitous as the fraction of space covered by spheres approaches 29%. When the 29% threshold is crossed, the average size of a cluster literally becomes infinite.

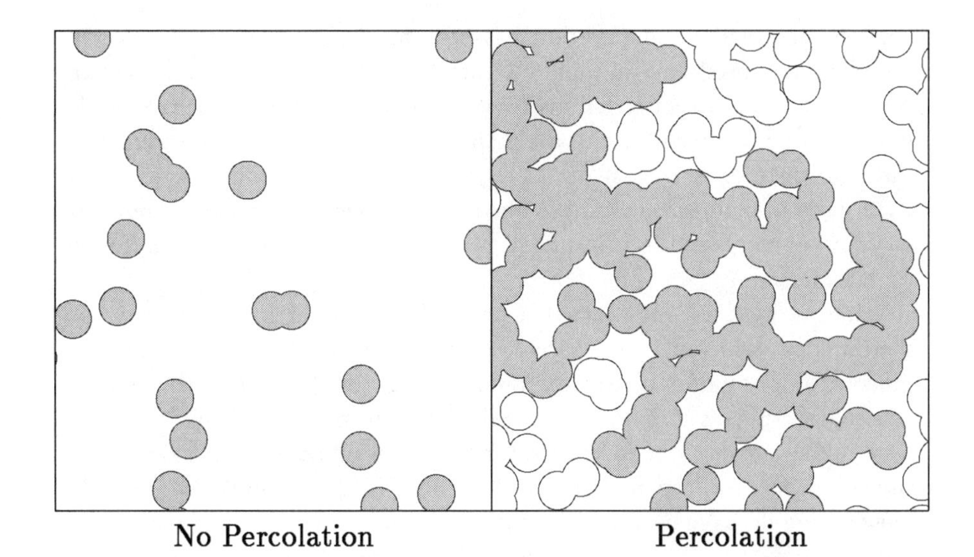

Figure 11.2 **The percolation of randomly placed equal-sized spheres.** When less than 29% of the volume is filled, the spheres will form only finite clusters, as shown on the left. For larger densities, however, an infinite cluster will form, as illustrated on the right by the shaded cluster. While the diagram shows only a finite region, the infinite cluster would extend beyond the diagram no matter how large a volume it depicted.

Some fraction of the spheres link together to form an infinite cluster, extending throughout all of space, as illustrated on the right side of Figure 11.2.* The mathematical physicists call this phenomenon *percolation*, even though it seems to have nothing to do with making coffee. In truth, there is a relation to brewing coffee, since the passage of a liquid through a porous material can be studied by similar techniques: if the randomly situated holes in the porous material are sufficiently dense, they will link together into an infinite network through which a liquid might flow.

The bubbles that form during inflation materialize randomly at different times, and then start to grow. At any instant, however, the distribution of bubbles is a collection of randomly placed spheres, so we could ask

* Since the spheres are placed randomly, it seemed strange at first to learn that one can know with certainty that an infinite cluster will form if more than 29% of space is filled with spheres. The formation of an infinite cluster, however, depends on an infinite number of overlaps between bubbles. Even though any one overlap may or may not occur when the spheres are randomly placed into position, the general behavior of an *infinite* number of overlaps can be predicted unequivocally.

whether or not the configuration has percolated, i.e., whether or not the bubbles have fused into an infinite cluster. If the answer were yes, then we would still have further questions. We would have to think harder to decide whether the bubble collisions would suffice to spread the energy uniformly through space. However, if the answer were no—if the bubbles remained exclusively in finite-sized clusters—then it would seem clear that the bubbles could never merge to form the huge region of uniformity needed for a theory of our universe.

It was hard to guess whether the bubbles would percolate, because there were arguments in both directions. On the positive side, we knew that percolation occurs for equal-sized bubbles when the fraction of the volume covered by spheres is 29%. Since the bubbles in the early universe would very quickly cover a fraction of 99.9999999% or more, it seemed very plausible that they would percolate. On the other hand, the bubbles of the early universe would have been very far from equal-sized. The bubbles that materialized earliest would start to grow rapidly, so they would be spectacularly larger than those that formed later. We did not know whether this wide diversity of sizes could change the result.

While I was traveling on my eastward excursion, I sought out mathematicians and physicists who were interested in percolation. One physicist at Princeton invented two proofs that the bubbles would *not* percolate, but I was able to find flaws in both of them.

My feeling of relief, however, was only short-lived. A week later, while I was visiting Cornell, a friend put me in touch with Harry Kesten of the Cornell math department. Harry invited me to his home on a Sunday afternoon, and I explained the problem over tea. Harry's mind began clicking. Like many veteran problem-solvers, he began by seeking the simplest problem he could find that was "like" the problem in question. Instead of spheres in three dimensions, he decided to think about circles in two dimensions. Instead of randomly placed circles, he chose to consider squares on a checkerboard, some fraction of which were blackened at random. Since the inflationary universe problem involved spheres with a wide range of sizes, Harry divided each square of his hypothetical checkerboard into smaller and smaller squares, at each stage asking what would happen if some fraction of the squares were blackened. By the end of the afternoon Harry had a plausible argument—but not yet a proof—that his checkerboard would not percolate.

The next day Harry telephoned with a solid proof, and a few weeks later Erick Weinberg and I checked that Harry's argument applied to the spherical bubbles of inflation as well as to Harry's checkerboard. My hopes for percolation were dashed. The bubbles that would form in an inflating universe would remain forever in finite clusters, as illustrated in Figure 11.3.

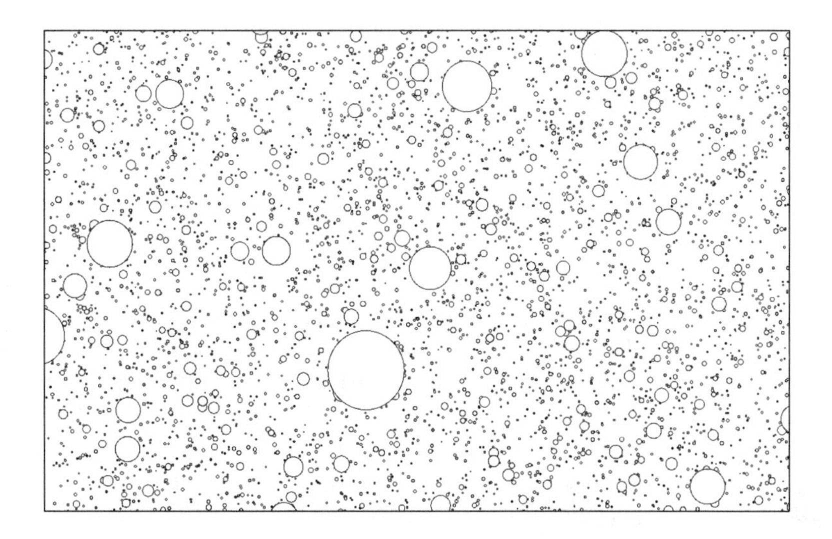

Figure 11.3 **Bubbles forming at the end of inflation.**

Even as the fraction of the volume filled with bubbles exceeds 99.9999999%, the clusters never merge into an infinite block. The problem is that the newly materializing bubbles are much smaller than those that have already been growing. As these tiny bubbles appear randomly in the gaps between previous bubbles, they cover an ever-increasing fraction of the volume without ever closing the gaps.

Each cluster, on average, is dominated by a single, largest bubble—the bubble that formed first, and had the most time to grow. As the large bubble grows it collides with other bubbles, but these are typically much smaller, creating a thin fuzz around the boundaries of the large bubble. Since the bubble wall gains energy as it grows, the wall of the larger, older bubble is far more energetic than the walls of the small bubbles with which it typically collides. The wall of the large bubble rips almost undistorted through the smaller bubbles, just as a jet airplane is unscathed by collisions with mosquitoes. The bulk of the energy in the universe, therefore, remains locked in the walls of the these large bubbles. There would be no energy available to produce the hot soup of particles needed to start the big bang. The absence of an acceptable ending to inflation soon became known as the *graceful exit problem.*

Although disappointed, I still thought that inflation looked too good to be wrong. The graceful exit problem could perhaps be solved if, at the end of inflation, there was a sudden rise in the rate at which bubbles materialize. Such a rapid burst of bubble formation could fill in the gaps between bubbles, leading to percolation, bubble collisions, and the necessary hot soup of particles. Unfortunately, I could not think of any process that might trigger such a burst of bubble formation.

At the beginning of August, 1980, I submitted my paper on inflation to the *Physical Review.* I titled it "The Inflationary Universe: A Possible Solution to the Horizon and Flatness Problems" [1]. The monopole problem was included in the paper, but I omitted it from the title because it seemed less persuasive. The monopole problem exists only if grand unified theories are correct, while the horizon and flatness problems depend only on well-accepted aspects of cosmology. In addition to discussing the successes of inflation, I also summarized the graceful exit problem, expressing the hope that "some variation can be found which avoids these undesirable features but maintains the desirable ones." (The details of the graceful exit problem were published later in a joint paper with Erick Weinberg [2], and a paper reaching similar conclusions was also published by Stephen Hawking, Ian Moss, and John Stewart, all of Cambridge University [3].)

So, as I headed East to assume my new position at MIT, my feelings were mixed. Inflation seemed to be *almost* the perfect theory for describing how our universe began. It accounted for the origin of virtually all the matter in the universe, and also explained why the early universe had a density so close to critical, why the cosmic background radiation is so uniform, and why the universe is not inundated with magnetic monopoles. But how, I kept asking myself, could inflation end without destroying the exquisite uniformity that it creates?

CHAPTER 12

THE NEW INFLATIONARY UNIVERSE

In the summer of 1980 I had three main concerns: writing an article on the inflationary universe, moving to Massachusetts, and discovering a workable ending for inflation. I was preoccupied with the first of these goals, my wife Susan took primary responsibility for the second, and no progress was made on the third.

At the beginning of July, Susan and our son Larry, now $2^1/_2$ years old, flew to Boston to live with Susan's sister while they looked for a place for us to live. Meanwhile I remained in California, working hard on the inflationary universe paper. I completed the manuscript by the end of July, and left it to be typed while I took off for a one-week conference at Santa Barbara. Returning to SLAC for a few days, I proofread and submitted the paper, and packed our belongings for shipping. I also copied computer programs to punch cards, which at that time provided the easiest method to transport the information. After loading our car with the delicate items that we chose not to risk in shipping, I set off across the country a day after Susan had negotiated the purchase of a three-bedroom condominium in Brookline, about three miles from MIT.

The condominium was in a building that was undergoing renovation and conversion from being a rental property. We expected to have the apartment at the beginning of September, but due to delays in the construction we were not able to move in until October 9. Meanwhile we lived as vagabonds, staying for one or two week intervals in various apartments belonging to Boston area physicists who were temporarily out of town. The hardest moments were the several occasions when Larry became upset and cried that he wanted to "go home." When we were finally settled in the condominium, however, all of us were very happy with our new home.

During the following year I continued to work on inflation, but my life was complicated by two problems, one medical and the other legal.

The medical problem began in November, when I developed a serious case of gastroenteritis. After several recurrences, my doctor discovered a tumor that had apparently been growing for many years in my colon. He advised that my colon be removed, and I checked with two other doctors who agreed with the recommendation. The operation was performed at the end of June by a superb surgeon (William Silen). I was in the hospital for two weeks as expected, and then gradually developed the strength to resume a normal life.

My legal problem, which centered on the condominium purchase, took somewhat longer to resolve. Although Susan and I had hired one of Boston's most prestigious law firms to handle the seemingly straightforward transaction, we found ourselves in conflict with recently adopted regulations concerning the conversion of rental units into condominiums. We became the test case for the new procedures, leading to a prolonged and indecisive morass of civil and criminal litigation. The issue was not settled until May 1983, when I missed the first day of an important physics conference to speak at Brookline's Town Meeting in favor of a grandfather clause—a provision that would exempt condominium buyers who purchased their property during the early, chaotic phase of the new legislation. The clause passed overwhelmingly, so my family and I finally became the legal residents of our own condominium.

During my first year and a half at MIT, when I was not busy talking to lawyers or talking to doctors or teaching classes, I continued to work with Erick Weinberg on the question of how inflation might end. While the ultimate goal was to find a way to modify the proposal so that it would have a successful ending, we had no ideas along those lines. We worked hard, however, writing a definitive paper on the failure of the inflationary universe model as I had originally proposed it. The proof that the bubbles which form in the phase transition at the end of inflation would not merge into an infinite cluster was rather complicated, so we tried to simplify the proof as much as possible.

In addition to our work on bubble merging, we also discussed the question of whether our universe might be contained within a single bubble. If it could, then the question of bubble merging would become irrelevant. When we tried to calculate the properties of the inside of an isolated bubble, we found that the calculation depended on the properties of the Higgs fields that cause the phase transition. These Higgs fields, however, are only a theoretical proposal, and if they exist at all their detailed properties are unknown. So the best that Erick and I could do was to assume what we considered typical properties for the Higgs fields, and then to proceed with the calculation. Using this approach, we found that a universe that might exist inside an isolated bubble would be an extraordinarily empty one. For example, when such a universe reached the age of 10 billion years, roughly the age of our universe, the temperature of the cosmic background radiation would not match the value in our universe, 2.7°K, but instead would be about 10^{-29}°K. Omega, the ratio of the actual mass density to the critical density, would not be in the vicinity of one as in our universe, but would instead be in the vicinity of 10^{-86}. The one-bubble universe, we concluded, would be far too empty to bear any resemblance to the real universe.

On December 15, 1981, I finished putting in the last footnotes on the paper with Erick about the failures of the original inflationary model. Much more important, however, on the same day I read a paper that I had received a few days earlier from Andrei D. Linde (pronounced LEEN-deh), a physicist at the Lebedev Physical Institute in Moscow. (I learned a few weeks later that Paul Steinhardt and Andreas Albrecht at the University of Pennsylvania had independently reached similar conclusions.) Linde's paper claimed to solve the graceful exit problem by exactly the method that Erick Weinberg and I had thought we had excluded. "The whole observable part of the universe," Linde wrote, "is contained *inside one bubble,* so we see no inhomogeneities caused by the wall collisions." "Nonsense," I thought, as I skimmed through the paper, "this calculation can't possibly be right."

Then I read through the paper a little more carefully, expecting to find the obvious blunder that Linde must have committed. But there was no blunder. Erick and I had analyzed the ending of inflation for "typical" Higgs fields, and we correctly concluded that the resulting bubbles would be far too empty to allow a one-bubble universe. Linde had apparently reached the same conclusion, but he went on to consider Higgs fields that are not so "typical." In particular, Linde considered a Higgs field with an energy density graph that resembles neither the Mexican hat of Figure 8.3 nor the dented Mexican hat of Figure 10.1. Instead he considered an energy graph that was first proposed in 1973 by none other than Sidney Coleman and Erick Weinberg. (The graph will be shown later as Figure 12.1.) On the second reading I realized that Linde had discovered a possibility that completely avoids the problems of the one-bubble universe that Erick and I had

Andrei Linde

found. I noticed that Jeffrey Goldstone had his door open in the office across the hall from mine, so I immediately ran to tell him that inflation was alive again. "Damn it," I wrote in my diary that night, "how did I miss it?"

At the time, Linde and I had read each other's papers and had exchanged letters, but we had not yet met. Linde was born in 1948, to parents who were both physicists, and received his first diploma in physics from Moscow State University in 1971. He attended graduate school at the Lebedev Physical Institute, receiving his Ph.D. in 1974. Motivated by an analogy with superconductivity and its disappearance at high temperatures, Linde and his thesis advisor, D.A. Kirzhnits, studied the high temperature behavior

Paul Steinhardt

of particle theories with spontaneous symmetry breaking. Their work focused on the unified electroweak theory, but their methods were later applied to grand unified theories. In 1972 Kirzhnits and Linde became the first to predict that theories with spontaneous symmetry breaking would lead to phase transitions, as was discussed in Chapter 8.

Kirzhnits continued to work on superconductivity and various problems in particle theory, but Linde's main interest was in the electroweak theory and its application to the early universe. In 1976 Linde showed that the mass of the electroweak Higgs particle must be greater than 6 GeV, or else the theory would not produce the spontaneous symmetry breaking that is essential for its agreement with observation. (The same result was derived independently by Steven Weinberg.) The following year Linde used cosmological arguments to conclude that the mass must be greater than 9 GeV. If it were not, Linde calculated, the universe would have been caught in a very long-lived false vacuum state from which it would not have had time to emerge. (Today we know from particle accelerator experiments that the mass of the electroweak Higgs particle is at least 58 GeV, but in 1977 Linde's result represented important new information.)

My initial contact with Linde stemmed from his paper on the Higgs particle mass, a topic which Erick Weinberg and I reconsidered while we were at SLAC. Disagreeing with Linde's logic, we repeated the calculation our own way and found exactly the same result. Our paper properly cited Linde's previous work, acknowledging that he had described similar calculations, but we impolitely neglected to mention that our conclusion for the Higgs particle mass was identical to his. Linde responded with a letter saying that our paper "is rather interesting for me, but I think that your comments to my papers are improper." Linde has taken a conciliatory attitude about this affair, however, and has not let it stand in the way of a friendship that has since grown between us.

In the late 1970s Linde independently invented much of the inflationary universe theory. He and Gennady Chibisov had realized that the universe might supercool at the grand unified theory phase transition, and they realized that this would lead to exponential expansion. They also noticed that the bubble collisions under these circumstances would cause a problem, leading to gross nonuniformities in the universe. However, since they failed to realize that the exponential expansion would solve the horizon and flatness problems, they concluded that "there was no reason to publish such garbage" [1].

At the end of September 1980, I sent a copy of my inflationary universe paper to Linde, but apparently he had learned about it earlier. Since he had already discovered most of the ideas on his own, he instantly understood what inflation was about, and became convinced that inflation must be responsible for the universe in which we live. As he puts it, "It was very difficult to abandon this simple explanation of many different cosmological problems. I just had the feeling that it was impossible for God not to use such a good possibility to simplify His work, the creation of the universe."

But just as he appreciated the value of solving the horizon, flatness, and monopole problems, Linde also understood the severity of the graceful exit problem. If inflation was the key to our universe, there had to be some crucial element that was different from what was described in my paper. Linde worried intensely about this problem. While I was being treated for ulcerative colitis, Linde developed an ulcer. "I don't know the reason," Linde has said, but "maybe this [worry about inflation] was one of them."

In addition to developing an ulcer, however, Linde developed the new inflationary universe. He had been studying the various ways that Higgs fields might behave, and had gotten some feeling for the different possibilities by solving the equations on a computer. By late spring of 1981, Linde discovered what seemed to be a way to salvage inflation. To make sure that he was not overlooking some subtlety, he placed a late-night telephone call to another theoretical physicist, Valery A. Rubakov. Speaking from his bathroom so as not to wake his sleeping family, Linde exchanged thoughts with Rubakov about phase transitions, bubbles, and Higgs fields. By the end of the conversation, Linde became convinced that his new ideas would work [2]. In great excitement he woke his wife—theoretical physicist Renata Kallosh—and told her, "It seems that I know how the universe originated."

Linde still worried whether his new idea was correct, and also whether anyone would believe it. At the beginning of October 1981, he spoke about it at a quantum gravity meeting in Moscow, which was attended by Stephen Hawking, of Cambridge University. Hawking raised some questions about Linde's calculations, so Linde refined his arguments and then submitted the paper at the end of the month.*

Although Erick and I had been miserly in the credit we gave to Linde in our Higgs mass paper, Linde was generous in giving credit to my work. He called his theory "new inflation," and even used the title of my paper as a prototype for his: "A New Inflationary Universe Scenario: A Possible Solution of the Horizon, Flatness, Homogeneity, Isotropy, and Primordial

* Linde informs me that some of Hawking's questions came out the following day, when Hawking lectured at the Sternberg Astronomical Institute of Moscow State University. Linde attended the talk, and was asked to translate. It was a slow-moving and complicated procedure, since Hawking's speech could be understood only by his close associates. Hawking spoke in English to a student traveling with him, the student repeated the words, and then Linde translated into Russian. Hawking began by describing the problems of the original inflationary theory, so Linde tried to speed things up by elaborating on his own. "Steve would say something, his student would say few words, and then I was speaking for about five minutes." Hawking said that there was an interesting idea by Linde for solving these problems, and Linde continued to happily translate. Then, however, Hawking's direction changed, as he explained the problems that he saw in the new theory. Linde was trapped, and for a half hour he had no choice but to relay to the whole institute arguments that he did not believe about why new inflation could not work! After the talk Linde seized the chance to tell the audience his own point of view, and he and Hawking continued the conversation for several hours at Hawking's hotel. Was the controversy resolved? In any case, Hawking invited Linde to the Nuffield Workshop on the Very Early Universe, which he was organizing for the following summer, and Hawking soon started to write papers about new inflation himself.

Monopole Problems" [3]. The increase by 150% in the number of problems solved indicates a difference between Linde's style of presentation and mine.

While Linde was inventing new inflation in Russia, an independent development occurred in the United States, spearheaded by Paul Steinhardt. Like Linde, Steinhardt was a particle theorist whose work in cosmology was partly motivated by condensed matter physics. Like me, Steinhardt's work began with the consideration of a problem involving magnetic monopoles.

Paul was born in 1952, and received his Ph.D. from Harvard in 1978, working with Sidney Coleman. He then joined the Harvard staff for three years as a "Junior Fellow." At Harvard he pursued mainly particle theory and cosmology, but at the same time he was a consultant at the IBM Research Laboratories, Yorktown Heights, New York, where his activities centered on condensed matter physics. Paul's first papers on the early universe concerned the phase transition in the electroweak theory, but he then turned to the study of phase transitions in grand unified theories. While others had already studied the production of magnetic monopoles in GUT phase transitions, Steinhardt asked what would happen to monopoles that were already present at the start of a phase transition.

When it became known that the original inflationary theory failed because bubble nucleation would destroy the uniformity of the universe, Paul was in an ideal position to see the alternative. While the vast majority of phase transitions proceed by bubble nucleation, Paul had learned about an obscure type of condensed matter phase transition called *spinodal decomposition*, which takes place without bubbles. Armed with this background, he was able to see how a similar phase transition could occur in a grand unified theory.

In the fall of 1981 Paul began an assistant professorship at the University of Pennsylvania. He suggested to a graduate student, Andreas (Andy) Albrecht, that he work out the consequences of the new idea. The basic properties fell into place, but Paul and Andy continued to worry about subtleties. When copies of Linde's paper on new inflation began to circulate in the United States, Paul and Andy realized the necessity of putting their ideas to paper immediately. By the end of January, their description of the new inflationary theory was submitted to *Physical Review Letters* [4].

The Higgs fields that drive the inflation are a theoretical invention, so the nature of these fields cannot be deduced from known physics. While the qualitative properties of the fields are patterned after those of the Higgs field of the electroweak theory, the detailed properties have to be hypothesized. The secret of new inflation is to choose a special shape for the energy density graph of the inflationary Higgs fields. The desired shape, which might be called a flattened Mexican hat, is illustrated in Figure 12.1. The general shape is similar to that of the standard Mexican hat of Figure 8.3 (on page 140), except that the central peak is a flat, gentle, plateau, rather than a rounded mountain.

The evolution of the Higgs field can again be visualized as the motion of a ball rolling on the hill of the energy diagram. For the steep hill of the standard Mexican hat, the ball tends to rapidly barrel to the bottom, starting to oscillate about the vacuum circle. For the flattened Mexican hat, however, the rolling is very slow. If the plateau is nearly level and the ball starts near the center, the ball will hover sluggishly before it finally begins to wander toward the vacuum circle.

As long as the ball is near the top of the hill, the energy density of the Higgs field remains high. Although this state is not nearly so stable as the

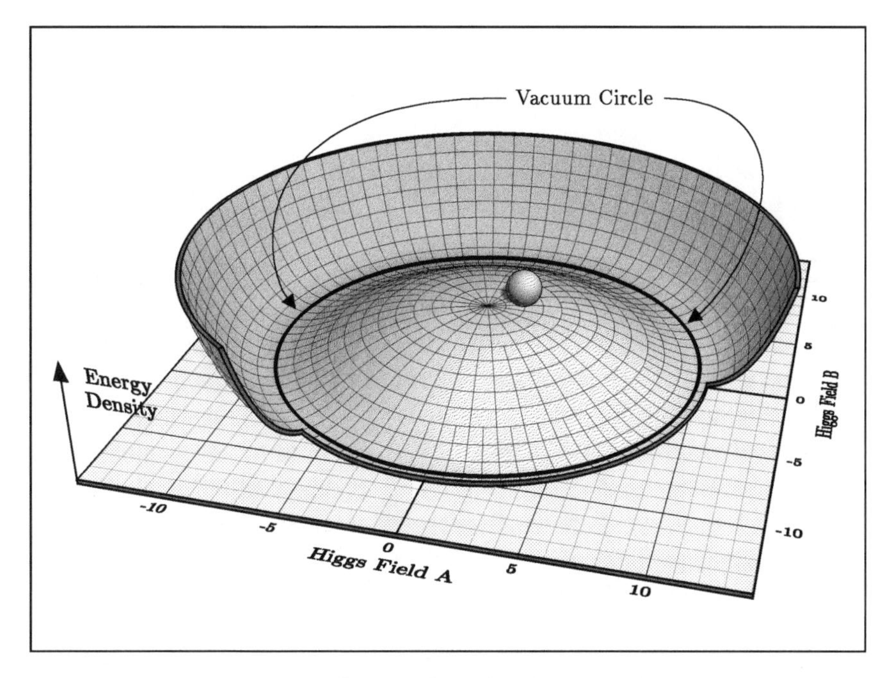

Figure 12.1 **Energy density of Higgs fields for the new inflationary theory.** The new inflationary theory assumes that the energy density of the Higgs fields is described not by the Mexican hat diagram of Figure 8.3, or the dented Mexican hat of Figure 10.1, but instead by a flattened Mexican hat. The evolution of the Higgs fields is similar to that of a ball that very gradually rolls away from the center of the plateau. The bubble continues to inflate while the ball remains near the center, allowing a single bubble to become easily large enough to encompass the observed universe.

false vacuum state envisioned in the original inflationary theory, it has essentially the same properties. The false vacuum state shown at the center of the dented Mexican hat in Figure 10.1 (on page 168) was actually much more stable than necessary, so the ball shown dawdling near the center of the plateau in Figure 12.1 is stable enough. This state can also be called a false vacuum,* and it can also drive inflation.

In the original inflationary theory, with the dented Mexican hat energy diagram, the phase transition follows the paradigm of boiling water. The Higgs field in a small, spherical region undergoes a process called quantum tunneling, which takes it from the false vacuum value to a point on the other side of the energy barrier, as shown in Figure 12.2(b). After the tunneling, the Higgs field lies on a steep part of the energy diagram hill. In the central region of the bubble, the Higgs field plummets to the bottom of the hill, quickly terminating the period of inflation. The field oscillates back and forth about the trough of the vacuum circle, but the oscillations are soon arrested, as shown in Figure 12.2(c), by a process similar to friction. The energy of the Higgs field is converted into a gas of many kinds of particles, but their density is rapidly diluted by the growth of the bubble. More energy is released as the bubble grows, since the Higgs field near the edge is plunging from the high energy value it has outside the bubble to the low energy value it has inside. The energy produced, however, moves outward with the bubble wall, so the interior of the bubble remains essentially barren.

If the energy density of the Higgs fields is described by the flattened Mexican hat diagram, however, then the scenario is totally different. It is a much gentler, gradual phase transition, more like the congealing of Jell-O than the boiling of water. Inflation continues as the Higgs field begins to slowly drift away from the center of the plateau, so the energy density remains high while the bubble enlarges by many orders of magnitude, as shown in Figure 12.2(e). When the Higgs field finally slips off the plateau, the central region of the bubble has become large enough to easily encompass the observed universe. As shown in Figure 12.2(f), the Higgs field throughout this huge region oscillates and converts its energy to a hot soup of particles, exactly as required for the standard hot big bang model.

All of the successes of the original theory are preserved by the new inflationary universe, and the graceful exit problem disappears completely.

The discovery of new inflation happened before there had been time for interest in old inflation to die down, so there was never a lull in my lec-

* Strictly speaking the phrase "false vacuum" is reserved for a local minimum of the energy density graph, but the center of a broad plateau behaves essentially the same way.

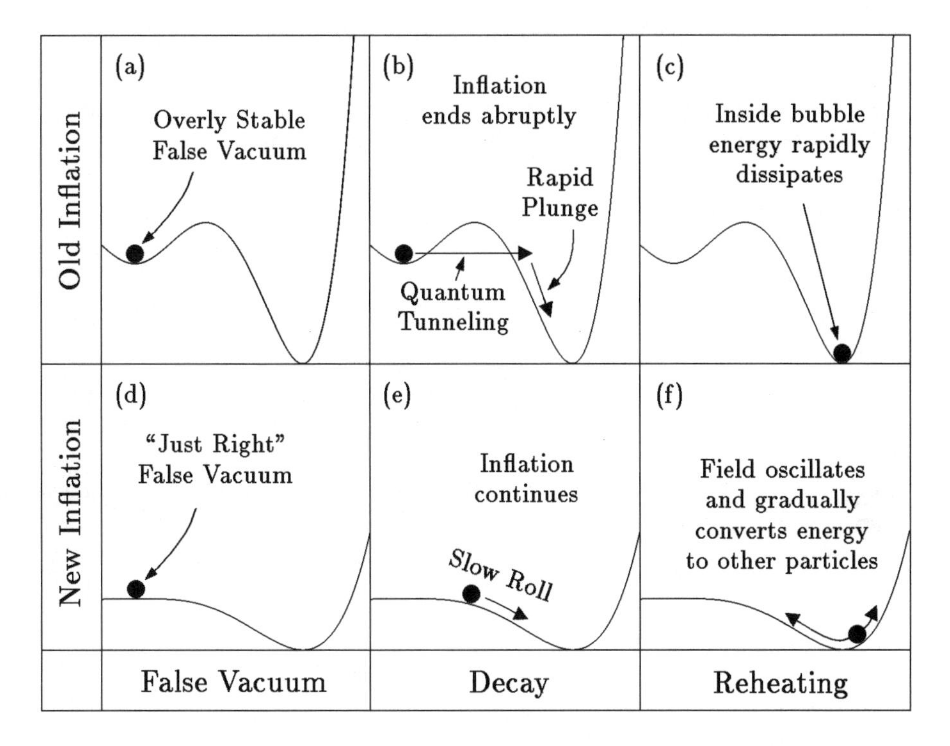

Figure 12.2 **Comparison of old and new inflationary theories.** The diagram illustrates the nature of the false vacuum, the mechanism for the decay of the false vacuum, and the mode of reheating the universe after inflation, for both the old and new inflationary theories. Each box shows a plot of energy density vs. the value of the Higgs field, with a ball to indicate the evolution of the Higgs field. In the new inflationary theory there is continued inflation followed by a significant energy release deep inside the bubble, making possible a one-bubble universe.

---

turing schedule. When new inflation was invented, my calendar for the spring term already included lectures at IBM, Brookhaven National Laboratory, and Columbia University, as well as a European tour that included a conference at the Royal Society in London, a meeting in the French Alps, a lecture at CERN, and another conference in England organized by the Rutherford Laboratory. The European trip was very interesting, since I became part of a "traveling circus" of cosmologists who were attending these meetings, a group that included David Schramm, Michael Turner, Paul Steinhardt, and a number of others. All of us were excited about new inflation, which became a dominant theme in our lectures.

In May of 1982 I was asked to give the Harvard-MIT Joint Theoretical Seminar, so I naturally chose to talk about the new inflationary universe. On my title transparency I included the informal subtitle, "How Linde and Steinhardt Solved the Problems of Cosmology, While I Was Asleep."

CHAPTER 13

WRINKLES ON A SMOOTH BACKGROUND

One of the great virtues of the inflationary theory is its explanation of the large-scale uniformity of the universe. This uniformity is seen most clearly in the high precision measurements that have been made of the cosmic background radiation, which is found to have a constant temperature across the sky to an accuracy of about one part in 100,000. As discussed in Chapter 10, any attempt to understand this large-scale uniformity in the standard big bang theory without inflation is thwarted by the horizon problem—that is, the theory implies that the early universe evolved so quickly that there was no time for information, matter, or energy to be exchanged between one region and another. Thus, the large-scale uniformity can be "explained" in the standard theory only by *assuming* that the universe began that way. The question of *why* the universe began in a state of such exquisite uniformity is not addressed.

In the inflationary theory, on the other hand, the large-scale uniformity is a natural consequence of the evolution of the universe. If inflation is right, then the observed universe emerged from a much smaller region than had been previously thought, so there was plenty of time for cosmic uniformity to be established before inflation began. Inflation then magnified this submicroscopic region, enlarging it enough to easily encompass the observed universe.

However, although the universe appears remarkably featureless if we blur our vision and average over large volumes, a look at the nearby universe shows that the matter is clumped in a complicated, irregular pattern. The Earth, for example, has a mass density roughly 10^{30} times higher than the cosmic average, and even the "thin" air that we breathe has a density

10^{26} times higher. (To reduce a quart of air to the average density of the universe, you would have to spread it over a sphere with a radius 5 times larger than the Earth!) Our solar system lies in the Milky Way galaxy, where the average density in the galactic disk is approximately 10^6 times the cosmic average. The Milky Way, in turn, belongs to a conglomeration of about twenty galaxies known as the *Local Group*. The group, with an average mass density roughly 200 times the cosmic average, is dominated by the Milky Way and its sister galaxy, Andromeda, $2^1/_2$ million light-years away. The matter distribution surrounding the Local Group is also concentrated in clumps, the largest of which is the *Virgo Cluster*, an aggregation including 200 bright galaxies which is centered about 60 million light-years away. Virgo is also the center of the *Local Supercluster*, a loosely knit assemblage of some 100 clusters of galaxies, including the Local Group. Looking further into space, at a distance of perhaps 350 million light-years, Margaret Geller and John Huchra of the Harvard-Smithsonian Center for Astrophysics have discovered a giant sheet of galaxies—the *Great Wall*—which stretches more than 500 million light-years across the sky. The Geller-Huchra data also show that the universe has a frothy texture, with nearly empty bubble-like voids surrounded by sheets of galaxies. The voids are typically 150 million light-years across.

Cosmologists attribute the evolution of this rich tapestry to the influence of gravity. The universe is said to be *gravitationally unstable*, which means that gravity strongly amplifies any nonuniformities in the distribution of matter. That is, if any region has a mass density that is slightly higher than average, the excess mass will produce a slightly stronger-than-average gravitational field, pulling extra matter into the region. The extra matter creates a still stronger gravitational field, resulting in yet more matter being drawn into the region. However, for this process of gravitational clumping to start, the initial distribution of matter must have had small irregularities, which would serve as seeds for the development of cosmic structure. If the universe had started out completely uniform, then it would still be uniform today. If inflation is to describe the real universe, it must explain the large-scale uniformity in a way that can also accommodate the lumpiness that is seen on smaller scales.

In the context of the original form of the inflationary universe theory, there was no problem finding a possible source for these primordial nonuniformities. In this theory the inflationary period was assumed to end with a first order phase transition, in which bubbles of the new phase would form randomly and collide. The randomness of the bubble formation would obviously lead to nonuniformities, so one could hope that this randomness would account for the structures observed in the universe today. Henry Tye and I expressed this optimistic hope in the paper that we wrote on supercooled phase transitions, but we made no attempt to calculate the consequences. However, when the effects of the random bubble formation and bubble collisions were later understood (as described in Chapter 11), it was found that the resulting irregularities were far too gross to describe galaxies. Instead, the nonuniformities were so severe that the original inflationary universe theory had to be abandoned.

In the new inflationary theory, on the other hand, the randomness of bubble formation is not a factor. In this version of the theory, the entire observed universe is assumed to lie deep inside a single bubble, so any bubble collisions are far too remote to have any observable effects. Within our bubble, the colossal expansion of inflation would have diluted any normal matter to a negligible density. The only form of matter that would remain significant after the enormous expansion is the false vacuum itself, the mass density of which does not decrease as the universe expands. Since the mass density of the false vacuum is controlled by the underlying laws of physics, it would have the same value everywhere. Inflation would produce a smooth universe, even if the matter was extremely lumpy before inflation began. For a while, those of us working on the new inflationary universe theory were seriously worried that it would lead to an impeccably smooth distribution of matter. Where could we hope to find the initial nonuniformities that are needed to seed the evolution of cosmic structure?

During my travels in Europe in March of 1982, just a few months after the new inflationary theory was proposed, the question of structure formation in this theory was beginning to attract attention. At the *Rencontre de Moriond* Astrophysics Meeting in the French Alps, for example, I had hot chocolate late one afternoon with Michael Turner of the University of Chicago. The conference was organized with lectures in the mornings, late afternoons, and evenings, with the early afternoons free for skiing. This was my first experience on skis, so as I was recovering from the bruises of the day's falls, Michael told me about some recent ideas of his collaborator, James Bardeen of the University of Washington. Jim, the son of the twice Nobel physics laureate John Bardeen, had two years earlier written the definitive paper on the evolution of density nonuniformities in cosmology. Now Jim was worrying about how these nonuniformities could possibly survive a period of inflation. He was trying to devise a version of the theory in which the phase transition ending inflation is extraordinarily sensitive to local variations in the temperature, providing an amplification mechanism that counteracts the smoothing effect of the inflation itself. A week later at the Rutherford Laboratory I chatted with Graham Ross, who was thinking about similar ideas. At the Rutherford meeting I spoke to David Schramm of the University of Chicago, who suggested that we might need to learn about turbulence to resolve this problem. None of these early ideas about structure formation led to successful theories, but they set in motion a period of international brainstorming that culminated in the discovery of one of inflation's most revolutionary consequences: If inflation is right, the intricate pattern of galaxies and clusters of galaxies may be the product of quantum processes in the early universe. The same Heisenberg uncertainty principle that governs the behavior of electrons and quarks may also be responsible for Andromeda and the Great Wall!

The first time I heard about this extraordinary possibility was in the middle of May, when I had a phone conversation with Paul Steinhardt, co-inventor of the new inflationary universe. He told me that he and Michael Turner were exploring the peculiar effects that arise when quantum theory is combined with general relativity. Although no complete unification of these theories existed, Stephen Hawking had shown in 1975 that if our general understanding of quantum theory and general relativity was correct, then black holes are not really black after all. Although classically the gravitational field of a black hole is so strong that not even light can escape from its interior, Hawking discovered that quantum effects cause black holes to emit radiation with the same spectrum as a hot object, the black-body spectrum discussed in Chapter 4. The temperature of the black hole depends on its mass, with the temperature becoming higher as the mass becomes smaller.*

In 1977 Hawking had joined with Gary Gibbons, also of Cambridge University, to explore the consequences of quantum theory for an exponentially expanding universe [1], precisely the kind of spacetime that later became the central feature of the inflationary universe. Using logic similar to that of the black hole case, they found that an exponentially expanding universe would become filled with radiation like that of a hot object, with the temperature proportional to the expansion rate.† With thermal radiation

* The temperature is found by dividing 6.17×10^{-8}°K by the mass, measured in solar masses, where 1 solar mass = 1.99×10^{33} grams. A black hole formed by the collapse of a star must weigh at least 3 solar masses, so the Hawking radiation is totally negligible. On the other hand, 1-gram black holes can theoretically exist, although nobody knows how such exotica might be produced. If a 1-gram black hole did exist, its temperature would be 10^{26}°K, and the Hawking formula indicates that it would explode by radiating all its mass in 10^{-27} seconds! Scientists are still debating whether it would totally self-destruct, or whether some permanent remnant might survive.

† Precisely, they found that the temperature of the radiation is given by $T = hH/(4\pi^2 k)$, where h is Planck's constant, H is the Hubble constant of the exponentially expanding space, and k is Boltzmann's constant. Numerically, if H is measured in kilometers per second per million light-years, then the temperature in degrees Kelvin is 1.285×10^{-31} times H.

comes thermal fluctuations, the relentless, unpredictable jittering of hot matter. While the temperature and the accompanying fluctuations would be minuscule for the expansion rate of the universe today, it could have been significant for an inflationary period in the very early history of the universe. Paul and Michael were trying to calculate whether these thermal fluctuations could produce the nonuniformities, also called *density perturbations*, that evolved to produce the structure we see today.

I learned more about Steinhardt and Turner's work a few weeks later, at the end of May, when Steinhardt visited Cambridge, Massachusetts. I biked to Harvard to meet Paul, who spent a morning explaining his calculations to me and a few other physicists. Typically cosmologists would quote the magnitude of the density perturbations at the time of recombination, about 300,000 years after the big bang, the time when the plasma of free electrons and nuclei condensed to form a neutral gas. According to the best estimates, the present structure of the universe indicates that the magnitude of the density perturbations at the time of recombination must have been about 10^{-3} or 10^{-4}. That is, a typical region of high density was about one part in 10^3 or 10^4 more dense than average, while a typical low density region was a similar amount below average. (Today, based on measurements of nonuniformities in the cosmic background radiation, cosmologists estimate the density perturbations at recombination to be only about 10^{-5}, somewhat less than the estimates in 1982.)

In their preliminary calculation of the effect of the Gibbons-Hawking fluctuations, Paul and Michael estimated a density perturbation magnitude of only 10^{-16}. There was a huge gulf between this estimate and the value of 10^{-3} or 10^{-4} desired for galaxy formation, but they were not ready to give up hope. They recognized that their work was very preliminary, and the Gibbons-Hawking quantum fluctuations seemed the only possibility for obtaining structure from the new inflationary universe theory.

Although the numbers did not seem right, the underlying idea was breathtaking. Physicists are very familiar with the effects of quantum randomness on the scale of atoms or molecules, but on the scale of baseballs or planets, the deterministic laws of Newton have always proved accurate. Paul and Michael were suggesting, however, that the intricate astronomical landscape of galaxies, clusters, and voids is in fact the direct result of quantum processes! The key to this odd proposal is the colossal expansion rate in the inflationary theory. For the physics of the very small, quantum indeterminacy is always the governing principle. In the inflationary theory, perturbations created by quantum processes on subatomic distance scales are rapidly stretched to the size of galaxies or even larger than the observed universe. The inflationary theory therefore offers the possibility of linking the texture of the cosmos to the physics of elementary particles.

Meanwhile, on the other side of the Atlantic, Stephen Hawking had become interested in the same questions. While Paul, Michael, and I were neophytes to general relativity, Hawking was one of the world's experts, with years of experience studying problems such as the evolution of density perturbations in an expanding universe.

In April, Stephen had visited the University of Chicago, by coincidence at the same time that Paul was there to give a colloquium. Since Michael is based at Chicago, the three were soon engaged in a trialogue. Not much information was exchanged, as all three had just begun to think about the problem, but at least the contact was established. On June 7, Hawking discussed his work in a seminar at Princeton, so Paul drove from Philadelphia to hear it. The next day Paul gave me a detailed report by telephone.

Since his first year of graduate school, Hawking has suffered from amyotrophic lateral sclerosis, also known as ALS, or Lou Gehrig's disease, after the baseball hero who died from it in 1941. The progressive motor neuron disease has confined Hawking to a wheelchair since the early 1970s. Hawking, however, has never allowed his physical condition to cripple his life: He is the father of three children, and he writes papers, travels around the globe, and delivers seminars as actively as any physicist I know. Since 1986, Hawking has communicated with a voice synthesizer and computer, selecting words with one finger from a menu. In 1982, however, he still had a voice, but his speech was so impaired that only his close associates could consistently understand him. He lectured with the help of an interpreter (often one of his graduate students), who would turn the transparencies and repeat his sentences. The process was slow, so frequently results were given with only the sketchiest description of how the calculation was performed. Nonetheless, Paul had been able to extract at least a qualitative understanding of Hawking's approach.

In attacking any problem, the crucial first step is to identify the important elements. In this case, Hawking singled out the process by which inflation ends, the rolling of the Higgs field down the hill of the energy density diagram, Figure 12.1. Classically this rolling would be uniform, but quantum mechanics implies that there is randomness in the motion. In some places the Higgs field will roll down the hill a tiny amount faster than the classical prediction, ending inflation a bit prematurely. In other places the Higgs field will roll a smidgen slower, prolonging inflation and its associated production of matter. The result would be a slightly nonuniform distribution of matter at the end of inflation, as illustrated in Figure 13.1. In a few lines Hawking summarized how he calculated the evolution of these nonuniformities through the time of recombination, reaching two startling conclusions. First, the magnitude of the perturbations would be 10^{-4}, exactly what was wanted to explain the evolution of cosmic structure. Second, the

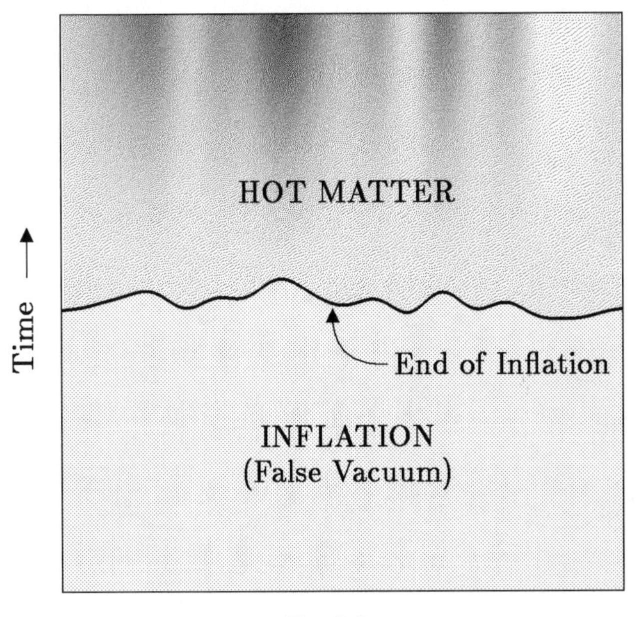

Figure 13.1 **The quantum generation of density perturbations in inflation.** Due to quantum effects, inflation ends at different times in different places, as illustrated by the wavy line in this space-time diagram. Since matter is created by inflation, the hot gas of particles resulting from inflation is densest in those regions that inflated the longest. The initial variations are minute, but gravity causes further clumping, producing the concentrated lumps indicated by the dark shading near the top of the diagram.

---

spectrum of the density perturbations would also have exactly the form desired for the seeds of galaxy formation.

The meaning of the word "spectrum" in the context of density perturbations may seem unfamiliar, but it is closely related to the use of the same word for light waves or sound waves. No matter how complicated a light wave is, it is always possible to decompose it into waves of a standard form, each with a definite wavelength. The same is true for density perturbations, the variation of density with position. The top curve in Figure 13.2 shows a sample graph of how the density of matter might vary with position. The curve is complicated, and cannot be said to have any particular wavelength. Nonetheless, the curve is precisely the sum of the curves shown below it as (a), (b), and (c), each of which has a definite wavelength. The spectrum of this density perturbation is a description of how strongly each wavelength

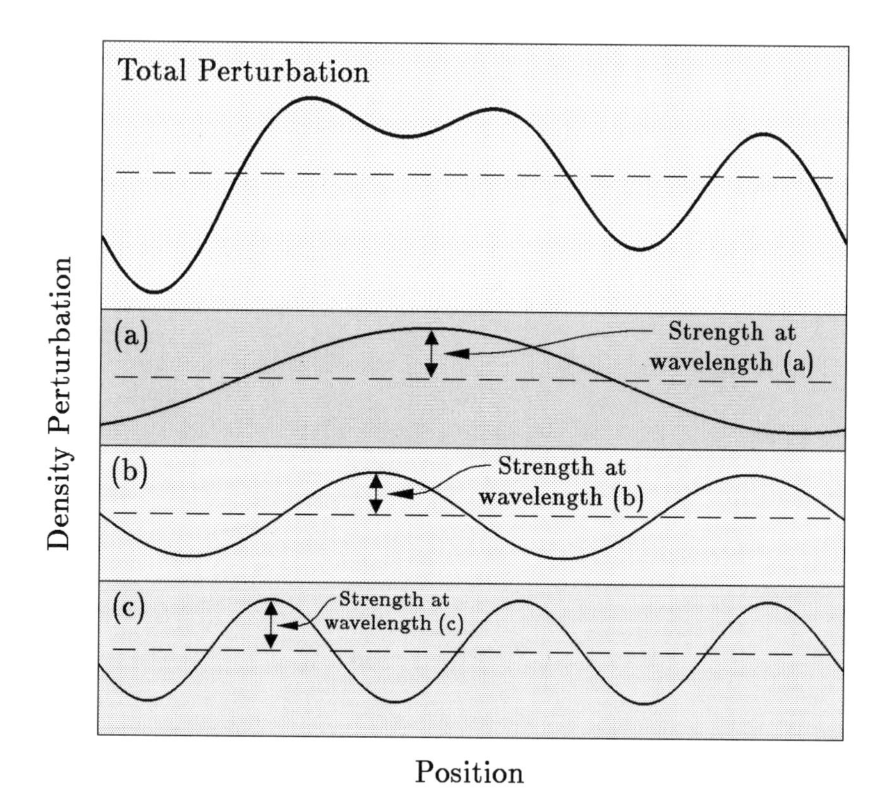

Figure 13.2 **The spectrum of a density perturbation.** The top curve shows a sample density perturbation, the variation of the density of matter with position. Although complex, the curve is exactly equal to the sum of the three standard waveforms (a), (b), and (c), each of which has a definite wavelength. The spectrum is a description of the strength of each contribution, as measured by the peak of each curve.

---

contributes to the sum. The example was of course contrived to be simple, so we need to ask what would happen if the curve at the top of Figure 13.2 were replaced by an arbitrary curve. According to a famous mathematical procedure developed by Jean-Baptiste-Joseph Fourier in the early 1800s, it would still be possible to decompose the curve into a sum of standard waveforms, although an infinite number of terms would usually be needed.* The

* Fourier, the son of a tailor, was among the most creative intellectuals of his time. Accompanying Napoleon on his expedition to Egypt from 1798 to 1801, he engaged in extensive research on Egyptian antiquities. Fourier's twenty-one volume *Description de l'Égypte* is often regarded as the origin of Egyptology as a separate discipline. When Fourier died in 1830, he was the perpetual secretary of the Académie des Sciences, and also a member of the Académie Française and the Académie de Médecine.

spectrum would again be a description, usually in the form of a graph, of how strongly each wavelength contributed to the sum.

Hawking found that the spectrum of density perturbations predicted by inflation has a simple form: it is *scale-invariant,* meaning essentially that each wavelength has the same strength.* As Hawking pointed out, this is exactly the form that was proposed in the early 1970s by Edward Harrison of the University of Massachusetts, and independently by the Soviet cosmologist Yakov B. Zeldovich. While cosmic structure formation is actually a difficult and unsolved problem, cosmologists had believed for about a decade that the scale-invariant spectrum was the simplest assumption consistent with our knowledge of the universe.

If Hawking was right, the new inflationary universe had achieved an incredible triumph. The theory was invented without any consideration of the question of density perturbations, yet Hawking was claiming that the theory correctly explained both the magnitude and the spectrum of these perturbations.

Paul and I still did not understand Hawking's calculations, but several fragments were described in enough detail for us to follow. The answer, Hawking found, depends crucially on the rate at which the Higgs field changes, as it rolls down the hill of the energy density diagram. Just as a high-speed bullet might be unperturbed by a gust of wind that would drive an arrow far off course, the Higgs field is less sensitive to quantum fluctuations if it is evolving swiftly. While the evolution of density perturbations was beyond my depth at the time, both Paul and I thought that we knew how to calculate the rolling of the Higgs field. When we compared our answer to Hawking's, however, we found an outrageous disagreement. Our answer was about 10^8 times smaller than his! This would make the perturbations 10^8 times larger than Hawking found, which means 10^8 times larger than what was wanted for galaxy formation. Since we did not understand the rest of Hawking's calculation, however, we were not sure what to conclude. Both Hawking's answer and our revision of Hawking's answer were in glaring contradiction with the answer that Paul and Michael Turner had found, which was 10^{12} times smaller than Hawking's answer.

The problem was obviously very important, and in fact the week before I had begun to work on it in earnest, in collaboration with So-Young Pi, a Korean-born physicist who was at that time a postdoc visiting Harvard, while on leave from SLAC. She and I had been planning for some time to

* For the reader looking for precision, I mention that the cosmologists' definition of scale-invariance is complicated by the fact that waves with different wavelengths vary with time in different ways. At any given time, waves with shorter wavelengths are actually stronger than those with longer wavelengths. However, if the strength of each wave is measured at the time when its wavelength is equal to the Hubble length (defined as the speed of light divided by the Hubble constant), then all wavelengths would be measured with the same strength.

So-Young Pi

start a collaboration, and the quantum generation of density perturbations seemed like an exciting choice of topic. With Steinhardt, Turner, and Hawking already well along in their work, however, we knew that we would have to rapidly come up to speed if we were going to make a useful contribution. Furthermore, the major cosmology meeting of the year, the Nuffield Workshop on the Very Early Universe, was only a few weeks away. The meeting, funded by the Nuffield Foundation and organized by Stephen Hawking and Gary Gibbons, would bring together about thirty of the most active early universe researchers in the world. The goals of the conference, which highlighted the exciting unsolved problems in cosmology, were spelled out clearly in the letter of invitation, shown in Figure 13.3.

Luckily, So-Young and I benefited from a jump-start supplied by Jemal Guven, an Irish graduate student of Turkish descent who was working with me at MIT. For several months he had been studying the evolution of density perturbations in an exponentially expanding universe, although the motivation of his work had been essentially the opposite of our new interests. His research, which neglected all quantum effects, had been aimed at understanding how any pre-existing density perturbations would be smoothed by inflation. The mathematics was the same, however, so we eagerly listened to his mini-lectures on the techniques of performing these calculations.

UNIVERSITY OF CAMBRIDGE
Department of Applied Mathematics and Theoretical Physics
Silver Street, Cambridge CB3 9EW
Telephone: Cambridge (0223) 51645

Dr. A. H. Guth,
Center for Theoretical Physics,
Laboratory for Nuclear Science and
Department of Physics,
M.I.T.
Cambridge - Massachusetts, 02139,
United States of America.

14 October, 1981.

Dear Dr. Guth,

As you may know, the Nuffield Foundation has provided funds for a series of workshops on Quantum Gravity and related topics. Next year we are planning to hold a workshop here in Cambridge on the very early Universe (< 1 sec). The dates envisaged are June 21st to July 9th 1982.

The standard model seems to provide a satisfactory account of the evolution of the universe after 1 sec but it assumes certain initial conditions such as thermal equilibrium, spatial homogeneity and isotropy with small fluctuations, spatial flatness, and the baryon to entropy ratio. The aim of the workshop would be to discuss how these conditions could have arisen from physical processes in the very early universe on the basis of grand unified theories and quantum gravity. Topics covered would include phase transitions, the generation of baryon number, the production of monopoles, primordial black holes and other long lived particles, the existence and nature of the initial singularity, particle creation and the origin of fluctuations. As in previous years the aim would be to limit the formal programme to about two seminars a day to leave time for informal discussion.

We would like to invite you to take part in the workshop. We could pay your airfare and subsistence, but we would be grateful if you could take advantage of any cheap fares that are convenient to enable us to stretch our limited funds as far as possible.

Please let us know whether you will be able to come. We would be grateful for any comments you may have on the organization of the workshop or of topics to be discussed.

Yours sincerely,

Stephen W. Hawking
Gary W. Gibbons

Figure 13.3 The invitation to the Nuffield Workshop on the Very Early Universe.

So-Young and I decided to pursue Hawking's approach, focusing on the variation from place to place in the time at which inflation ends, as illustrated by Figure 13.1. Meanwhile Paul and Michael were pursuing a more ambitious program, trying to track the detailed evolution of the Higgs field, taking into account even the gravitational effects associated with the perturbations of the Higgs field during the period of inflation. We had no way of knowing for sure that Hawking's simplified picture was accurate, but we thought it made sense to start with the simplest calculation and worry later about possible complications. We further assumed, again without any real justification, that the ending of inflation was sudden, so that the false vacuum almost instantaneously decayed to produce a gas of extremely hot particles. In the context of Figure 13.1, this assumption is the statement that the "inflation" and "hot matter" regions of the diagram are separated by a sharp line rather than a fuzzy transition region.

So-Young and I knew how to use the inflationary theory to describe the state of the universe—the properties of the matter and the form of the gravitational field—for the lower half of Figure 13.1, during the inflationary era. And we could at least estimate the variation from place to place in the time at which inflation ends, shown by the undulations of the line in the middle of the diagram. The big hurdle was to learn how to use the equations of general relativity to follow the evolution forward in time. Our goal was to determine the density perturbations at the time of recombination, which could then be compared with estimates based on galaxy formation. Before I left for the Nuffield meeting, which So-Young did not attend, she and I had made significant progress, calculating the evolution through the ending of inflation. In terms of Figure 13.1, we had learned how to determine the state of the universe just above the wavy line. Following advice from Jemal, we had also found a 1976 paper by D.W. Olson, then at Cornell, which seemed to contain all the equations that we would need to trace the evolution of these perturbations to the time of recombination. The program of calculation seemed clear, but we were not yet in a position to foresee the answer.

Paul Steinhardt and Michael Turner were continuing their work, and two days before leaving for England I received a telephone update from Paul. They had discovered subtleties in their calculation, which they were sure would make their answer much larger—but they did not know by how much.

The days before a long trip are usually hectic, and this trip was no exception. At dinner on Friday, June 18, my last full day at home, my frenzy was compounded by a phone call from one of my colleagues at MIT. Our group had just been informed, he told me, of a benefactor who planned to donate a half million dollars. To be considered, however, we needed to submit a proposal immediately. The group had decided that the early universe

Michael Turner in his office at the University of Chicago.

would be an enticing theme, so they hoped that I could draft something before I left! In addition, there were a host of small errands to complete, including the installation of two air conditioners and the fixing of my son's scooter. It was to be a working trip, so I had to pack not only clothes, but also the notes and references that I would need to continue the calculations. I did not start packing until 2:30 A.M., finishing at 5:00. I worked on the proposal until 8:30, and finally fell into bed, exhausted.

I slept until early afternoon, and then spent some relaxed hours with my family, mostly playing with my $4^1/_2$ year-old son. We had dinner at our favorite restaurant, and then my family took me to the airport for my 9:00 P.M. flight. The departure was delayed for two hours, however, so I pulled out my clipboard and continued the calculations of density perturbations. Airports are actually good places to work—no visitors and no telephone calls. Before the plane was ready for boarding, I had derived one of the key formulas that I needed to complete the calculation of the density perturbations generated by inflation.

The flight was crowded and uncomfortable, and I slept very little. Landing at Heathrow Airport at 10:30 A.M. British time, I took a bus to Victoria

Station, a cab to King's Cross, and a train to Cambridge. At 3:00 I finally reached Sidney Sussex College at Cambridge University, where the conference participants were to be housed. Due to a mix-up I was not on the room list, but within an hour they arranged a room for me and I had a chance to nap. I awoke to have dinner with another American physicist, and then continued work on the calculation of density perturbations until 2:30 in the morning.

At breakfast the next day, I met for the first time the effervescent Andrei Linde, the Russian co-inventor of the new inflationary universe. He was part of a delegation of five Soviet physicists, headed by the prominent, elderly member of the Soviet Academy of Sciences, M.A. Markov, from the Institute for Nuclear Research in Moscow. Andrei was vibrant and enthusiastic, and was anxious to show me the four new preprints that he had brought to the conference for circulation. I invited him back to my room after breakfast, so we could continue our chat. Along with much discussion of physics, I learned that Andrei's wife, Renata Kallosh, is also a theoretical physicist, working on supergravity. They had two sons, ages 7 and 4; when the boys were not in school, they had a woman who came "not quite regularly" to help take care of them. Andrei also told me about the procedures for publishing papers that were then in effect in the Soviet Union. Although papers were often published in English in Western journals, they had to first be written and typed in Russian for internal review. After a lengthy process, the paper would hopefully receive formal approval. Then it would be translated into English, retyped, and finally submitted to the journal, where the editors would begin their review process. If the referee for the journal requested changes in the paper, then the entire internal Soviet review process would have to start again. After hearing this account, I understood for the first time why Russian papers tended to be so short and often cryptic. For Andrei, the conference provided a valuable opportunity to circulate his preprints* without waiting for bureaucratic approval.

Still obsessed by the density perturbation calculation, I returned to it as soon as Andrei left. I finally arrived at the result that I had been seeking: a formula that related the density perturbations at the time of recombination to the nonuniformities in the Higgs field at the end of inflation, as it rolled down the hill of the energy density diagram. So-Young and I had not calculated the nonuniformities of the Higgs field, but if I adopted the estimate used by Hawking (which seemed very plausible), I obtained a result very different from his. While Hawking had found perturbations with the desired

* It is common practice in many fields of science to circulate research papers as preprints, so that other scientists can learn of the work without waiting the several months or even a year required for publication in a scientific journal. Preprints are still used in the 1990s, although many scientists submit their publications to computerized archives, with instant worldwide access provided by the Internet.

magnitude of about 10^{-4}, I found the magnitude to be larger than one! But it was already Monday afternoon and I did not want to miss the opening talk of the conference, so I grabbed a hamburger lunch at a fast food restaurant and raced to the Department of Applied Math and Theoretical Physics, where the lectures were to be held.

From what I recall, I was much more exhilarated than discouraged by the result on the magnitude of the density fluctuations. It had been two weeks since Paul Steinhardt and I had decided that Hawking's answer looked wrong, so my calculation that afternoon was not a major surprise. At this point I still did not appreciate the impact of inflation, so I still did not feel like an established physicist. (In reality, I was not an established physicist, since I still did not hold a tenured position.) In this frame of mind, I was delighted to convince myself that I was right where the renowned Stephen Hawking had been wrong. As for the impact on inflation, there was some good news along with the bad: The approach by So-Young and me led to a scale-invariant spectrum, as Hawking had found. It was extremely exciting to see that quantum physics could lead so naturally to the spectrum favored by cosmologists. The magnitude was wrong, but it was clear from the beginning that the predicted magnitude depended very much on the details of the assumed particle theory. We were all working in the context of what was then the most promising grand unified theory, the original theory proposed by Georgi and Glashow in 1974. This theory had a strong virtue in simplicity, but many other successful grand unified theories had been constructed. By complicating the grand unified theory, one could arrange for the magnitude of density perturbations to turn out small.

That night I wanted to think more about density perturbations, but I was scheduled to speak the next day, and had not yet finished preparing my overhead transparencies. My talk, titled "Phase Transitions in the Very Early Universe," was a general review of phase transitions, the original inflationary theory and the problems that it solved, the failure of the original inflationary theory, and the success of the new inflationary universe. I was nervous about speaking in front of such a high-powered audience of international experts, but the presentation seemed to go well.

Immediately before my lecture I had a brief conversation with Stephen Hawking. By this time I had also received a preprint describing Stephen's calculations, but the preprint was not much more detailed than the summary of Stephen's Princeton talk that I had learned from Paul. I told Stephen that I would be very happy if his calculation of density perturbations were correct, but that I was finding that the perturbations were much larger. Our discussion came to an abrupt end, however, when the clock struck 2:30, and it was time for my lecture to begin. Our differences remained unresolved.

Andrei Linde spoke the next day, Wednesday, presenting a whirlwind tour of calculations related to the new inflationary universe, focusing almost entirely on results obtained since his original paper. While the equations went by too quickly to ponder, the key points were accentuated by a series of beautifully drawn cartoons, two of which are shown as Figure 13.4. The first shows how the inflationary universe solves the monopole and several related problems by sweeping the dirt out of the house. That is, although inflation does not destroy monopoles or other unwanted relics from the big bang, the enormous expansion can spread these hobgoblins over such enormous distances that they cannot be seen. The second cartoon portrays the struggle to build a successful, detailed inflationary scenario. While the artist's moral commitment to inflation is unambiguous, the drawing suggests that the forces of evil are continuing to present thorny problems that require resolution. Do realistic particle theories lead to a flattened Mexican hat energy density diagram, like Figure 12.1, as needed for new inflation? Are the predictions for density perturbations in accord with observation?

The speakers on Thursday included Jim Bardeen, who spoke on the key topic, "Fluctuations in the Very Early Universe." He discussed the mathematical formalism for describing the evolution of density perturbations, focusing particularly on the inflationary universe. He did not yet, however,

Figure 13.4 Cartoons by Andrei Linde.

have any results on the magnitude or spectrum of the density perturbations produced by quantum processes. I had lunch with Jim after his talk, and I gave him a summary of how So-Young and I had been attacking the problem. From what I recall, he was very skeptical of our approximations, which caused me a bit of anxiety.

There were no talks on the weekend, so on Saturday Michael Turner and I went sightseeing in London. In the afternoon we made a quick visit to the British Museum, seeing the Rosetta Stone and the Magna Carta, and in the evening we saw a superb production of *Children of a Lesser God*. Returning to Sidney Sussex College at 1:00 A.M., we had to wake the night porter to let us through the gate. We talked some physics during the day, and I learned that Turner and Steinhardt were still in a state of uncertainty over the density perturbations. They had been assuming that the density perturbations do not change much during the rollover of the Higgs field, but Michael now felt that their approximations needed to be re-examined. The next day at lunch, however, he told me that he still did not find significant growth of perturbations during the rollover.

On Sunday afternoon there was a tea at the Hawkings', where I met Stephen's wife Jane, his son Timmy (age 3), and his daughter Lucy (age 12). Stephen's 15-year-old son, Robert, was not there, but I did have the opportunity to meet him before the conference ended.

In addition to the social activities of the weekend, I was able to spend some time working on the calculation of density perturbations. At this point my main goal was to rederive the equations that I had taken from papers in the literature. I wanted to be sure that the equations were correct, and that I understood what they meant. I made progress, but I was still baffled by several contradictions in my formulas.

The first speaker on Monday (the second week of the workshop) was Paul Steinhardt, who spoke about a new version of inflation developed by him and five other physicists, describing inflation in the context of a specific, particularly attractive version of a grand unified theory. Later on Monday we heard Alexei Starobinsky* of the Landau Institute for Theoretical Physics in Moscow. The combination of a stutter with a Russian accent made his talk particularly difficult to follow, but the effort was rewarded. Starobinsky had also studied the density perturbations arising from quantum fluctuations in

* On a visit to Cambridge $2^1/_2$ years earlier, Starobinsky had proposed a form of inflation that preceded my own work on cosmology. His paper, however, had not yet attracted a great deal of attention in the West, since it did not discuss the success of the scenario in solving the horizon, flatness, or monopole problems. It focused instead on the avoidance of an initial singularity, suggesting a solution that is probably invalid. Nonetheless, the mechanism that Starobinsky proposed to drive the exponential expansion is today considered a viable alternative to the Higgs field mechanism described in this book.

the new inflationary universe, and like me he was finding perturbations that were too large. The magnitude that he mentioned in the talk was 10^{-2}. Otherwise I did not understand his calculation, except that he did say clearly that the main source of perturbations was the non-simultaneity of the end of the inflationary phase. I recall that in the discussion that followed Starobinsky's talk, I announced that I was also finding perturbations that were too large, even larger than those described by Starobinsky.

On Tuesday Stephen Hawking presented his lecture, impishly titled "The End of Inflation." The title presumably referred to the ending of inflation in the early history of the universe, but it was clearly devised to alternatively suggest that the inflationary theory was dead. In any case, Paul Steinhardt and I (among others) expected Stephen to describe the calculation of density perturbations from his preprint, and we were poised to challenge him when he came to the crucial, controversial step. But it never happened! Hawking's lecture followed the preprint until the end, but then at the last step he substituted a new calculation for the rate of change of the Higgs field, and concluded that the density perturbations would be much too large. He was a bit vague about the answer, but he was clearly finding something in the neighborhood of one. True to Hawking's style, he did not mention that he was correcting his preprint, or that his new answer was basically in agreement with Starobinsky or me. I suspect that when I was discussing this issue with Stephen the week before, he had already realized that his preprint was wrong. But instead of just saying so, Stephen preferred to defend the incorrect result until he could announce his new calculation in a public forum. Stephen did not claim that this result killed the inflationary theory, but he pointed out that inflation would have to be abandoned if some change were not found.

Shortly before Stephen's talk, on Tuesday morning, I had finished my project of rederiving the equations that I had taken from the literature. The timing was right, because by Tuesday night I had a new surprise to think about. Jim Bardeen knocked on my door at about 10:00 P.M., and told me that he disagreed with my picture of inflation ending at different times in different places. His latest calculations showed that as the Higgs field started rolling down the hill of the energy diagram, it influenced the local expansion rate of the universe in a way that caused a feedback effect on the Higgs field. The final result: Inflation ended at almost exactly the same time everyplace, invalidating the calculations by Hawking and Starobinsky, and those by So-Young Pi and me. "Amazing if true," I wrote in my diary, as I was thinking about how to check Jim's claims. Until this time So-Young and I had completely ignored the details of the Higgs field, treating it only as a mechanism that would at some random moment trigger the decay of the false vacuum. Jim was claiming that this approximation was invalid, and it was certainly true that we had not justified it. To find out if Jim was right, I would have to derive the equations for the evolution of the Higgs field and its interaction with the nonuniformities of the universe.

At breakfast the next morning Jim had an updated report. Although his calculations still showed that inflation ended simultaneously, he nonetheless found that the density perturbations resulting from the quantum fluctuations of the Higgs field were large—in the neighborhood of one. So, while neither of us really understood the other's calculations, we were at least finding approximately the same answer.

The first speaker that day was Michael Turner, who gave a progress report on the work on density perturbations by himself, Paul Steinhardt, and Jim Bardeen. In a frank seminar which made no attempt to conceal the confusion, Michael began by summarizing the original calculation by Steinhardt and himself, which led to a magnitude of 10^{-16}. He pointed out that this calculation included only an approximate treatment of the rollover of the Higgs field, but it seemed hard to believe that a more accurate calculation could change the answer by a large factor. Then he went on to report the late-breaking news of Jim Bardeen's calculations of the previous night. Proceeding humorously in the style of a TV newscaster, he described the 10:00 P.M. report that the inflationary phase ends simultaneously throughout the universe, followed by the 11:40 P.M. report that the perturbations were in any case large. The final summary suggested that more work would be needed to resolve the discrepancies. Turner's talk was followed by a mini-talk by Bardeen, which was in turn followed by a free-for-all discussion in which everyone involved reiterated his results. Turner, Steinhardt, Bardeen, and I all went for lunch together after the seminar, but we made no further progress in understanding the differences in our calculations.

That night I continued with my own calculations of the evolving Higgs field, and decided that my original answer still looked right. I was able to follow the Higgs field as it rolled down the hill of the energy density diagram, although I did not yet include the mechanisms which transformed the energy of the Higgs field into ordinary particles. At least up to this stage, the calculation agreed perfectly with the simplified approximations that So-Young and I had used.*

* Although Bardeen was claiming that the inflationary era ended simultaneously throughout the universe, while the calculations of Hawking, Starobinsky, Pi, and myself were based on the non-simultaneity of the ending of inflation, it now appears that all of us were correct! In general relativity there are many different ways to define exactly what one means by "time," so events at different places might occur at the same time by one definition, but not by another. In fact, one of the big communication problems at Nuffield was that we were all using slightly different definitions of time.

On Thursday night, Michael Turner told me that he had improved his approximations, and his estimate of the density perturbations had increased dramatically. He was no longer finding a magnitude of 10^{-16}, but instead about 10^{-1}. On Friday night Jim Bardeen knocked on my door and showed me a new formula he had obtained, which at first looked very complicated. Only later did I understand it, and I decided that it was very similar to my answer.

On Saturday, Jim Bardeen and I went to London together, heading directly for the British Museum. After seeing an Egyptian excavation exhibit, the Elgin marbles, and an ancient India exhibit, we had lunch at the museum and then saw a performance of *Amadeus*. We returned to Cambridge by 9:30 P.M., just before the start of a scheduled train strike.

Sunday was devoted to work, as I still wanted to make sure that the density perturbations were not modified when the energy of the Higgs field was transformed to ordinary particles. The third and final week of the conference was about to begin, and I had arranged to speak that Thursday on density perturbations. I wanted to be sure that I understood what I was talking about. On Sunday evening I had dinner with Alexei Starobinsky, and learned a good deal more about his method of calculation. The next day Michael Turner told me that he had figured out how to avoid some key approximations, and was now getting a much larger answer.

My calculations were finally completed on Tuesday night. I convinced myself, by very clumsy and complicated arguments, that the energy transfer would not change the answer that I had obtained. I also calculated the details of the quantum fluctuations of the Higgs field as it rolled down the hill in the energy density diagram. The result was similar to the estimate that So-Young and I had borrowed from Hawking, but at least the calculation eliminated one source of feelings of insecurity.

Wednesday night was devoted to preparing my talk. I did not even realize until the next day that I had missed the conference banquet. My calculations had been done with many changes in notation, so much work was needed to mold the equations into a coherent story. I spent an hour and a half struggling to track down an irritating factor of 3 discrepancy between two methods of doing the same calculation. I discovered that the answer from the night before was in fact wrong by a factor of 2. It was 6:00 A.M. before I finished writing the talk, and then I worked until 9:00 making the transparencies. Exhausted, I fell into bed for $3^1/_2$ hours of badly needed sleep.

My talk that afternoon went smoothly, and I was rather proud of the way that I had pulled all the parts of the rather complicated calculation together. Before the talk I learned that Bardeen, Steinhardt, and Turner now agreed with the answer that I had been finding.* In the discussion that followed, Starobinsky said that he had rechecked his earlier calculations, and that his estimate of 10^{-2} had been revised to approximately 1. While this was a little smaller than the answer found by the rest of us, the discrepancy did not seem important. After three weeks of spirited disagreements, the workshop had led to a consensus [2].

The spectrum of density perturbations predicted by inflation was found to depend only very slightly on the assumptions about the underlying particle physics, and was predicted to be scale-invariant. While the perturbation spectrum of the real universe was not known, this result was in agreement with earlier speculations, starting with Harrison and Zeldovich, that were based on the phenomenology of cosmic structure formation. Although the scale-invariant spectrum was not known to be right, we could easily imagine a wide variety of other spectra, which we might have found, that would have been obviously wrong. Those of us at the workshop viewed the spectrum calculation as a major success for inflation.

The magnitude of the density perturbations, however, was found to depend very sensitively on the underlying particle theory. For the very attractive grand unified theory of Georgi and Glashow, the predicted amplitude was of order 10^2, five or six orders of magnitude too large.

According to our calculations, the magnitude of the predicted density perturbations depends crucially on the shape of the energy density diagram for the Higgs field. To get the magnitude right, the diagram must resemble Figure 12.1, except that the plateau must be extraordinarily flat—about 10^{13} times flatter than one might expect.† Recall, however, that the Higgs fields of grand unified theories were originally introduced not to drive inflation, but to create the spontaneous symmetry breaking responsible for establishing the differences between quarks, electrons, and neutrinos. To accomplish this purpose, the Higgs field cannot have so flat an energy density diagram. So, a key conclusion of the Nuffield calculations is that the field which drives inflation cannot be the same field that is responsible for symmetry breaking. For the density perturbations to be small, the underlying particle theory must contain a new field, now often called the *inflaton*

* While agreement on the approximate numerical value could easily have been a coincidence, our agreement was actually much more detailed. We agreed on the precise relation between the density perturbations and the shape of the Higgs energy density diagram.

† It is hard to describe flatness quantitatively, but one way is to imagine that we start with an energy density diagram such as Figure 10.1 or 12.1, typical of what one expects for the Higgs field in the simplest grand unified theory. To make this diagram flat enough to be consistent with inflation and the density perturbations of the universe, the entire diagram has to be squashed in the vertical direction by a factor of about 10^{13}.

(pronounced IN-flah-tahn), which resembles the Higgs field except that its energy density diagram is much flatter.

In assessing this mixed success—a good spectrum but a magnitude that would be wrong in the favored grand unified theory—one must remember that the new inflationary universe was the first believable theory that gave any prediction at all for the properties of the density perturbations. In standard cosmology without inflation, an entire spectrum of density perturbations had to be included arbitrarily in the hypothesized initial conditions, or else galaxies would never form. Because of the horizon problem of standard cosmology—the fact that different parts of the early universe did not have enough time to exchange information or mass—there was not even a hope of finding a mechanism to control the perturbations. If two regions had no time to exchange mass, a higher mass density in one region than the other could only be attributed to the initial conditions, which in standard cosmology are specified by fiat. Thus, the mere fact that inflation provides a context in which density perturbations can be calculated, rather than assumed, was in itself a step forward.

It was a tremendous feeling of relief to be done with my talk, after having worked so hard. On Thursday night I went out for dinner with Jim Bardeen, and afterward I spent an hour and a half wandering through the streets of Cambridge. The next morning, before the first talk, I went on a brief shopping spree. I bought four or five physics books at Heffers, and also rounded up some toys for my son Larry—two Lego space stations and an inflatable spaceship capsule large enough for one or two kids to play inside. The space capsule was a big hit, for several years filling a sizable fraction of the floor space in my son's room.

On Friday afternoon Frank Wilczek* of the University of California at Santa Barbara gave the conference summary talk, formally ending the three-week workshop. Quoting from the written version of that talk,

> By my count seventeen of the thirty-six lectures at the workshop were mainly about the idea of an inflationary Universe, and many of the other talks were heavily influenced by this idea. By supposing that the Universe inflates by an enormous factor, we explain at one stroke why it is observed to be so nearly flat, so nearly homogeneous, and perhaps why it is so nearly free of various topological monstrosities such as magnetic monopoles, axion domain walls, primordial black holes . . . . The idea is so simple, and yet it provides a qualitative understanding of some of the deepest puzzles of cosmology!

* Wilczek's role in the development of QCD, the theory of the strong interactions, was described in Chapter 7.

He then went on to discuss some of the unsolved problems, such as finding a convincing particle theory that leads to the right magnitude of density perturbations. He concluded with a statement that I consider still valid today:

> I think it is fair to say that while the general idea of an inflationary universe is extremely attractive, the specific models so far put forward do not inspire confidence in detail.

By this time I was finally starting to appreciate the enormous impact that inflation was beginning to have. Before the conference, I knew that inflation was exciting to me and to Paul Steinhardt and to someone in Moscow named Andrei Linde. But it was not so clear how much anyone else cared. I had just witnessed, however, an international conference on cosmology, organized by Stephen Hawking and Gary Gibbons. At the conference it became clear that inflation was not only an idea that people had heard about, but it had in fact become the primary focus of discussion. An official conference summary, which was written by John Barrow and Michael Turner for the British Journal *Nature*, was titled "The Inflationary Universe—Birth, Death and Transfiguration" [3].

CHAPTER 14

OBSERVATIONAL CLUES FROM DEEP BELOW AND FAR BEYOND

While those of us working on inflation all knew that no particular grand unified theory was well established, there was still a feeling of uneasiness in discovering, at the Nuffield Workshop, that inflation was incompatible with the simplest grand unified theory. Inflation, we discovered, must be driven by a field with an extremely flat energy density diagram, or else the resulting density perturbations would be too large. The simplest grand unified theory—the original theory of Georgi and Glashow—unfortunately contained no field of this description.

While more complicated theories were certainly possible, we worried that the situation was uncomfortably similar to the one that existed a decade earlier. At that time the unification of the weak and electromagnetic forces was the major open question in particle physics, and theorists had thrashed out a bewildering array of theories that accomplished this goal. Nonetheless, when the experiments finally became clear enough to settle the issue, we found that nature had opted for economy. The right theory was the simplest, the one that had been first proposed by Weinberg and Salam. It was natural, therefore, to believe that the pattern would repeat, and that the original grand unified theory of Georgi and Glashow would prove to be correct. If so, inflation would have to be scrapped. Although one could imagine modifying the Georgi-Glashow theory by adding the new field needed for inflation, such an addition would obliterate the simplicity that made the theory attractive in the first place.

In the next few years, however, our beliefs about grand unified theories were reshaped by a series of heroic experiments on the lifetime of the proton. Most grand unified theories predict that the proton is not completely stable, and the Georgi-Glashow theory in particular predicted a half-life between 10^{27} and 10^{31} years. At the time of the Nuffield meeting this prediction looked very plausible, as an Indian-Japanese collaboration had already reported six proton decay "candidates" seen in a detector located 2.3 kilometers underground in the Kolar Gold Field near Bangalore, India. Describing these events as "very probably nucleon decays barring hitherto unknown phenomena," the authors estimated a half-life of 5×10^{30} years [1]. The credibility of the Kolar discovery was bolstered by rumors that another candidate event had been seen by a CERN group working in a tunnel 3 kilometers under Mont Blanc.

By 1985, however, the experimental status of proton decay had changed dramatically. The Irvine-Michigan-Brookhaven detector, 600 meters deep in the Morton-Thiokol salt mine in Ohio, had completed 417 days of observing a volume of water more than 50 times more massive than either the Kolar or Mont Blanc experiments. No certifiable proton decays had been seen, indicating that the proton half-life must be longer than 2×10^{32} years [2]. The death knell of the simplest grand unified theory had been sounded. The contradiction discovered at Nuffield between inflation and the Georgi-Glashow theory was no longer a cause for concern.

Once the Georgi-Glashow theory was excluded, there was no natural front-runner. Even today there are a wide variety of grand unified theories under consideration, with no clear favorite. An inflaton field—a field with an energy density diagram suitably flat for driving inflation—can be included in any of them. It must be admitted, however, that the ad hoc addition of such a field makes the theory look a bit contrived. To be honest, a theory of this sort *is* contrived, with the goal of arranging for the density perturbations to come out right. We still appear to be a long way from pinning down the details of the particle physics that underlies inflation.

While the need to tailor the particle physics to match the observed level of density perturbations might be viewed as evidence against the inflationary theory, I would argue that it is not nearly as strong as the evidence in favor. Since inflation is so successful in explaining the origin of matter, the flatness of the early universe, and the large scale uniformity of the universe, it seems reasonable to just accept the smallness of the observed density perturbations as a measurement of the energy density diagram of the inflaton.

While most particle physicists, including myself, believe that the correct theory of nature will be simple and elegant, we have to admit that our best theory at present, with or without an inflaton field, is neither. In fact, some physicists have commented disparagingly that the standard model of particle physics looks like it was designed by a committee. For example, the masses of the W^+ particle and the electron arise in essentially the same way, so the fact that the mass of the electron is 160,000 times smaller is built into the theory only by rigging the parameters to make it happen. Similarly, the fact that the gravitational force between two protons is 10^{36} times weaker than the electrostatic force is not explained in any fundamental way, but is incorporated into the theory only by fixing the parameters to reflect this monumental mismatch. Since we do not understand why the electron is so light or why gravity is so weak, we need not be overly concerned that we do not understand why the inflaton energy density diagram is so flat. Clearly we do not yet understand the principles that determine the values of the fundamental parameters of our theories. For now we can only hope that the correct fundamental theory, when it is finally understood, will lead naturally to an inflaton field with the right energy density diagram, along with an explanation of the other parameters, such as the mass of the electron and the strength of gravity.*

Although our ignorance of the inflaton energy density diagram prevents us from predicting the magnitude of the density perturbations, inflation nonetheless makes a fairly definite prediction for the shape of the density perturbation spectrum. This shape can be changed slightly by modifying the energy density diagram, but it is usually very close to the scale-invariant spectrum proposed by Harrison and Zeldovich.

In 1982, when theoretical physicists at the Nuffield meeting were working furiously to calculate this spectrum, the possibility of ever measuring it seemed extremely remote. I do not recall any discussion of this possibility, nor can I find any mention of it in the published proceedings. Probably none of us at Nuffield were aware that while we were theorizing in Cambridge, England, NASA officials on the other side of the Atlantic were pushing through the final approval of the *Cosmic Background Explorer* satellite, known as COBE. One of the three instruments aboard the satellite, the Differential Microwave Radiometer (DMR), would be dedicated to measuring the nonuniformities in the temperature of the cosmic background radiation.

* As mentioned in a footnote at the start of Chapter 8, a serious candidate for the correct fundamental theory, called superstring theory, has already been found. The consequences of superstring theory, however, are difficult to extract, so it is still too early to tell whether or not superstring theory contains a suitable inflaton field. The possibility of inflation in the context of superstring theory has been an active area of study. One of my own recent papers, in collaboration with Lisa Randall and Marin Soljačić, showed that inflation can occur naturally as a consequence of properties which superstring theory is believed to possess.

David Wilkinson

Since the photons of the background radiation are believed to have been moving in straight lines since about 300,000 years after the big bang, the radiation provides an image of the universe at this extremely young age. In principle, it should be possible to extract the primordial density spectrum from this kind of data.

Although full-scale funding for COBE was approved in 1982, the same year as the Nuffield meeting, the satellite was not launched until the end of 1989. And then it took a year to gather the data and a little more than a year to analyze it.

The first I learned about the measurement was at a cosmology conference in Irvine, California, organized by the National Academy of Sciences during the last weekend in March 1992. In the cafeteria line I asked David Wilkinson, a member of the COBE team (and formerly a member of the

Princeton team that originally identified the cosmic background radiation), if COBE had found nonuniformities. He declined to answer while we were among other people, but once we were alone he told me that COBE had indeed seen nonuniformities, but that the group was still trying to refine its estimates of the experimental uncertainties. They were planning to make an announcement at the upcoming Washington meeting of the American Physical Society (APS), on April 23. I asked him if I would consider the announcement good news or not, and he smiled and said without hesitation, "Good news." A week later I received a phone call from George Smoot, the Principal Investigator for the Differential Microwave Radiometer (DMR) experiment on the COBE satellite, who gave me more of the details.

Although the results had been described to me, I did not see a graph of the data until April 23, at the COBE session of the APS meeting. I arrived just as the session was starting at 8:00 A.M., and took a seat next to Paul Steinhardt. Paul handed me a piece of paper that he had been given by the COBE team. "This says it all," he commented, with a strong tone of satisfaction in his voice.

The graph on the piece of paper is shown as Figure 14.1 [3].* The precise meaning of the graph is a little complicated, and is explained in the caption. The key point, however, is simple: The agreement between the data and the predictions of inflation was nearly perfect. The undulations of the small triangles indicating the data points followed faithfully the shaded gray band that showed the consequences of the scale-invariant density perturbation spectrum predicted by inflation. The prediction was shown as a band, rather than a line, because the density perturbations are created entirely by quantum processes. As with most quantum experiments, the precise outcome is unpredictable, but one can predict the most probable range of outcomes.

There were six speakers in the COBE session, beginning with George Smoot. The talks described the measurements, the estimates of instrumental uncertainties, the subtraction of astronomical backgrounds, and a discussion of the implications for theories of galaxy formation. The measurements were so precise that in their analysis, the COBE team had to take into account the effect of the Earth's magnetic field on the electrical properties of their instruments. The presentations placed strong emphasis on the agreement between the observations and the inflationary theory.

The conference talks were followed almost immediately by a well-attended press conference, at which a bank of television cameras and a

* The graph shown in Figure 14.1 is not precisely the same as the original COBE graph, but has been modified to make it easier to explain. The original graph showed a quantity that can be obtained from the average of $(T_1-T_2)^2$ by multiplying by $-\frac{1}{2}$ and then adding a constant.

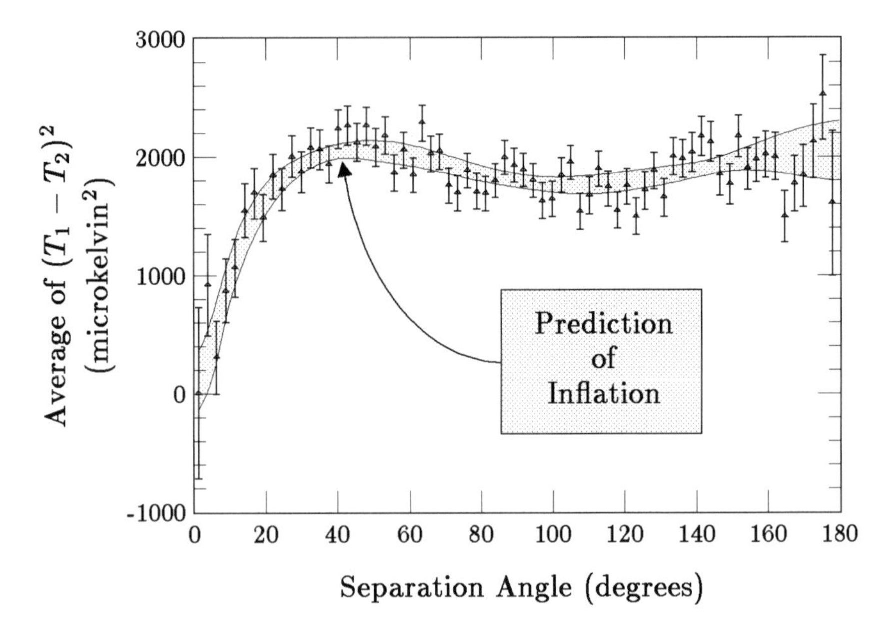

Figure 14.1 **COBE nonuniformity data.** The data from the COBE satellite gives the temperature T of the background radiation for any direction in the sky. The experimental uncertainties for any one direction are large, but statistically meaningful quantities can be obtained by averaging. The COBE team considered two directions separated by some specific angle, say 15°, and computed the square of the temperature difference, $(T_1 - T_2)^2$, measured in microkelvins (10^{-6}°K). By averaging this quantity over all pairs of directions separated by 15°, they obtained a statistically reliable number for that angle. The process was repeated for angles between 0° and 180°. The computed points are shown as small triangles, with the estimated uncertainty shown as a vertical line extending above and below the point. The gray band shows the theoretically predicted range of values corresponding to the scale-invariant spectrum of density perturbations arising from inflation. The effect of the Earth's motion through the background radiation has been subtracted from both the data and the prediction, as has a specific angular pattern (called a quadrupole) which is believed to be contaminated by interference from our own galaxy. Since inflation determines the shape of the spectrum but not the magnitude of density perturbations, the magnitude of the predicted gray band was adjusted to fit the data.

---

horde of science journalists recorded comments by Smoot and three other COBE team members. Again there was strong emphasis on the agreement with inflation, and Smoot announced that within six months, he expected "everyone" to believe in inflation. The COBE team released three papers that day, one of which [4] was devoted to the interpretation of the results. The concluding sentence of that paper was:

> But the results of this paper have an even greater significance, since the . . . [temperature variation] observed by the DMR experiment is a direct measure of the fluctuations produced during the inflationary epoch, . . . and thus provide the earliest observational information about the origin of the universe, going back to 10^{-35} seconds after the big bang.

Staring at the COBE graph, I think that Paul and I had the same sense of awe. Almost ten years had elapsed since we had both returned from the Nuffield meeting, but we still remembered the thrill of wrestling with the equations of general relativity until they finally yielded the predictions of inflation for the density perturbations of the universe. We remembered the long nights of painstaking calculations, and the arguments over breakfast about why we were getting different answers. It was hard to believe that the intellectual frenzy in which we were so passionately involved at Nuffield had really turned out to be more than a game. A massive fifteen-year NASA project had finally succeeded in measuring the nonuniformities of the cosmic background radiation. And the agreement was gorgeous! To a theoretical physicist, there is no greater joy than to see that this curious activity that we call calculation—the depositing of ink on paper, followed by throwing away the paper and depositing new ink on more paper—can actually tell us something about reality!

The 1992 Washington APS meeting was an amazing experience for me, made even more amazing by the fact that I was chosen that year for a major APS award, the Julius Edgar Lilienfeld Prize. In honor of this award, I was asked to present a lecture to the entire conference, as part of the APS annual "Unity Day" meeting. The lecture was on Wednesday, the day before the COBE announcement.

Human experience, however, is always in the end a mixture of both sad and happy occasions. If we are fortunate, the sad occasions can at least be tempered by happy ones. Since my parents were very proud of the APS award, they came to Washington to hear my lecture. The following night my elderly father slipped on a slick hotel bathroom floor and fractured his hip. He never fully recovered from this accident, and was never again able to get out of bed without the help of a walker. He died of a brain tumor on October 27, 1993, and I am saddened by the fact that I did not finish this book in time for him to read it.

CHAPTER 15

THE ETERNALLY EXISTING, SELF-REPRODUCING INFLATIONARY UNIVERSE

A cosmological theory lives or dies on the basis of the description it predicts for the *observable* universe. Most of this book, therefore, has been concerned with the observable properties of inflation and the standard big bang theory. Human curiosity, however, extends beyond all boundaries, even that of the visible universe. The predictions of inflation for the universe beyond the visible have been studied by Andrei Linde, Alexander Vilenkin of Tufts University, and others. In this chapter, the title of which was borrowed from a paper by Linde [1], I will describe the rather extraordinary picture that inflation suggests for the universe on the largest scales.

If the inflationary theory is correct, then the unbridled expansion of the inflationary era almost certainly stretched our region of space to a size much larger than the range of our telescopes; the universe that we can see is no more than a minute fraction of all that exists. While we can never expect to test the predictions of inflation for the region beyond the observable universe, there is no way that we can prevent ourselves from thinking about it. And if we can develop confidence in the inflationary theory by testing its observable predictions, we can give at least some credence to the predictions that it makes about the unobservable.

The driving force behind inflation, you will recall, is a peculiar state of matter called a false vacuum. In microscopic terms the false vacuum consists of a Higgs-like field, often called an inflaton field, perched in some precarious position, such as the center of the energy density diagrams of Figures 10.1 or 12.1. Such a state is called *metastable*, meaning that it is stable for short times, but an observer who waits long enough will see that it is unstable. Eventually the inflaton field will find its way to the bottom of the hill of the energy density diagram, and the false vacuum will be said to have decayed.

The manner in which the false vacuum decays is very similar to the decay of a radioactive substance. In both cases, the decay is exponential, meaning that it is described by a half-life. The half-life of the false vacuum that drives inflation is unknown, but something like 10^{-30} seconds or 10^{-35} seconds would be reasonable for inflation in the context of grand unified theories. If one imagines filling the volume of false vacuum with a dense network of test points, then at the end of one half-life only half these points, on average, will be in regions that have remained false vacuum. The other half will be in regions that have "decayed," in the sense that the inflaton field has begun to roll down the hill of the energy density diagram. At the end of the second half-life one quarter of the test points will remain in false vacuum, and so on.

But now we come to the key difference between the false vacuum and a radioactive substance: while the false vacuum is decaying, those regions that have not decayed continue to exponentially expand. For the numbers typical of most inflationary theories, the rate of expansion is much faster than the rate of decay! So, if we wait one half-life, only half of the test points will remain in the false vacuum. However, the volume of the region remaining as false vacuum will be far larger than the volume of the entire region at the start. Therefore, even though the false vacuum is decaying, its volume is dramatically growing. A region of false vacuum will grow forever [2]: once inflation begins, it never stops!

The fate of a region of false vacuum is illustrated schematically in Figure 15.1. This drawing is not intended to be realistic, but instead is meant only to give a clear illustration of the important features. The uppermost horizontal bar represents a region of false vacuum, and the lower bars represent the same region, after three equal intervals of time. The physical size of the region increases by a factor of 3 with each successive bar, but this increase in size is not shown; if it were, the diagram would not fit on the page. By the time of the second bar, a third of the region has decayed to produce a localized hot big bang. Unfortunately, such a localized big bang region does not yet have an accepted name, even among inflationary enthusiasts. I will call such a region a "pocket universe." The use of the word "pocket," however, is not intended to mean that the region is small by conventional standards. On the contrary, our entire observed universe is believed to be only a minute fraction of one of these pocket universes. The pocket universe, on the other hand, is only a minute fraction of all that exists, so in that sense it is only a "pocket."

In addition to the pocket universe, the second bar shows two regions that remain as false vacuum; each of these false vacuum regions is physically

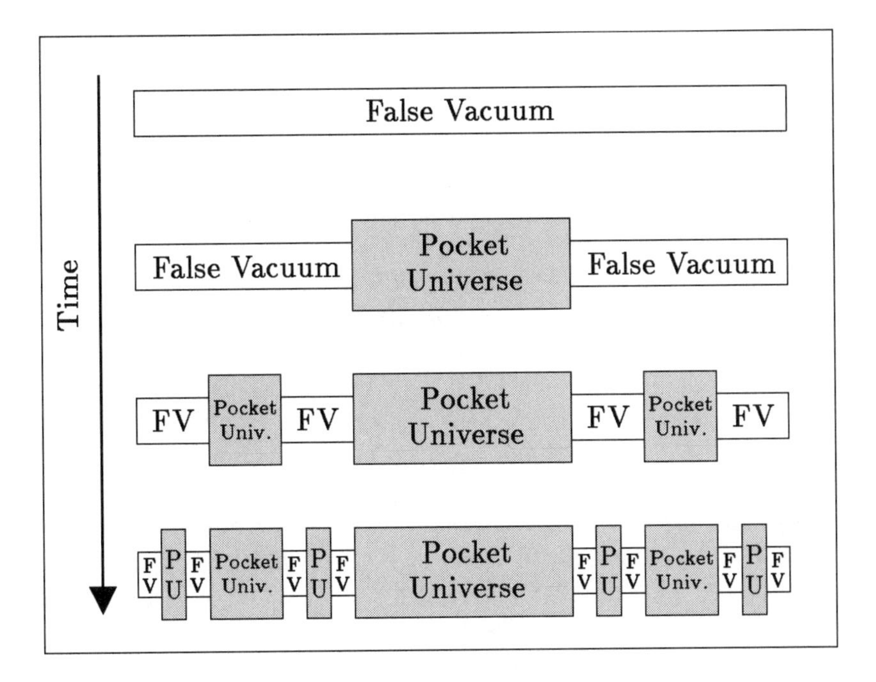

Figure 15.1 **Schematic drawing of eternal inflation.** The four bars represent a part of the universe at successive, evenly spaced times. Each bar is actually three times the length of its predecessor, although the expansion is not shown.

the same size as the entire region of the first bar. In going from the second bar to the third, which represents another factor of 3 increase in size, the central region continues to evolve as a standard big bang universe. Each of the two regions of false vacuum, however, evolve in the same way as the original false vacuum region shown in the top bar. By the time of the third bar, one third of each false vacuum region in the second bar has evolved into a pocket universe, and the remaining regions of false vacuum have each evolved to become the same physical size as the starting region shown in the first bar. By the fourth bar, we have seven pocket universes of varying ages, and eight regions of false vacuum, each as large as the region with which we started. The process does not stop here, but goes on forever, producing an infinite number of pocket universes at an ever-increasing rate. A *fractal* pattern is created, meaning that the sequence of false vacuum, pocket universe, and false vacuum is replicated on smaller and smaller scales. Thus, a region of false vacuum does not produce merely one universe, but instead produces an infinite number of universes! In the cosmic shopping mall, an infinity of pocket universes can be purchased for the price of one.

Each pocket universe undergoes a big bang history, just as we believe that the observed universe is doing. Since each pocket universe goes through the process of inflation, it will become almost exactly flat. For a period far longer than the 10 to 15 billion year history of our universe since the big bang, the evolution of each pocket universe will be indistinguishable from that of a flat universe. Ultimately, however, the deviations from flatness will grow. Parts of each pocket universe will recollapse into a crunch, while other parts dwindle off toward negligible mass density. On the very large scale, however, from a view that shows all the pocket universes, the evolution will strongly resemble the old steady state model of the universe. As the pocket universes live out their lives and recollapse or dwindle away, new universes are generated to take their place. Although the ultimate fate of our own pocket universe is no more appealing in the inflationary scenario than in a simple big bang theory, the universe as a whole will regenerate eternally, forever producing new pocket universes. While life in our pocket universe will presumably die out, life in the universe as a whole will thrive for eternity. Although eternal inflation predicts that each pocket universe will evolve in agreement with our observations, it implies a panoramic view of the whole universe that captures the same emotional appeal as the steady-state theory.

While the infinite pocket universe production shown in Figure 15.1 is an unavoidable feature of almost all versions of inflation, two important aspects of Figure 15.1 have been oversimplified for purposes of illustration. First, the real universe has three space dimensions, while the universes in the figure are shown as one dimensional. Second, the decay of the false vacuum, like the decay of a radioactive substance, is fundamentally a random process. The pocket universes, therefore, are not spaced regularly as shown in the diagram, but instead are sprinkled randomly in space and time.*

One fascinating question, debated for a number of years, is whether eternal evolution into the future can be accompanied by eternal evolution into the past. That is, must an inflationary universe have had a beginning, or might it have existed forever?

The idea of a truly eternal universe—one that has always existed and will always exist—is very appealing, since it frees us from all questions

* Since the formation of pocket universes distorts the spacetime in a complicated way that cannot be shown on the diagram, the resulting spacetime structure is very difficult to visualize. This very large scale fractal structure of an inflationary universe has been studied by Andrei Linde, Alexander Vilenkin, and a number of other physicists [3], but the structure is so intricate that its properties are still not well-understood.

about how the universe was created, or what existed before the universe was created. In recent history there have been two versions of eternal universe theories that have been widely discussed: the steady-state and oscillating theories. The steady state theory, proposed by Bondi, Gold, and Hoyle, held that the universe is infinitely large and infinitely old, and that new matter is continuously created to maintain a constant density as the universe expands. The oscillating universe theory accepted the big bang description of an evolving universe, but hypothesized that the universe will recollapse, and that expansion and collapse are just one of an infinite series of cycles. Today, however, the steady state theory is considered by almost all to be excluded by observation, and the oscillating theory is viewed with skepticism (see footnote in Chapter 2). Inflation, therefore, is seen by some as a new hope for a universe without a beginning. Is it possible that there was never a first patch of false vacuum, but instead false vacuum has existed forever, creating pocket universes indefinitely far into the past?

The debate on this question is probably not over, but it is pretty clear which side is winning. At the beginning of 1994, Arvind Borde of the Brookhaven National Laboratory and Alexander Vilenkin [4] proved that if a certain set of technical hypotheses is accepted, then the universe cannot avoid having a beginning, even if the universe is eternal into the future. The set of hypotheses includes the assumption that the universe is open, so if the universe is closed then the question remains unresolved.* The other hypotheses of the theorem are plausible, but not beyond dispute.

Since the other pocket universes would be far too remote to ever observe, the concept of eternal inflation does not lead to directly testable predictions. Nonetheless, I believe that eternal inflation has important implications for anyone who speculates on future directions in cosmology.

First, although eternal inflation apparently does not avoid the question of how the universe originated, it certainly pushes the question to an obscure corner of the theoretical physicist's map. In particular, any hypotheses about cosmic origins become totally divorced from observational cosmology. Even with only a single episode of inflation, essentially all evidence of what came before inflation is erased. But if inflation has been happening for an arbitrarily long time, and our universe is not the first but perhaps the

* Since the universes under discussion are highly irregular on the largest scales, the terms "open" and "closed" do not have quite the same meaning that they had for Friedmann universes. The precise definition is very technical (as is typical for proofs in general relativity), but roughly speaking Borde and Vilenkin call a universe closed if it contains at least one spatial region with finite volume and no boundary; otherwise it is open.

10^{1000}th pocket universe to be created, then any hope of learning about how it all began by observing our universe seems totally futile. The beauty of inflation—that it can make predictions independent of the details of the initial conditions—becomes the cosmologist's ultimate barrier. Because the properties of the observed universe are determined by the physics of inflation, they tell us nothing about what existed before inflation. The question of how it all began remains philosophically important, and it might still be possible to make persuasive arguments to justify some particular theory. However, if eternal inflation is correct, it will unfortunately be impossible to ever test these arguments by observation.

Second, eternal inflation lays to rest the difficult question of deciding how plausible it is for inflation to have started. In the early days of inflation we assumed that the universe began as a uniform hot soup of particles, much as in the standard hot big bang theory. In that case one can argue, with some assumptions about the underlying particle physics, that inflation is likely to start. Now, however, it is clear that inflation can create a uniform hot soup of particles with just the properties that were previously assumed. Since the initial state is no longer dictated by necessity, theorists today prefer to explore a much wider range of assumptions about what may have preceded inflation. A number of physicists have pursued the consequences of "random" initial conditions, but this line of inquiry is hindered by the fact that "randomness" is ill-defined. The word "random" does not become meaningful until the probability rules are stated.* The onset of inflation has been found plausible under some sets of reasonable assumptions, but no one knows which are the right ones. With eternal inflation, however, the question becomes irrelevant. No matter how unlikely it is for inflation to start, the eternal exponential growth can easily make up for it.

To see why the onset of inflation becomes irrelevant, it is useful to consider a numerical example. Imagine that the correct theory of nature has two different fields, either one of which could play the role of the inflaton field that drives inflation. In that case the theory can support two alternative types of inflation, depending on which of these two fields is driving it. Suppose that the first type of inflation, which I will call type *A*, has a probability of starting that is vastly larger than inflation of the other type, called *B*. For my example, I will assume that type *A* inflation is 10^{1000} times more likely to start than type *B*. (This number is far larger than the number of elementary particles in the visible universe, which is only about 10^{89}.) Thus, at the

* For example, the outcome of rolling a pair of dice is a random number between 2 and 12, but the numbers do not have equal probability. The probability of rolling a 3 is twice that of rolling a 2, since a 3 can be achieved by rolling either ⚀⚁ or ⚁⚀, while a 2 can be achieved only by rolling ⚀⚀. For a single die all outcomes are equally probable, but it is hard to know what properties of the nascent universe, if any, should be taken to be analogous to the roll of a single die.

beginning of the thought-experiment universe, the volume undergoing type *A* inflation will be 10^{1000} times larger than the volume undergoing type *B* inflation. Type *B* inflation, on the other hand, will be assumed to have an exponential expansion rate just a tiny bit larger than that of *A*—say 0.001% larger. For the expansion and decay rates, I will take numbers that are plausible for inflation driven by the physics of grand unified theories. The region undergoing type *A* inflation will double in volume every 10^{-37} seconds, and the half-life for decay for both types *A* and *B* will be a hundred times longer, or 10^{-35} seconds.

To decide whether type *A* or type *B* will dominate the universe, in principle we should wait an eternity, since the inflation never ends. However, to make my point it is sufficient to wait just one second. For the numbers given, the volume of type *B* inflation will overtake the volume of type *A* inflation in 3×10^{-29} seconds, and by the end of one second the volume of type *B* will be $10^{3 \times 10^{31}}$ times larger than that of type *A*! And if we had assumed an additional factor of $10^{1,000,000}$ in the starting probability for type *A*, the volume of type *B* at the end of one second would still be $10^{3 \times 10^{31}}$ times larger than that of type *A*! The factor of $10^{1,000,000}$ is so unimportant, compared to the exponential expansion, that its effect disappears when the answer is rounded off. The effect of the exponential expansion is so dramatic that even an unbelievably large difference in initial probabilities can be compensated in less than a second. If multiple forms of inflation are possible (which is very likely), then the form that will dominate is the one with the fastest rate of exponential expansion.

The third implication of eternal inflation concerns the plausibility of cosmological theories that do not lead to reproduction. I claim that once the viability of eternal inflation fully penetrates our psyches, our mindset concerning theories of the universe will be radically altered. Today, the dominant view of the origin of the universe, in both Judeo-Christian and scientific contexts, portrays it as a unique event. In the fifth century A.D. Saint Augustine described in his *Confessions* how time itself began with the creation of the universe, and modern scientists frequently refer to the big bang as the beginning of time. While scientists generally agree that we do not really know what came before the big bang, the tendency to think of the big bang as the origin of time remains habitual. However, if the ideas of eternal inflation are correct, then the big bang was not a singular act of creation, but was more like the biological process of cell division. If a biologist discovered a bacterium that belonged to no known species, she would presumably invent a new species in which to classify it. However, even though only a single specimen of the new species had been found, she would undoubtedly assume that it was the offspring of a bacterial parent cell. Although she believes firmly that life on earth originated from nonliving materials, the possibility that this particular cell is the result of such an improbable occurrence would be too preposterous to even consider. Given the plausibility of eternal inflation, I believe that soon any cosmological theory that does not lead to the eternal reproduction of universes will be considered as unimaginable as a species of bacteria that cannot reproduce.

CHAPTER 16

WORMHOLES AND THE CREATION OF UNIVERSES IN THE LABORATORY

If meringue is made by beating egg whites and sugar, how do you make a universe? This is clearly a question that has gripped the human imagination since the dawn of civilization.

According to the legends of the Fulani, a nomadic people living in the Republic of Mali in Western Africa, the universe was fashioned by the god Doondari from a huge drop of milk. Doondari first turned the milk to stone, which in turn created fire, then water, and then air. From these elements Doondari created humanity. The ancient Babylonian myth *Enuma elish* ascribes the universe to a much more violent incident, in which the thunder and rain deity Marduk slew the sea goddess Tiamat. Cleaving Tiamat's body in two, Marduk used half to forge the sun, moon, and constellations of the heavens, and then molded the other half into the mountains, plains, and rivers of the Earth. Tiamat's consort was also executed, and his blood was used by Marduk's father to create humans, so that the gods would not have to toil in constructing the city of Babylon.

Since the inflationary theory implies that the entire observed universe can evolve from a tiny speck, it is hard to stop oneself from asking whether a universe can in principle be created in a laboratory. Given what we know of the laws of physics, would it be possible for an extraordinarily advanced civilization to play the role of Doondari or Marduk, creating new universes at will? In collaboration with several graduate students and colleagues at MIT, I began to study this question in the mid-1980s. In this chapter I will describe what we have learned.

The first thing to think about is the list of ingredients needed to make a universe. Do we start with milk, the body of a sea goddess, or something else? Curiously, scientific theories in the twentieth century continue to offer an enormous range of answers to the question of what the universe was made from. One of the most dramatic differences between the inflationary theory and the standard (non-inflationary) big bang theory is the answer that each gives to this fundamental question.

For the standard big bang picture, the detailed answer depends on the starting time that one chooses. We cannot start at the instant of the big bang, because at that moment the temperature, pressure, and mass density are all infinite, so the description is ill-defined. A very reasonable starting time might be one second after the big bang, as this marks the beginning of the nucleosynthesis processes that led to the production of helium and the other light elements. Since the prediction of the abundances of these elements is a major success of the big bang theory, we are quite confident that the theory is reliable back to this time. While most cosmologists believe that the theory is reliable even at much earlier times, there is no direct observational evidence for times earlier than one second.

If the recipe for the standard big bang universe were written in a *Cosmic Cookbook*, how would it read? To begin the universe at an age of one second, the ingredient list would include 10^{89} photons, 10^{89} electrons, 10^{89} positrons, 10^{89} neutrinos, 10^{89} antineutrinos, 10^{79} protons, and 10^{79} neutrons. The ingredients should be stirred vigorously to produce a uniform batter, which should then be heated to a temperature of 10^{10}°K. After heating, the total mass/energy of the mix would be about 10^{65} grams, or 10^{32} solar masses. This number, by the way, is about 10 billion times larger than the total mass in the visible universe today. The mismatch between the mass of the ingredients and the mass of the present universe is caused by the cooling that takes place as the universe expands. The cooling robs the particles of most of their energy, transferring that energy to the gravitational field. So, to produce a universe by the standard big-bang description, one must start with the energy of 10 billion universes! Since a chef's first task is to assemble the ingredients, this recipe looks formidable enough to discourage anybody.

The *Cosmic Cookbook* entry for an inflationary universe, on the other hand, looks as simple as meringue. In this case the natural starting time would be the onset of inflation. In contrast to the standard big bang recipe, the inflationary version calls for only a single ingredient: a region of false vacuum. And the region need not be very large. If inflation is driven by the physics of grand unified theories, a patch of false vacuum 10^{-26} centimeters across is all the recipe demands. While the mass required for the previous recipe was 10^{32} solar masses, the mass in this case is only 10^{-32} solar masses. The sign of an exponent can make a big difference: in more easily recognizable units, the required mass is about 25 grams, or roughly one ounce! So, in the inflationary theory the universe evolves from essentially nothing at all, which is why I frequently refer to it as the ultimate free lunch.

In contrast to the earlier recipe, the ounce-of-false-vacuum prescription for producing an inflationary universe sounds appropriate for a beginners cooking class. Does this mean that the laws of physics truly enable us to create a new universe at will? If we tried to carry out this recipe, unfortunately, we would immediately encounter an annoying snag: since a sphere of false vacuum 10^{-26} centimeters across has a mass of one ounce, its density is a phenomenal 10^{80} grams per cubic centimeter! For comparison, the density of water is 1 gram per cubic centimeter, and even the density of an atomic nucleus is only 10^{15} grams per cubic centimeter. To reach the mass density of the false vacuum, one can imagine starting with water, and then compressing it to the density of a nucleus. Even with four more increases in density by the same factor, the density would still be 100,000 times lower than that of the false vacuum! If the mass of the entire observed universe were compressed to false-vacuum density, it would fit in a volume smaller than an atom!

The mass density of a grand unified theory false vacuum is not only beyond the range of present technology, it is beyond the range of any *conceivable* technology. As a practical matter, therefore, I would not recommend buying stock in a company that intends to market do-it-yourself universe kits. Nevertheless, in this chapter I will dismiss the gargantuan mass density of the false vacuum as a mere engineering problem, boldly assuming that some civilization in the distant, unforeseeable future will be capable of creating such densities. In this context, I will discuss whether the laws of physics, in principle, allow such an advanced civilization to create a universe. Is it possible, given what we know of the laws of physics, that someday our descendants might play the role of Marduk, producing new universes by slicing pieces of false vacuum the way the legendary god sliced the body of Tiamat? On the darker side, does the physics of the false vacuum create the possibility of an ultimate doomsday machine? Is our universe imperiled by the threat that a super-advanced civilization in some remote galaxy might create a cancerous patch of false vacuum that would engulf us all?

These questions sound very speculative, and of course they are. Some readers might complain that it is idle nonsense to hypothesize a civilization so amazingly advanced, and perhaps they are right. Nonetheless, I consider these questions interesting for at least three reasons. First, the discussion will be based on the real laws of physics, as best we understand them, and so at the very least it will help to illustrate how these laws work. Second, the implications go beyond the possibility of laboratory universes. While I find the thought of a human-made universe particularly intriguing, the thrust of the analysis is to learn under what circumstances, if any, a new universe can be created from an old one. Situations like the one I will be discussing, but more complicated, are believed to arise in eternal inflation, without the intervention of intelligent beings. By imagining a laboratory experiment, however, we can focus our attention on the simplest possible situation, which is almost always the best way to attack a new problem. And third—who really knows?—somewhere in this inconceivably vast universe, maybe there is a civilization so advanced!

The first step in trying to fabricate a laboratory universe is to create a patch of false vacuum. How exactly this can be achieved depends on the details of the high energy physics, which at present we have no way of knowing. This part of the problem, therefore, will be left for our descendants to solve; they will presumably understand the physics of arbitrarily high energies. At present, we can say that our current theories offer several possibilities. In many theories, the desired false vacuum can be created by heating a region of space to enormous temperatures (perhaps 10^{29}°K), and then rapidly cooling it. The region would then supercool into the false vacuum, exactly as we imagine that the early universe may have done. Another possibility would take advantage of the fact that the inflaton field, which creates the false vacuum and drives the inflation, presumably interacts with ordinary particles such as quarks and electrons. By compressing ordinary matter to enormous densities, one could exert an influence on the inflaton field that would push it into the false-vacuum state. The presence of the ordinary particles, however, would interfere with the properties of the false vacuum, preventing inflation from occurring. So, to complete the construction, it would be necessary to rapidly sweep away the ordinary matter, leaving only the inflaton field in its false-vacuum state.

Once a patch of false vacuum is created, the evolution is independent of how it was created. The false vacuum is characterized by having a huge energy density which cannot decrease as the volume expands, and a huge but negative pressure. Through the equations of general relativity, these properties alone determine how the spacetime is distorted by a region of false vacuum. In 1986 I wrote a paper, with two MIT graduate students, Steven Blau and Eduardo Guendelman [1], in which we calculated how such a patch of false vacuum would evolve. We discovered that we were not the first to investigate this question, but much of the work had already been done by groups in Japan, Italy, the Soviet Union, and Canada.

Figure 16.1 shows a *false vacuum bubble*—a sphere of false vacuum, surrounded by ordinary vacuum, or empty space. In this context, the ordinary vacuum is often called the *true vacuum.* The boundary between the true vacuum and the false vacuum is called the *bubble wall.* In microscopic terms the false vacuum consists of a region of space within which the inflaton field is perched at a metastable value, such as the center of the energy density diagrams of Figures 10.1 or 12.1. In the true vacuum the inflaton

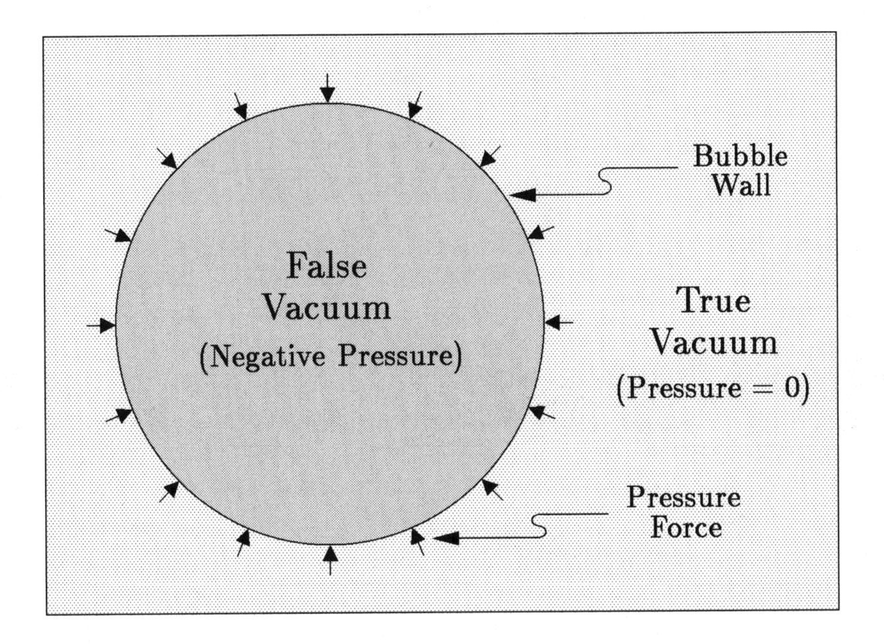

Figure 16.1 **A false vacuum bubble.** A sphere of false vacuum, which has negative pressure, is surrounded by true vacuum, with zero pressure. Since the pressure on the outside is higher, the pressure force pushes inward on the bubble wall. How, then, can the bubble expand?

field has the lowest possible energy, corresponding to any of the points on diagrams 10.1 or 12.1 labeled as "vacuum circle." In the bubble wall the inflaton varies continuously between the true-vacuum and false-vacuum values. The thickness of the bubble wall is usually small compared to the radius of the bubble, so the calculations that will be described in this chapter were based on the approximation that the bubble wall could be treated as an infinitely thin sheet.

Since the false vacuum creates a strong gravitational repulsion, we expect that the region of false vacuum will grow. However, if the bubble wall is to move outward, there must be a force pushing it that way. The diagram shows that the pressure outside the bubble is zero, and the pressure inside is negative. The pressure is therefore higher outside than inside, so the pressure difference will push inward on the bubble wall.

One might guess that the gravitational repulsion of the false vacuum would push outward on the bubble wall, so if this repulsion were strong enough, the bubble would start to grow. Not so, however, say the equations of general relativity. The repulsion of the false vacuum is an internal property only. The gravitational repulsion causes the false vacuum to swell, but the repulsion does not extend beyond the false vacuum. From the outside, the gravitational field of the false vacuum is attractive, just like anything else! If an object were placed in the true vacuum region of Figure 16.1, it would fall inward toward the center of the diagram, not outward. If one calculates the gravitational force on the bubble wall, it too is inward.

Since both pressure and gravity pull inward on the bubble wall, a bubble that is started from rest will start to contract. With nothing to halt the contraction, it will collapse to a black hole. While a black hole is interesting, it would certainly be viewed as a disappointment by our would-be universe creator.

Suppose, however, that the bubble was not started from rest, but instead was given an outward push. At first this does not seem to make much difference, since the forces on the bubble wall are inward. No matter how large the outward velocity of the wall, one would think, eventually these forces would stop it and cause the bubble to collapse. If the picture in Figure 16.1 is meaningful, it looks as if there is no way that a false vacuum bubble can become a universe.

Figure 16.1, however, is very oversimplified—it is flat! According to general relativity, the geometry of space is not flat, but is warped by a gravitational field—that is, it becomes non-Euclidean, no longer obeying the geometrical relations with which we are all familiar. For the case of a false vacuum bubble on the verge of creating a universe, the geometry becomes highly distorted.

Now we come to an important challenge: how can we visualize the distortion of space? It is not easy, since we have no direct experience with non-Euclidean space. Although general relativity implies that the space in which we live bends and twists with the gravitational fields of matter, in practice this distortion is far too small to notice. The ratio of the circumference of the Earth to its diameter, for example, differs from Euclidean geometry by less than one part in a billion. To the accuracy that we can see, the space in which we live is relentlessly Euclidean, denying us the opportunity to explore the intricacies of curved space.

General relativists, however, have invented a clever way to visualize curved space, which works at least for simple cases. The first step is to realize that we often do not need to visualize all three dimensions at once. For example, the false vacuum bubble that we are trying to visualize is spherically symmetric, or, in other words, perfectly round. It looks exactly the same from one angle as it does from another. For that reason it is sufficient to study a two-dimensional slice through the middle of the false vacuum bubble, as shown in Figure 16.2. If we know what is happening at all points on this slice, then we know everything. So now all we have to do is find a way to visualize the curvature of a two-dimensional surface.

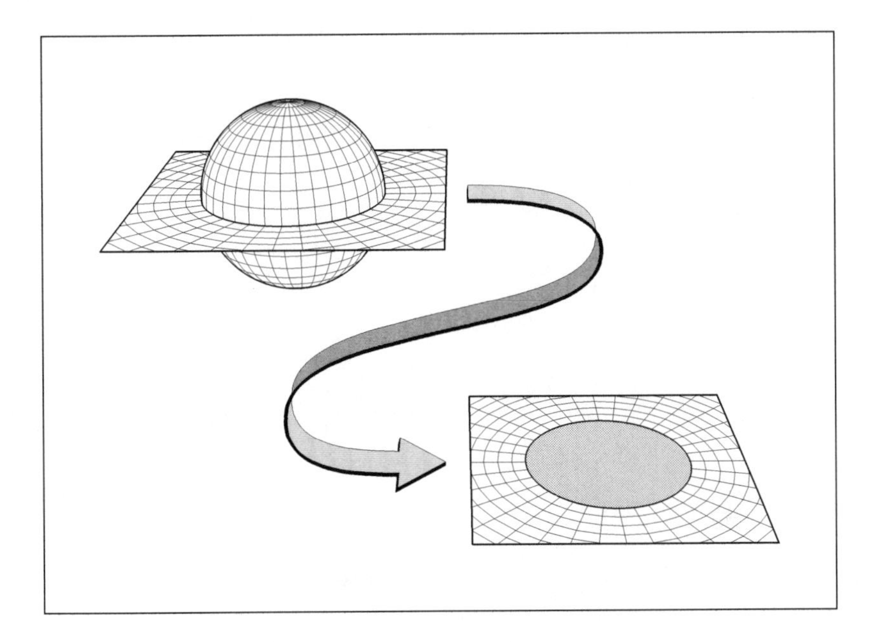

Figure 16.2 **A two-dimensional slice through the center of a false vacuum bubble.** Since the bubble is spherical, the full three-dimensional geometry can be determined from the geometry of the two-dimensional slice.

If the two-dimensional surface is diagrammed flat on a page, then we have gained nothing—it looks flat, and we have not made it easier to visualize the possibility that it might be non-Euclidean. However, there is no need to diagram the surface flat on a page. We live in a three-dimensional space, so we can use all three dimensions for purposes of the display. If a two-dimensional surface is shown in three dimensions, then curvature can be made visible. One must remember, however, that the third dimension has nothing to do with the original space, but is introduced *only* to help visualize the geometry. The only points in the diagram that have meaning are those on the two-dimensional surface. The three-dimensional picture, however, can be used to accurately display the geometry of the two-dimensional surface. Such a three-dimensional picture of a two-dimensional surface is called an *embedding diagram.*

An embedding diagram of the false vacuum bubble is shown in Figure 16.3. The region of false vacuum at the center causes a visible indentation in the space, much as a heavy stone would cause an indentation in a rubber sheet. The shape of the surface, however, was constructed using the equations of general relativity, not the equations of a rubber sheet. The diagram

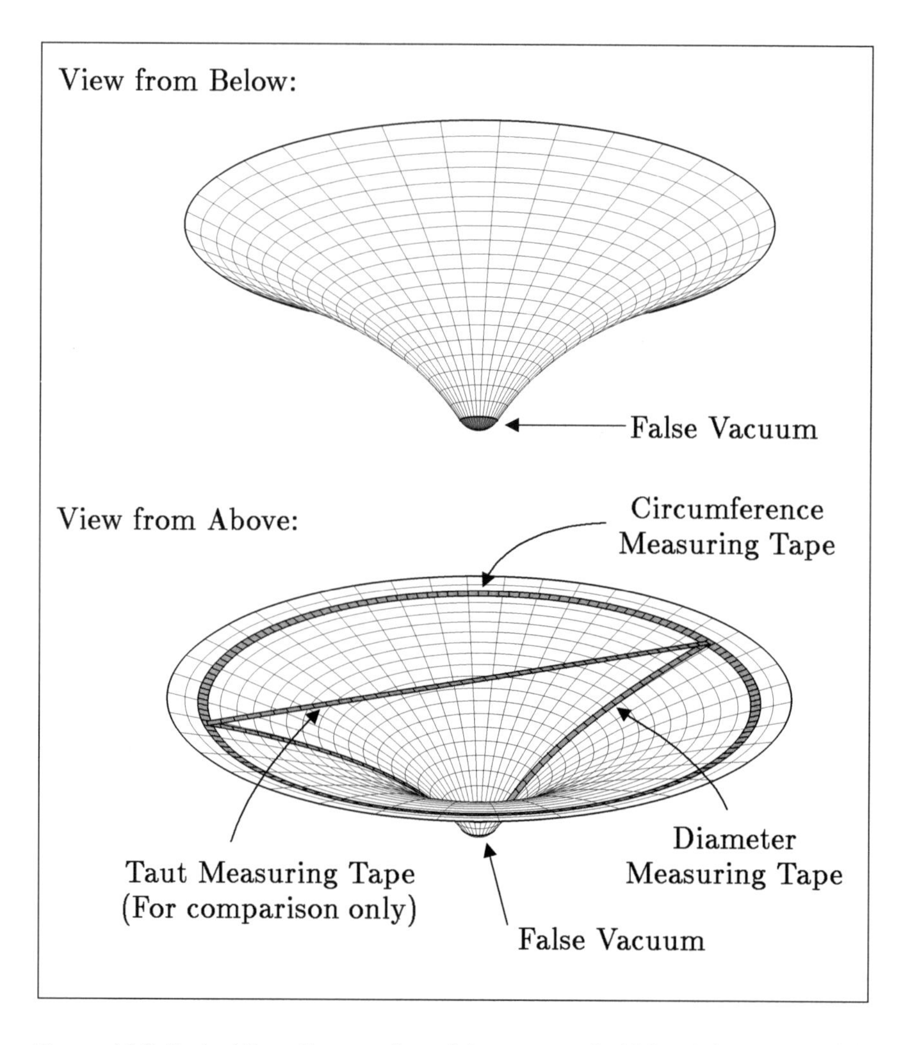

Figure 16.3 **Embedding diagram for a false vacuum bubble.** The non-Euclidean geometry of a slice through the center of a false vacuum bubble can be exhibited by showing it as a curved surface in three dimensions. Only points in the two-dimensional surface correspond to points in the false vacuum bubble; the rest of three-dimensional space is used only for visualization. The diagram shows two views, one from a little below the surface, and one from a little above. The view from above includes measuring tapes on the surface that could be used to determine the circumference and diameter of a circle. The taut measuring tape does not lie in the surface, so it does not correspond to a measurement of the false vacuum bubble. The length of the taut measuring tape is equal to the circumference divided by π, so the diameter of the circle, measured by the tape in the surface, is larger than that.

is not merely a suggestive image, but rather shows precisely the predictions of general relativity for the geometry of a false vacuum bubble. The geometry is completely described by distance relationships, which can be measured from the diagram.

To understand exactly what the embedding diagram is telling us, imagine that it has been constructed as a three-dimensional sculpture. Consider, for example, a circle centered on the false vacuum bubble. If you were to wrap a measuring tape around the circle on the sculpture, as shown, you would measure the same distance as would be measured on the corresponding circle around the false vacuum bubble. Now imagine measuring the diameter by the same method, by pulling the tape measure along the diameter curve shown on the diagram. In Euclidean geometry, the diameter of a circle is always equal to the circumference divided by π (3.14159265. . .). We can see from the three-dimensional sculpture that if the tape were pulled taut, then the tape and circle would lie in a flat plane, and we would find the Euclidean answer. This answer, however, does not apply to the false vacuum bubble, since the tape is stretched across points that are not in the surface, and which have no correspondence to points in the real object. The true diameter is the length of the diameter tape shown in the surface, which is larger than the Euclidean answer. In the non-Euclidean geometry of a false vacuum bubble, the diameter of a circle is longer than the circumference divided by π. You could find out exactly how much longer, for a circle of any size, by making accurate measurements on the sculpture.*

With the help of embedding diagrams, we can now visualize how an expanding false vacuum bubble will evolve. If its initial outward velocity is too slow, then its fate will be the same as a false vacuum bubble that starts from rest. The pressure and gravity forces will quickly stop the expansion, and the bubble will collapse to a black hole. However, if the initial outward velocity is large enough, then the bubble will follow the sequence shown in Figure 16.4. As the bubble grows the indentation will become deeper, as shown in part (a). The indentation will continue to deepen, developing a neck, or *wormhole*, as shown in part (b). Once the wormhole develops, a dramatic change takes place—the bubble has turned inside out. Now the region of false vacuum can grow larger and larger without encroaching on

* No matter how the surface is curved in the embedding diagram, the diameter measured along the surface will always be greater than or equal to the Euclidean diameter, which is measured along a straight line. Surfaces with this property are called *positively curved*. A surface could alternatively be *negatively curved*, with the diameter of a circle less than the circumference divided by π, but such surfaces cannot be displayed by embedding diagrams. Fortunately, the non-Euclidean surfaces for false vacuum bubbles are positively curved, at least during the time intervals that are illustrated in this chapter.

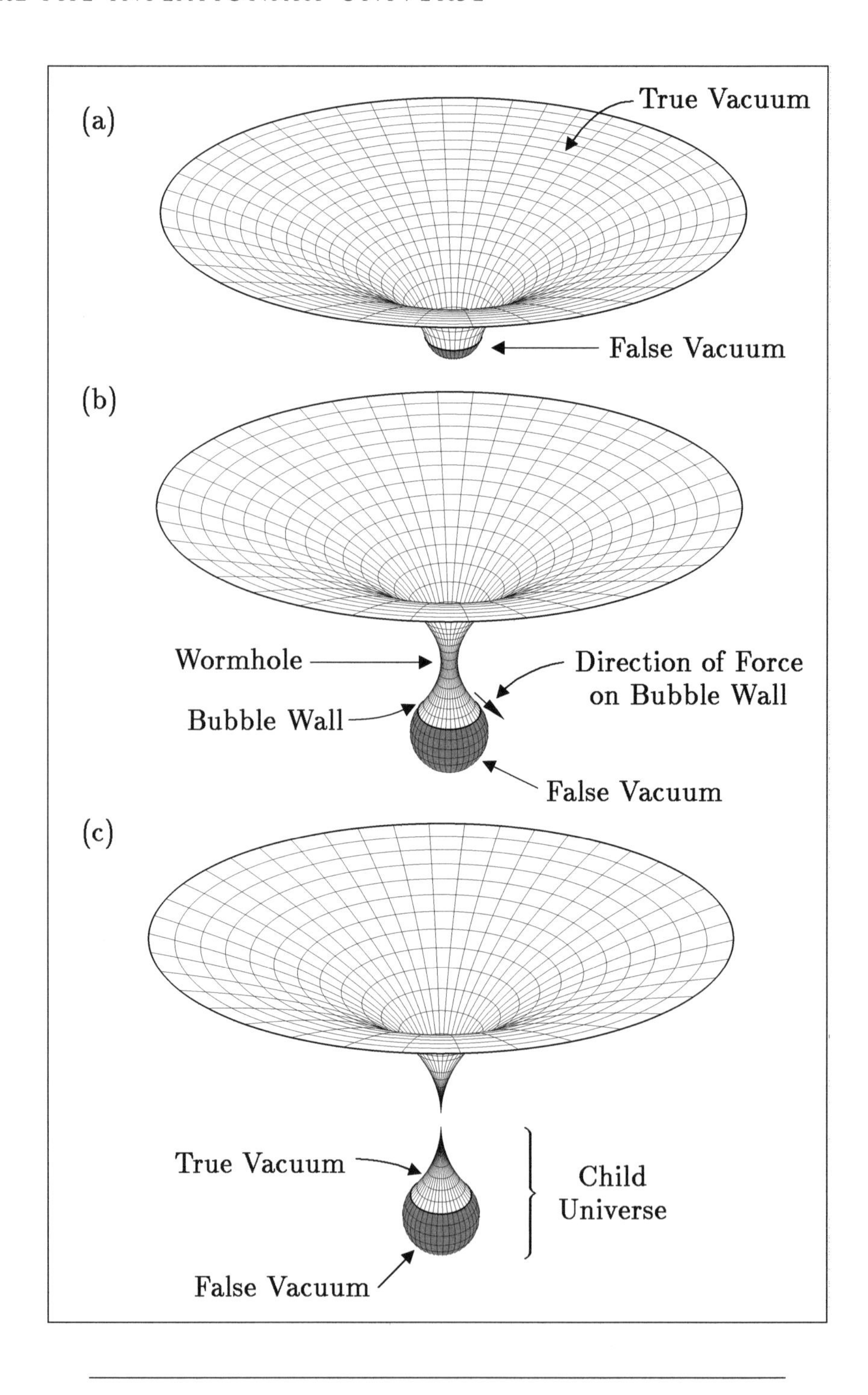
(a)
True Vacuum
False Vacuum
(b)
Wormhole
Direction of Force on Bubble Wall
Bubble Wall
False Vacuum
(c)
True Vacuum
Child Universe
False Vacuum

the original space. It creates new space as it expands, resembling an inflating balloon. Before the formation of the wormhole, the pressure difference between the true and false vacuum resulted in a force pushing the bubble wall inward, toward a smaller radius. After the formation of the wormhole, the pressure difference continues to exert a force from the true vacuum toward the false vacuum. Now, however, this force acts to cause the bubble wall to enlarge! The climax of the evolution is shown in part (c): the region of false vacuum, with a region of true vacuum attached, disconnects from the parent space, forming a new, completely isolated closed universe. It will then continue to enlarge, going through the usual evolution of an inflationary universe. A new universe has been created, and the parent universe is unharmed—universe creation is not a doomsday machine. From the point of view of the parent universe, the "umbilical cord" of the child universe is indistinguishable from a black hole. The umbilical connection in the child universe would similarly look like a black hole.

If we assume that the false vacuum driving the inflation has a mass density of about 10^{80} grams per cubic centimeter, characteristic of a typical grand unified theory, then the time that it takes the child universe to disconnect is roughly 10^{-37} seconds. After this time there will be no contact between the parent universe and child. While Marduk had the opportunity to rule the world that he created, a false-vacuum-bubble universe creator would watch helplessly as his new universe slipped inexorably through the wormhole and severed all contact. (The scenario might be viewed as an extreme form of Woody Allen's joke from *Love and Death*: "The important thing, I think, is not to be bitter. You know, if it turns out that there is a God, I don't think that He's evil. I think that the worst you can say about Him is that, basically, He's an underachiever.") Once the new universe has separated, on a much slower pace the black hole that remains in the parent universe would evaporate. It would disappear in roughly 10^{-23} seconds, releasing the energy equivalent of a 500 kiloton nuclear explosion. While

---

Figure 16.4 **(facing page) The creation of a child universe.** If a false vacuum bubble is caused to expand quickly enough, then the shape shown in Figure 16.3 will evolve through the stages shown here. The false-vacuum region will grow (a), and the space will distort to form a neck (b), also called a wormhole. The false-vacuum region will continue to expand like a balloon, creating new space as it does. In about 10^{-37} seconds the false-vacuum region will disconnect completely (c), forming a child universe, completely isolated from the parent universe. To observers in the original space, the event looks like the formation of a black hole. The region shown has a diameter of about 10^{-25} centimeters.

the parent universe would be in no danger of annihilation, the safety of the experimenters would require precautions similar to those used in hydrogen bomb tests.

The story of inflationary universe creation sounds complete at this point, but there is an important and surprising twist: It is not clear whether it is possible, even in principle, to attain the expansion velocity needed for the false vacuum bubble to evolve into a universe. I am not saying that a universe-creating false vacuum bubble cannot exist. The required expansion velocity is less than the speed of light, and the universe-creation scenario described in Figure 16.4 is completely consistent with general relativity and with all known laws of physics. The question is not whether such a universe-creating bubble is possible—as far as we know, it *is* possible. The question is whether the laws of physics allow us to get from here (the present universe) to there (a universe containing such a rapidly expanding false vacuum bubble).

The existence of this problem came as a surprise to me when I first learned it, while working on the question of child-universe creation in collaboration with Edward H. Farhi, a colleague at MIT [2]. I had been accustomed to believing that any situation could be constructed in principle, as long as it could be described within the language of physics, and as long as it needed only a finite amount of energy and other conserved quantities. If one wanted an object to move faster than it was moving, for example, one could always imagine attaching an extra rocket to propel it. The only barrier would be the speed of light, which is imposed by energy restrictions: the energy of a particle rises without limit as the speed approaches that of light, so no finite amount of energy can ever cause the particle to move faster than light.

In the context of general relativity, however, I learned that there are sometimes more stringent restrictions on what can be created from a given starting point. If one can ignore gravity, then the rigidity of Euclidean space provides the bedrock that any construction engineer would want. A larger building or a faster car can always be attained in principle by bringing in the necessary cranes, girders, or engines. When gravitational fields become strong, however, then space itself becomes malleable. As Figures 16.3 and 16.4 suggest, space no longer provides a bedrock, but instead begins to behave more like quicksand. If one tries to construct a larger building or a faster car by bringing in heavy machinery, the quicksand caves in all the quicker.

In the case of the false vacuum bubble, the possibility of attaining an expansion velocity fast enough for universe creation runs into conflict with a theorem proved in 1965 by Roger Penrose, of Oxford University [3]. The theorem is usually phrased as a statement about collapsing objects, and it is easiest to understand in this form first. (The connection between a collapsing object and an expanding bubble may not be clear, but I will come back to that.) While the precise statement of Penrose's theorem is very complicated, the gist can be stated simply: if an object is contracting fast enough to satisfy a specific technical criterion, then there is nothing that can stop it from collapsing to a black hole singularity. One can imagine trying to prevent the collapse by bringing in extra tools—ropes, rockets, explosives, or whatever. According to the theorem, however, nothing will help. As each additional tool is brought in to try to resist the collapse, the gravitational field of the new object only hastens the submersion into the quicksand.

While the theorem in the previous paragraph refers to contracting objects, not expanding objects, general relativity makes no distinction between the two directions of time. That is, if one filmed a movie of any event described by general relativity, then the movie shown backward in time would also describe a possible sequence of events. Similarly, if a movie shows a sequence of events that is impossible according to general relativity, then the movie shown backward must also be impossible. According to the Penrose theorem, rapid collapse without a subsequent singularity is impossible. Reversing the direction of time, it follows that rapid expansion without a preceding singularity is also impossible. In this form the theorem applies to the false vacuum bubble. For the bubble to expand fast enough to become a universe, Penrose tells us that it must begin from a singularity!

The kind of singularity that would be needed here is the time-reversed form of a black hole, often called a *white hole*. That is, a film of matter emerging from such a singularity would be identical to a film of matter collapsing into a black hole, but shown backward. While matter falls into a black hole and can never escape, matter emerges from a white hole but can never enter it. The white hole is exactly the kind of initial singularity that was hypothesized in the standard form of the big bang theory, but certainly there is no known way to create such a singularity in the laboratory. In fact, there is really no conceivable way, since such a singularity marks the beginning of time, at least at a point if not everywhere. Since the beginning of time cannot be caused by something that preceded it, there is no way to cause a white hole.

Since even a super-advanced civilization cannot be expected to create a white hole, the Penrose theorem implies that a false vacuum bubble cannot be caused to expand fast enough to produce a new universe. However, there are at least two conceivable ways to evade this uninspiring conclusion.

First, as one might guess, it is always possible to question the hypotheses on which a theorem is based. In the case of the Penrose theorem, a key hypothesis is that the energy density of any form of matter, viewed by any observer, is always positive. If one could find a form of matter with negative energy density, then the theorem would no longer be valid. If negative energy densities were available, one would be free to bring in extra rockets to propel the walls of the false vacuum bubble. To prevent the weight of the rockets from hastening the collapse of the space, one could bring in negative mass material to cancel out the effect. (Earlier in the book I described the negative energy density of the gravitational field, but the negative energy is always accompanied by the positive energy of the matter creating the gravity. A calculation is needed to see which effect is more important, but Penrose has already told us the answer. His theorem includes all the effects of gravity, so the negative energy density of gravity cannot alter the result.)

The positive-energy-density assumption sounds reasonable enough, but surprisingly it is known that in quantum theory, negative energy densities can happen. An important example is the radiation emitted by black holes. This result was unexpected when Stephen Hawking discovered it in 1975, because it directly contradicted a theorem of classical general relativity, proven by Hawking four years earlier, which implied that black holes could never shrink. The theorem, like Penrose's theorem, was based on the assumption that energy densities are always positive. In his calculation of black hole radiation, Hawking showed that quantum effects cause the energy density in the vicinity of a black hole to become negative. It is conceivable, therefore, that negative energy densities associated with quantum effects might also provide a way to skirt the implications of Penrose's theorem for false vacuum bubbles. So far, however, this possibility has not been converted into a reality. No one has found a way to create a region of negative energy density that is large enough, and endures long enough, to generate a universe-creating false vacuum bubble. On the other hand, no one has shown it to be impossible.

The second hope for evading Penrose's theorem involves another consequence of quantum theory, the process of quantum tunneling. A quantum system can make a discontinuous transition from one configuration to another, even when classically the system would not have enough energy to exist in the intermediate configurations. This "quantum leap" is the origin of a frequently used metaphor (and a successful American television series!), and it also plays a crucial role in the decay of the false vacuum. In 1990 I collaborated with Farhi, and also a former student of mine at the University of Mexico, Jemal Guven, to pursue the possibility of universe-creation by quantum tunneling [4].

To exploit the possibility of quantum tunneling, the would-be universe creator would start with a bubble of false vacuum, as discussed earlier in the chapter. To prevent the bubble from immediately collapsing, she would apply an outward force to the bubble wall to start it expanding. Since she has not started with a singularity, however, she cannot impart a high enough velocity to create a universe—at least not according to the classical equations of general relativity. According to these equations, the bubble will reach a maximum radius and then collapse to a black hole. However, the equations of classical general relativity also imply that there is another classical configuration, with a larger radius, which would grow to produce a universe. The second configuration has exactly the same total energy as the first—the larger region of false vacuum contributes additional energy, but this is canceled by the negative energy of the larger region of strong gravitational field. Embedding diagrams for the two configurations are shown in Figure 16.5. Classically there is no transition between the two configurations, since there are no

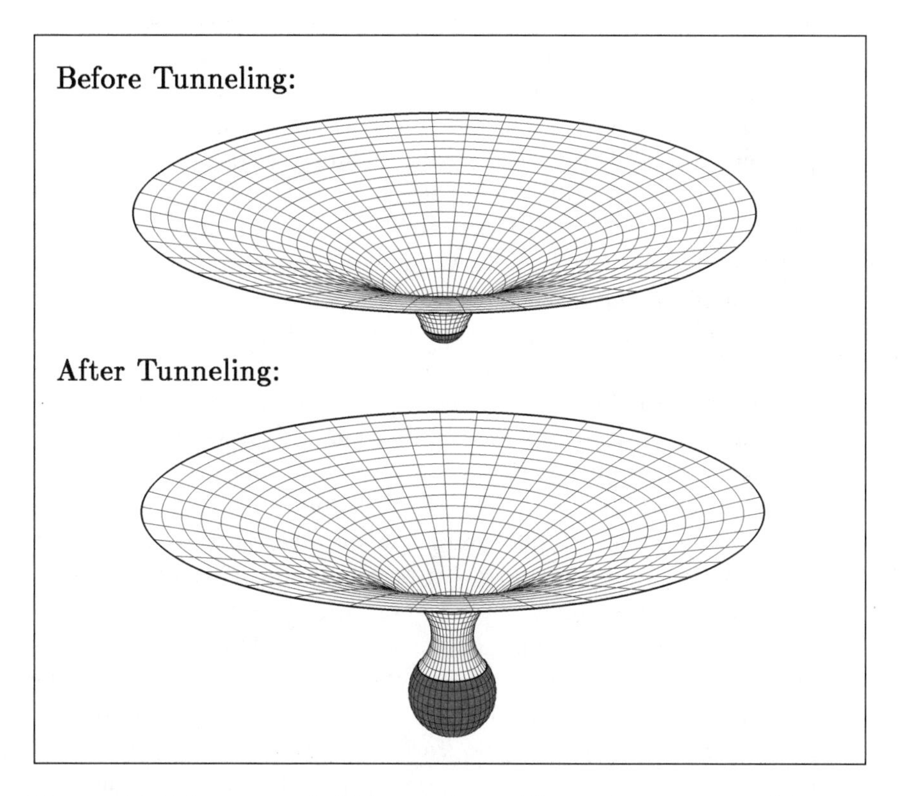

Figure 16.5 **False vacuum bubble tunneling.** The top figure shows a false vacuum bubble with a mass/energy about 35% smaller than the one shown in the previous figure. While the Penrose theorem implies that the bubble shown in the previous figure cannot be produced, the bubble shown here is small enough so that the theorem does not apply. Classically, this bubble would reach a maximum radius, shown in the top diagram, and would then collapse. Quantum mechanically, however, it can tunnel into the configuration shown below it. After the tunneling, it would grow classically, producing a child universe in a manner very similar to that shown in Figure 16.4.

intermediate configurations of the same energy. This is precisely the situation in which quantum tunneling can be called upon to bridge (or tunnel) the gap. If the bubble can tunnel from the first configuration to the second, then the subsequent classical evolution will lead to the creation of a new universe!

There is a shortfall in our understanding, however, because the tunneling process indicated in Figure 16.5 involves a significant change not only in the inflaton field, but also in the spatial geometry. This means that a reliable calculation would require a complete theory of quantum gravity, which does not at present exist. Nonetheless, I think it is fair to say that most physicists would expect that the process shown here is possible. We even have an approximate method for calculating the probability, which is based on analogies with other quantum systems that we do understand.* When this method is applied to false vacuum bubbles with a mass density of 10^{80} grams per cubic centimeter, a number typical for grand unified theories, the answer is found to be outrageously small. Each time a false vacuum bubble is set into expansion, the probability that it will tunnel to become a new universe is only about $10^{-10^{13}}$. To write out this number in decimal notation would require a decimal point, followed by 10^{13} zeros, followed by a one! So even if some super-advanced civilization develops the capacity to produce and manipulate regions of false vacuum, they would still require a fantastic amount of patience to produce a new universe.

It may, however, be premature for us to bemoan the difficulties that our super-advanced descendants will face. The calculation shows that the probability depends very sensitively on the mass density of the false vacuum. The probability becomes higher as the mass density of the false vacuum increases. Even if we knew that grand unified theories were correct, we would still be ignorant of the physics at higher energies. We really have no way of knowing whether there might be false vacuum states at mass densities much higher than 10^{80} grams per cubic centimeter. In particular, if there exists a false vacuum state associated with the unification of gravity with the other forces, expected to occur at about 10^{19} GeV, then the mass density would be about 10^{93} grams per cubic centimeter. For this density, the answer to our probability calculation would be approximately one—a new universe would be created with just about every attempt!

To put the story into perspective, one should remember that the process of eternal inflation, discussed in the previous chapter, leads to an exponential increase in the number of pocket universes on time scales as short as 10^{-37} seconds. Since the time needed for the development of a super-advanced civi-

* The probability calculation by Farhi, Guven, and me involved ambiguities, since it was based on analogy. There is no guarantee that it is right, but the credibility of the calculation was bolstered when a group at the University of Texas obtained the same answer by a different method [5].

lization is measured in billions of years or more, there appears to be no chance that laboratory production of universes could compete with the "natural" process of eternal inflation.

On the other hand, a child universe created in a laboratory by a super-advanced civilization would set into motion its own progression of eternal inflation. Could the super-advanced civilization find a way to enhance its efficiency? We may have to wait a few billion years to find out.

CHAPTER 17

A UNIVERSE EX NIHILO

If eternal inflation is correct, then the big bang which occurred 10 to 15 billion years ago was the beginning of our "pocket universe," but not the universe as a whole. The full universe existed long before our pocket universe, and will continue to exist for eternity. Nonetheless, since inflation appears to be eternal only into the future, but not the past, an important question remains open: How did it all start? Although eternal inflation pushes this question far into the past, and well beyond the range of observational tests, the question does not disappear. Since the universe exists, there is a human drive to learn how it began. The big bang theory does not really address this question, as it describes the universe from immediately after its creation onward. While there is no accepted scientific answer to the question of cosmic creation, speculative answers are now studied by respectable scientists who publish in respectable journals.

To my knowledge, the first serious suggestion that the creation of the universe from nothing could be described in scientific terms was the 1973 paper by Edward Tryon, "Is the Universe a Vacuum Fluctuation?" (which was mentioned in the first chapter of this book). The article was originally submitted to *Physical Review Letters,* which rejected it as being too speculative. Since Tryon was still enamored with the simplicity and beauty of the vacuum-fluctuation universe, he rewrote the article for a more general audience and submitted it to *Nature* as a "Letter to the Editor." There the editors not only published it, but they upgraded it to a feature article. Nevertheless, for the next decade the article was rarely cited, as no one had any good ideas for extending the proposal.

In discussing the creation of the universe, a key issue is the choice of a starting point. In Tryon's proposal, the universe was created from the vacuum, or empty space. According to quantum theory, the apparently quiescent vacuum is not really empty at all, but on a subatomic level is a perpetual tempest, seething with activity. For example, it is possible for an electron and its antiparticle, the positron, to materialize from the vacuum, exist for a brief flash of time, and then disappear into nothingness. Such *vacuum fluctuations* cannot be observed directly, as they typically last for only about 10^{-21} seconds, and the separation between the electron and positron is typically no larger than 10^{-10} centimeters. Nonetheless, physicists are convinced that these fluctuations are real, as they can be "seen" by indirect methods. For example, every electron acts like a tiny magnet, and atomic physicists can measure the magnetic strength of an electron to the extraordinary precision of 10 decimal places. Theorists can calculate this number using quantum theory, and they find that it is influenced by the materialization of electron-positron pairs. If the effect of this materialization is left out of the calculation, then the answer agrees with experiment for the first five decimal places, but the sixth, seventh, eighth, ninth, and tenth decimal places turn out wrong. But if the effect of electron-positron pairs is included, all ten decimal places agree!

Any object, in principle, might materialize briefly in the vacuum—even a refrigerator or a pocket calculator.* The probability for an object to materialize, however, decreases dramatically with the object's mass and complexity. Normally, only particle-antiparticle pairs need to be considered. In his controversial two-page paper, Tryon advanced the startling proposal that on rare occasions, whole universes might materialize from the vacuum, and that our universe may have begun in this way.

Since quantum fluctuations are normally very short-lived, it would seem preposterous for Tryon to claim that a ten-billion-year-old universe could arise by this mechanism. In general the lifetime of a fluctuation depends on its mass, becoming shorter as the mass becomes larger. For a quantum fluctuation to survive for ten billion years, the mass should be less than 10^{-65} grams, which is 10^{38} times smaller than the mass of one electron! Here Tryon invoked the fact, discussed in Chapter 1, that the energy of a gravitational field is negative. Furthermore, as Tryon learned from the well-known general relativist Peter Bergmann, in any closed universe the negative gravitational energy cancels the energy of matter exactly. The total energy,

* The conservation laws, however, cannot be violated. If the object has positive electrical charge, for example, then it can materialize only if it is accompanied by the production of an equal magnitude of negative charge. Energy never poses an absolute barrier, however, because it can be supplied temporarily by the stormy climate of the quantum vacuum.

or equivalently the total mass, is precisely equal to zero.* With zero mass, the lifetime of a quantum fluctuation can be infinite.

While the basic thrust of the idea looks attractive, today many theorists would question Tryon's starting point. If our universe was born from empty space, then where did the empty space come from?

In our everyday experience, we tend to equate empty space with "nothingness." Empty space has no mass, no color, no opacity, no texture, no hardness, no temperature—if that is not "nothing," what is? However, from the point of view of general relativity, empty space is unambiguously *something*. According to general relativity, space is not a passive background, but instead a flexible medium that can bend, twist, and flex. This bending of space is the way that a gravitational field is described. In this context, a proposal that the universe was created from empty space seems no more fundamental than a proposal that the universe was spawned by a piece of rubber. It might be true, but one would still want to ask where the piece of rubber came from.

In 1982, Alexander Vilenkin of Tufts University proposed an extension of Tryon's idea [1]. He suggested that the universe was created by quantum processes starting from "literally *nothing*," meaning not only the absence of matter, but the absence of space and time as well. This concept of absolute nothingness is hard to understand, because we are accustomed to thinking of space as an immutable background which could not possibly be removed. Just as a fish could not imagine the absence of water, we cannot imagine a situation devoid of space and time. At the risk of trying to illuminate the abstruse with the obscure, I mention that one way to understand absolute nothingness is to imagine a closed universe, which has a finite volume, and then imagine decreasing the volume to zero. In any case, whether one can visualize it or not, Vilenkin showed that the concept of absolute nothingness is at least mathematically well-defined, and can be used as a starting point for theories of creation.

As with Tryon's suggestion, Vilenkin's proposal was based on a quantum description of general relativity. While a completely successful merging of these two theories does not yet exist, we know enough about each theory

* Although Tryon was not aware of this source, the fact that the total energy of a closed universe is exactly zero had been explained some years earlier in the widely known textbook, *The Classical Theory of Fields,* Revised Second Edition, by L. D. Landau and E. M. Lifshitz (Addison-Wesley Press, Cambridge, Massachusetts, 1962), pp. 378–379. As far back as 1934, Richard C. Tolman of the California Institute of Technology showed how the energy of matter in a closed universe can grow without bound, with the energy "coming from the potential energy of the gravitational field." (*Relativity Thermodynamics and Cosmology*, Clarendon Press, Oxford, 1934; reprinted by Dover Publications, Inc., New York, 1987, p. 441). Tolman was actually discussing an oscillating universe model, but the conservation of energy issues are the same as in a one-cycle universe.

Alex Vilenkin at his computer terminal at Tufts University.

to develop a plausible scenario. From general relativity, Vilenkin took the idea that the geometry of space is not fixed, but that instead space is plastic and capable of distortion. There are many different possible geometries, including the closed universe, the open universe, and many less symmetric contortions of space and time. Among all possible geometries is the totally empty geometry, a space that contains no points whatever. From quantum theory, Vilenkin took the notion of quantum tunneling: A quantum system can suddenly and discontinuously make a transition from one configuration to another, as long as no conservation law makes the transformation impossible. Putting these ideas together, one can imagine that the universe started in the totally empty geometry—absolute nothingness—and then made a quantum tunneling transition to a nonempty state. Calculations show that a universe created this way would typically be subatomic in size, but that is no problem. While Tryon had fretted over the implausibility of a quantum fluctuation of cosmic proportions, Vilenkin was able to invoke inflation to enlarge the universe to its current size.

In the fall of 1981, Stephen Hawking introduced an alternative vision of a universe ex nihilo, presented to a conference at the Vatican organized by the Pontifical Academy of Sciences. Shortly afterward he developed this idea in a paper [2] with James Hartle of the University of California at Santa Barbara. Hawking's approach is quantitatively different from Vilenkin's, and in some ways qualitatively different as well. In particular, the concept of "absolute nothingness" does not appear. One could say that to Hawking, "nothing" is superfluous. The goal, after all, is to develop a theory of *something*, not *nothing*, so the invention of a prior history of nothingness might be unnecessary, or even misleading. Instead, Hawking proposed a description of the universe in its entirety, viewed as a self-contained entity, with no reference to anything that might have come before it. The description is timeless, in the sense that one set of equations delineates the universe for all time. As one looks to earlier and earlier times, one finds that the model universe is not eternal, but there is no creation event, either. Instead, at times of the order of 10^{-43} seconds, the approximation of a classical description of space and time breaks down completely, with the whole picture dissolving into quantum ambiguity. In Hawking's words,* the universe "would neither be created nor destroyed. It would just BE." As in Vilenkin's scenario, a form of inflation is used to enlarge a subatomic universe into a full-size cosmology.

Hartle and Hawking proposed a specific mathematical recipe for calculating the quantum description of the universe. The Hartle-Hawking description is often called the "no-boundary" proposal, because in the context of the rather abstract mathematical formalism used in its calculation, it is defined by the criterion that spacetime has no initial boundary. In principle, the full knowledge of such a quantum description, or *wave function*, would answer not only the basic questions of cosmology, but also all questions of science. While it would not uniquely define the history of the world,

* *A Brief History of Time: From the Big Bang to Black Holes,* by Stephen W. Hawking (Bantam Books, New York, 1988), p. 136.

since quantum randomness is important, it would determine the set of all possible histories with a probability assigned to each. "If you guys know the wave function of the universe," quipped the Nobel Prize-winner Murray Gell-Mann, "how come you aren't rich?" Answer: The recipe requires as input the knowledge of the full laws of physics, which Hartle and Hawking do not (yet) possess. In addition, the recipe is so complicated that in practice it can be implemented only in oversimplified model universes.

While the attempts to describe the materialization of the universe from nothing remain highly speculative, they represent an exciting enlargement of the boundaries of science. If someday this program can be completed, it would mean that the existence and history of the universe could be explained by the underlying laws of nature. That is, the laws of physics would imply the existence of the universe. We would have accomplished the spectacular goal of understanding why there is something rather than nothing—because, if this approach is right, perpetual "nothing" is impossible. If the creation of the universe can be described as a quantum process, we would be left with one deep mystery of existence: What is it that determined the laws of physics?

EPILOGUE

As this book goes to press at the start of 1997, inflation continues to be a vibrant area of study on both the theoretical and observational fronts.

On the theoretical side, cosmologists continue to explore new possibilities for fleshing out the details of the inflationary paradigm. In the body of the book I described only the two earliest proposals for inflation: the original inflationary model and the new inflationary model. The inflationary universe, however, is not really a theory, but rather a class of theories, and in fact it is a class that has been steadily growing. For the past decade there have been over two hundred scientific papers per year about inflation—much more than I have had time to read. And there is no sign yet of it starting to taper off.

Very shortly after new inflation was proposed, Andrei Linde proposed yet another variant, called *chaotic inflation* [1]. While new inflation depended on having a plateau in the energy density diagram, as in Figure 12.1 on page 209, the chaotic model showed that such a plateau is not necessary at all. Instead of a flattened Mexican hat diagram, chaotic inflation can work even if the energy density diagram has the shape of a simple bowl, with a unique minimum at the center. If the ball that represents the values of the fields starts high enough up the hill at the side of the bowl, then sufficient inflation can occur as the ball gently rolls to the point of minimum energy density. But why would the ball start high on the hill? Linde proposed that the pre-inflationary universe was "chaotic," meaning that the fields would take different values at different points in space in a complicated, random pattern. Most patches of space would presumably not experience much inflation, since the fields would tend to be near the bottom of the hill. There will be some patches, however, where the initial, random values of the fields would be far up the hill, and these patches would undergo enormous amounts of inflation. The patches which did not inflate would remain microscopic, so the universe at late times would be completely dominated by those regions which underwent large amounts of inflation, even if those regions were initially very rare.

In the late 1980s, Paul Steinhardt and Daile La invented yet another version of inflationary cosmology, called *extended inflation* [2], which differed from the previous successful models chiefly in the way that inflation ends. The ending of chaotic inflation was similar to that of new inflation, with a field rolling gently down a hill in the energy density diagram. The ending of extended inflation, however, closely resembled that of the original inflationary model: Many small bubbles of normal matter would form in the midst of the false vacuum, very much like the boiling of water. The original model failed, however, because the rapid expansion prevented the bubbles from ever joining to smoothly fill the space. The new trick, introduced by Steinhardt and La, was to propose that some new field interacts directly with the gravitational field, causing the strength of gravity to change with time. The changing strength of gravity could cause the expansion of the universe to slow down, giving the expanding bubbles a chance to catch up and fill the space.

The exploration of variations on the inflationary theme has continued, and a search of the scientific literature shows that there are roughly fifty different forms of inflation that have been named and studied. The list includes double, triple, and hybrid inflation, not to mention mutated hybrid inflation, tilted hybrid inflation, and hyperextended inflation. Cosmologists have also studied inflation that is gravity driven, spin driven, string driven, and vector field driven, as well as inflation that is warm, soft, tepid, and natural. I have recently been working with Lisa Randall, also at MIT, on a new version of inflation that she invented.* Since the model has some features of natural inflation and is based on the particle physics principle called supersymmetry, we decided to call it *supernatural inflation.* After circulating our paper, we learned from Andrei Linde that the name had in fact already been used—not just once, but twice!

From the many versions of inflation that have been developed, a few conclusions can be drawn. First, although inflation was invented in the context of grand unified theories, it does not really depend on them. Inflation can occur in the context of a wide variety of underlying particle theories. Inflation requires only that there is some state to play the role of a false vacuum, and that there is some mechanism to produce the baryons (e.g., the

* This work has also involved three MIT students: Marin Soljačić, Shufang Su, and Iosif Bena.

protons and neutrons) of the universe after inflation has ended. Thus inflation can survive even if grand unified theories are found to be wrong; even if grand unified theories are right, inflation may have occurred at energies so low that grand unified theories were not relevant.

Second, the abundance of inflationary models indicates emphatically that inflation is not a finished story. In my opinion the basic idea of inflation is almost certainly right, but the details of how inflation worked remain to be discovered. The inflationary paradigm leads to a tight connection between early universe physics and the high energy frontier of particle physics, so we can expect that the two fields will help to fertilize each other for some time to come.

While theorists have been sifting through the myriad variations of inflation, observers have continued the important task of pinning down the properties of the real universe. One particularly interesting—and troublesome—aspect of the recent improvements in cosmological measurements is the evidence suggesting an "age crisis": By some estimates, the age of the universe as determined from its expansion rate is younger than the age of the oldest stars.

The age crisis has clouded cosmology once before. As was discussed in Chapter 3, big bang cosmologists were plagued by a very severe form of this problem when the expansion rate was first measured by Edwin Hubble in 1929. Hubble's original value for the expansion rate was so high that the predicted age of the universe was less than 2 billion years, in gross conflict with evidence that the stars and even the earth are at least twice that old. This historical age crisis persisted for a quarter of a century, but it evaporated completely in the 1950s when astronomers discovered that Hubble had overestimated the expansion rate by at least a factor of 5.

Now, however, as astronomers develop newer and more reliable methods to measure the expansion rate, and to estimate the age of the oldest stars, it appears that a milder form of the age crisis has been rekindled. The problem reached the front page of *The New York Times* on October 27, 1994, when a team of astronomers known as the Hubble Space Telescope Key Project on the Extragalactic Distance Scale, led by Wendy Freedman of the Carnegie Observatories, announced what was regarded by many astronomers as the most reliable determination ever of the cosmic expansion rate. Their value for the expansion rate, or *Hubble constant*, was 25 kilometers per second per million light-years. This was much lower than Hubble's original measurements of 150 to 170, but was nonetheless uncomfortably high.

To convert the expansion rate to an age, one needs to assume a value for the mass density of the universe. A high mass density would mean that the expansion is rapidly slowing due to gravitational attraction, and therefore the average expansion rate over the history of the universe was much larger than the present rate. A large average expansion rate, in turn, means that the universe could reach its present condition quickly, and so the predicted age is on the low side. Unfortunately, it is very difficult to determine the mass density of the universe by direct observation. The problem is that at least 90% of the mass in the universe is invisible, and is detected only by its gravitational pull on visible matter. Even the composition of this *dark matter* remains unknown. Avishai Dekel of the Hebrew University in Jerusalem, David Burstein of Arizona State University, and Simon White of the Max Planck Institute for Astrophysics in Munich recently summarized [3] the ongoing attempts to measure the mass density, concluding only that different methods disagree, with numbers that range from about 0.2 times the critical density up to about 1.1 times the critical density.

The numerical relation between the age of the universe and the Hubble constant is shown in Figure E.1. For the value found by the Hubble Key

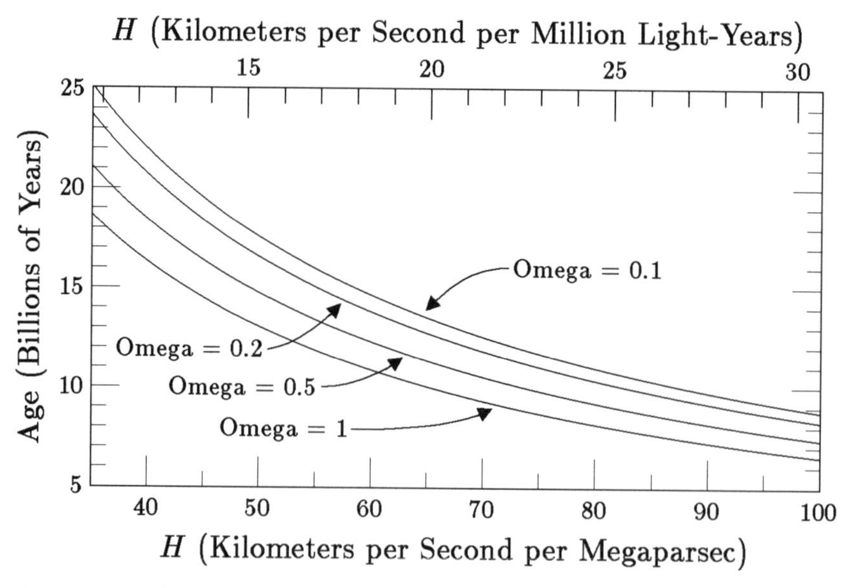

Figure E.1 **Relation between the age of the universe and the Hubble constant.** The Hubble constant is shown on the horizontal axis, with two different systems of units indicated at the top and bottom. The units at the bottom—kilometers per second per megaparsec—are the standard choice of astronomers. The relation is shown for four values of the mass density, indicated by the value of omega, the ratio of the average mass density to the critical density that would put the universe just on the border of eternal expansion and eventual collapse.

Project team, the age would be 11.0 billion years if the mass density is one tenth of the critical value (omega = 0.1), and only 8.1 billion years if the mass density is equal to the critical value (omega = 1).

How does this compare with the age of the oldest stars? The estimates of stellar ages have been rather consistent in recent years, with most researchers reaching conclusions very similar to those obtained in a recent study by Brian Chaboyer of the Canadian Institute for Theoretical Astrophysics, Peter Kernan and Lawrence Krauss of Case Western Reserve, and Pierre Demarque of Yale [4]. These authors concluded (with 95% confidence) that the age of the oldest stars is at least 12 billion years, in clear conflict with the age of the universe that is inferred from the Hubble Key Project measurement of the expansion rate.

Inflation plays an important role in this issue, because the simplest versions of inflation predict a critical density of matter, or omega = 1. This corresponds to the lowest curve shown in Figure E.1, aggravating the age crisis by forcing us to consider the most discrepant case. If the universe is really at least 12 billion years old, then the age calculation is consistent with a critical density of matter only if the Hubble constant is 17 kilometers per second per million light-years or less.

The age crisis is probably the most pressing unsolved problem in cosmology at this time. There are, however, several possible ways that the issue could conceivably be resolved in a manner that would be consistent with inflation. These are exciting times for cosmology, and it seems likely that within the next five years we will know the answer to this riddle.

One possible resolution would be for the actual value of the Hubble constant to turn out to be low. The Hubble constant is very difficult to measure, so everyone recognizes that there is considerable uncertainty in the reported values. When the Key Project team announced the value of 25, they estimated the uncertainty as plus or minus 5. The meaning of an uncertainty estimate is often difficult to interpret, but at least in this case the authors—although not the newspaper accounts—were very clear. They were talking about a *one-standard-deviation uncertainty*, which is very different from an estimate of the maximum possible error. By definition, a one-standard-deviation uncertainty is chosen so that the probability of the true result lying within the uncertainty is two out of three. That is, the Key Project results indicate two chances out of three that the true value of the Hubble constant lies between 20 and 30, but one chance out of three that it is either less than 20, or more than 30. Keeping this in mind, the possibility that the true value is less than 17 can hardly be ignored. Furthermore, while some other estimates have supported values around 25, many measurements have given lower values as well. The Key Project team [5] has in fact reduced its estimate to 22 (with a one-standard-deviation uncertainty of 3), so they have moved almost halfway toward the inflation-preferred value of 17. Meanwhile, Allan Sandage (also of the Carnegie Observatories) and Gustav Tammann of the University of Basel [6] have continued to find a value for the Hubble constant of about 17 plus or minus 3, well within the range preferred by the simplest inflationary models. And, to add spice to the field, a group of twenty-seven authors [7], many of whom had previously advocated a high value of the Hubble constant, recently used a new method and obtained a value of 13! The estimated observational uncertainty was only plus or minus 2, but there are also uncertainties in the method which are hard to estimate. Fortunately, the uncertainties in the method are expected to be significantly improved by Hubble Space Telescope observations in the near future.

A second conceivable resolution of the age crisis would be a revision in the estimates of the ages of the oldest stars. While the age estimates have been more stable over time than the Hubble constant measurements, one cannot consider them immune to revision. In fact, just as I was writing this section I picked up the current issue of *Science*, and noticed an article [8] about recent work by Allen Sweigart of NASA's Goddard Space Flight Center. Sweigart studied the effect of the mixing of material between the core of a star and its surface during the red giant phase of evolution. This effect has been well documented by observers, who have found that carbon, oxygen, and aluminum—products of the nuclear reactions at the center of a red giant—become increasingly abundant at the surface as the star becomes older. Yet the effect of this mixing has not been included in the standard calculations that have been used to estimate the age of the oldest stars. According to David Schramm of the University of Chicago, this work casts "a big question mark" over the age estimates. Sweigart has not calculated in detail the effect on stellar ages, but he noted that the ages could easily be reduced by as much as 1.5 billion years.

It would be simplest if the age crisis would simply disappear, as discussed in the previous two paragraphs, and I personally believe there is a reasonable chance that it will. However, it might not disappear. It is possible that further study will show the age crisis to be inescapable, and that we may be forced to modify our ideas about cosmology to accommodate it. If so, there are two possible ways that inflation could be modified to be consistent with an older universe.

One possible modification is called *open inflation*, proposed by Bharat Ratra and Jim Peebles* of Princeton University [9], and further developed by a number of other cosmologists. Open inflation is similar in many ways

* Peebles is the same cosmologist who played a major role in the early understanding of the cosmic background radiation, as discussed in Chapters 3 and 4.

to new inflation, except that one assumes that the amount of inflation that occurs as the inflaton field rolls down the hill in the energy density diagram is limited, and insufficient to drive the universe to flatness. Thus, in contrast to other inflationary theories which assume a large amount of inflation, open inflation can produce a universe with omega less than one. Our universe can then lie on one of the higher curves in Figure E.1, significantly alleviating the age crisis. Open inflation appears to be viable, but many cosmologists find it unattractive because we must assume that nature carefully arranged the energy density diagram of the inflation-driving field so that there is just enough inflation, but not too much. In conventional inflationary theories, by contrast, there is no such thing as too much inflation.

Another way to deal with the age crisis, within inflationary theory, is to resurrect the cosmological constant that Einstein had introduced and later discarded. The cosmological constant (if positive) produces a universal repulsion, which Einstein used originally to prevent his static theoretical universe from collapsing under the force of ordinary gravity. If our universe has a positive cosmological constant, the effect on the age question would be exactly the opposite of the effect of a large mass density. The cosmological constant would cause the expansion rate of the universe to increase, so its value now would be high compared to its value in the past. The relatively low value in the past would imply a longer time to reach the present state, so the universe would be older. This makes a bigger difference to the age than simply having a low mass density. Figure E.2 shows the age calculation for an inflationary universe with a cosmological constant.

The cosmological constant can handily solve the age problem, and it was in fact employed by Lemaître in the 1930s to cope with the original age crisis. Nonetheless, the cosmological constant has an inauspicious history—Einstein was wrong when he used it to explain a static universe, and Lemaître went astray when he used it to reconcile big bang cosmology with an erroneous value of the Hubble constant. Has the time come for it to play a proper role in physics, or is it once again merely a "fudge factor" that gives enough flexibility for invalid beliefs to persist?

From a modern particle physics point of view, there is yet one further suggestion that the cosmological constant is an unlikely solution to the age problem. While to Einstein the cosmological constant was an arbitrary term that he conjectured to belong to the equations of gravity, modern physicists have a clearer idea of what the cosmological constant really is: It is the energy density of the vacuum. Naively one might conclude that the cosmological constant must therefore be zero, since the vacuum is supposedly empty. The particle physicist, however, reaches the opposite conclusion. The vacuum is seen as an enormously complicated state, with nonzero values for Higgs fields, and with particle-antiparticle pairs which pop in and out of existence

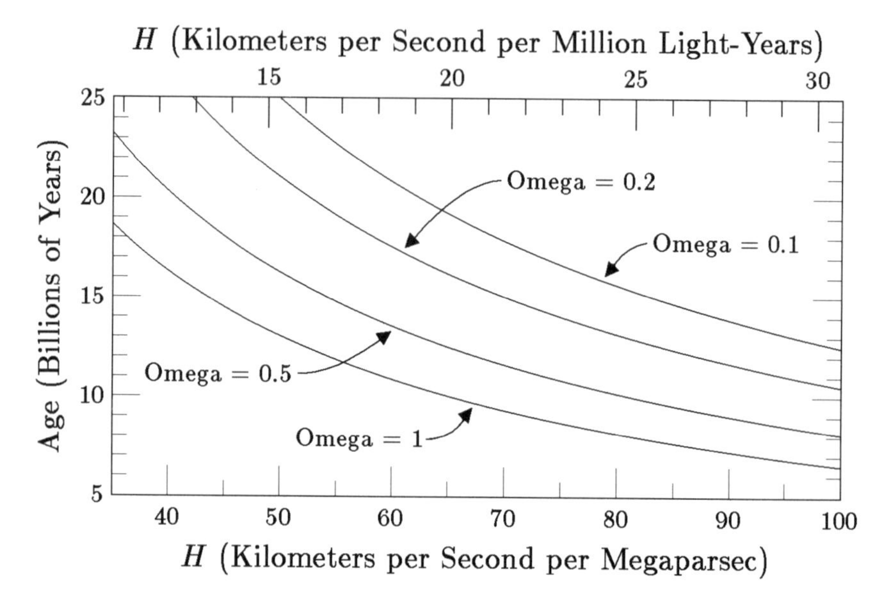

Figure E.2 **The age of an inflationary universe with a positive cosmological constant.** As in the previous diagram, the horizontal axis shows the Hubble constant, with different units indicated above and below. Curves are shown for four different values of omega. Each curve represents a geometrically flat universe, as expected from inflation, with a value of the cosmological constant that is fixed by the criterion of flatness.

at a fantastic rate. No one knows how to calculate the energy density of the vacuum, but when particle physicists estimate the energy associated with this tempest of activity, they come up with a number that is colossal: roughly 10^{120} times larger than the largest value consistent with observations. There are negative contributions to the energy density as well as positive ones, but no one knows why they should cancel. Something is happening that we do not understand, suppressing the cosmological constant by at least 120 orders of magnitude below our expectations. Our inability to understand this suppression is known as the *cosmological constant problem*, and is generally recognized as one of the outstanding problems [10] in particle theory. If the cosmological constant is to significantly affect the age of the universe, this mysterious suppression mechanism must by coincidence stop at almost exactly 120 orders of magnitude. If the cosmological constant were suppressed by 125 orders of magnitude, or 150 or 1000 orders of magnitude, then it would be too small to have any effect. Since there is no known reason why the suppression should be almost exactly 120 orders of magnitude, many particle physicists find it hard to believe that the cosmological constant is the right answer to the age riddle. This argument is not claimed to be compelling, but it gives us good reason to continue the search for alternative solutions.

At this point it is hard to say whether the age crisis will simply disappear with better measurements, or whether it will force us to accept the notion of open inflation, a cosmological constant, or some other modification of our theories. The present situation is confused, but cosmologists are very hopeful that we are on the threshold of settling all these questions.

There are a number of projects underway, but the brightest hopes for resolving the current uncertainties in cosmology are pinned on the new generation of space-based cosmic background radiation experiments—the successors to the spectacular Cosmic Background Explorer (COBE) mission, which was launched in 1989.

Two missions are currently being planned: a NASA project called MAP, under the leadership of the Goddard Space Flight Center and Princeton University, and a European Space Agency project called COBRAS/SAMBA.* MAP is scheduled for launch in about 2000, and COBRAS/SAMBA in about 2004. There are also at least a dozen balloon and ground-based experiments currently underway, which may shed significant light on these issues well before the new satellites begin to return data.

The technology available today for a space-based mission is vastly superior to the technology used in COBE, which was based on designs that were proposed in the mid-1970s. While COBE orbited at an altitude of 560 miles, each of the new satellites will observe the background radiation from an orbit about 900,000 miles from Earth, almost 4 times as far as the moon, well-protected from the Earth's microwave emissions and magnetic fields. While the COBE nonuniformity measurements used three different frequencies to help distinguish the cosmic background from local sources, MAP will use five frequencies, and COBRAS/SAMBA will use nine. The COBRAS/SAMBA experiment will be more than 10 times as sensitive as COBE. Most important of all is the gain in angular resolution—the sharpness of the image that the instruments will detect. The COBE images were very blurred, with an angular resolution of 7 degrees. The angular resolution of MAP, by contrast, will range from 0.3 to 0.9 degrees, depending on the frequency, and the angular resolution of COBRAS/SAMBA will range from .08 to 0.5 degrees.

With the improved technology, the expectations for the scientific payoff are staggering. Because of its blurry vision, COBE could not see any of the

* MAP is an acronym for *Microwave Anisotropy Probe*, while COBRAS/SAMBA is a double acronym for two proposals which were merged: *Cosmic Background Radiation Anisotropy Satellite* and *Satellite for Measurement of Background Anisotropies. Late news: COBRAS/SAMBA has been renamed PLANCK.*

fine detail in the cosmic background radiation temperature pattern. The broad patterns that COBE could see, however, gave us an exciting glimpse of the nonuniformities in the universe at very early times, confirming that they were in agreement with the predictions of inflation. Nonetheless, with the large degree of blurring, they were also in agreement with the predictions of theories in which the nonuniformities are generated by objects called *cosmic strings*, which could have formed in an early universe phase transition. The upcoming satellite experiments, however, are expected to be able to distinguish between these two possibilities.

Perhaps even more important, the small-scale details that will be visible in the new experiments are strongly affected by the evolution of the universe, and consequently carry the fingerprints of cosmic history. The observation of these nonuniformities, therefore, is expected to give a very precise determination of the Hubble constant, omega, and the cosmological constant. Assuming that the basic framework of our cosmological theories is correct, the MAP experiment should determine each of these parameters to better than 5%. The COBRAS/SAMBA experiment is expected to pin down these parameters with an astounding accuracy of 1 or 2%. If, on the other hand, the basic framework of our cosmological theories turns out not to be correct (which I think is very unlikely), these experiments should make it clear. So, although the basic features of the universe are largely unknown today, the prospects for the future are exhilarating. Within the next five to ten years we expect to know how much dark matter there is, whether the universe has the critical mass density, whether there is a cosmological constant, and whether there really is an age crisis. We will probably also know whether the complex fabric of cosmic structure can be explained, in detail, as the result of random quantum processes that took place during the first fraction of a second in the history of the universe.

Coming back to the inflationary universe and the present time, I would like to add a few words about where we stand.

It is fair to say that inflation is not proven, but I believe that it is rapidly making its way from working hypothesis to accepted fact. The universe in which we live has all the expected earmarks of inflation. It contains a vast quantity of matter and energy; it is undergoing uniform expansion; the early universe is known to have been extraordinarily close to the critical density; the universe is homogeneous on very large scales; and the nonuniformities observed on smaller scales are compatible with the nearly scale-invariant spectrum of perturbations that is predicted by inflation.

To me, the most impressive piece of evidence for inflation is the flatness problem—the closeness of the mass density of the early universe to the critical value. The quantitative precision of the agreement here is extraordinary. It may seem a bit hypothetical to extrapolate the universe backward to one second after the initial big bang, but the extrapolation requires only the physics of gravity and low-energy nuclear physics. The calculations are very simple, and among my fellow scientists I have heard no real skepticism about their validity. The extrapolations are validated, moreover, by the calculations of big bang nucleosynthesis, which give very good agreement with the observed abundances of the light chemical elements. By combining these calculations with the observation that today omega is certainly somewhere between 0.1 and 2, we can pinpoint the mass density at one second after the big bang to an accuracy of 15 decimal places. Such astounding precision is possible because the evolution of the universe has amplified any deviation from the critical value that the mass density might have had. I contend that this determination is far more reliable—and of course much more precise—than any measurement of an intergalactic distance or stellar age. And the prediction of inflation is in perfect agreement with this (slightly indirect) observation. Although it is technically a retrodiction, in that the observation was made before the theory was formulated, the mechanism of inflation inescapably leads to a critical density provided that there is sufficient inflation.

Of course, I must admit that a critical mass density for the early universe is such a bland proposition that one *might* suspect that an alternative explanation could be found. If the mass density could be predicted to 15 digits and the digits had no simple pattern, then everyone would certainly agree that inflation is established beyond reasonable doubt. Nonetheless, it has been fifteen years since inflation was invented, and so far the hypothetical alternative explanation has remained elusive.

Thus, inflation is at present the unique theory which allows us to calculate, to 15 decimal places, the mass density of the universe at one second after the big bang. One way to appreciate the significance of this agreement is to think about what you could learn if you had the ability to measure distances and time intervals to this accuracy. We usually think of special relativity, for example, as a theory that describes the behavior of objects moving at high speeds, near the speed of light. With precision measurements, however, you could detect the very small effects of relativity at lower speeds. With 15-decimal-place accuracy, you could see the effects of relativity at speeds that you could reach while riding a bicycle! Whether the prediction of a critical mass density for the early universe is said to be "bland" or not, this kind of precision is extraordinarily impressive. There is a strong suggestion that some form of inflation is responsible for the universe that we inhabit.

APPENDIX A

GRAVITATIONAL ENERGY

Since the negative energy of a gravitational field is crucial to the notion of a zero-energy universe, it is a subject worth examining carefully. In this appendix I will explain how the properties of gravity can be used to show that the energy of a gravitational field is unambiguously negative. The argument will be described in the context of Newton's theory of gravity, although the same conclusion can be reached using Einstein's theory of general relativity.

A simple way to demonstrate the sign of gravitational energy is to imagine a thin spherical shell of mass, as shown in Figure A.1(a). The shell will create a gravitational field, which at each point in space provides a measure of the force that would be experienced by a mass if it were located at that point. The gravitational field can be calculated by using Newton's methods. Newton first considers an ideal point mass—a mass concentration that is so small that it can be treated as if all the mass were located at a single point in space. For this case the gravitational field points directly toward the mass, with a strength that is described by the inverse square law—that is, the strength decreases as the square of the distance from the point mass. For a more complicated object such as the shell of Figure A.1(a), the gravitational field is in principle determined by mentally dividing the object into an infinite number of point masses, each with a mass that is infinitely small. A schematic illustration of this division into point masses is shown in part (b) of the figure. For each point mass one uses the simple inverse square law, and then one has to add the infinite number of contributions to obtain the final answer. The technique for handling such

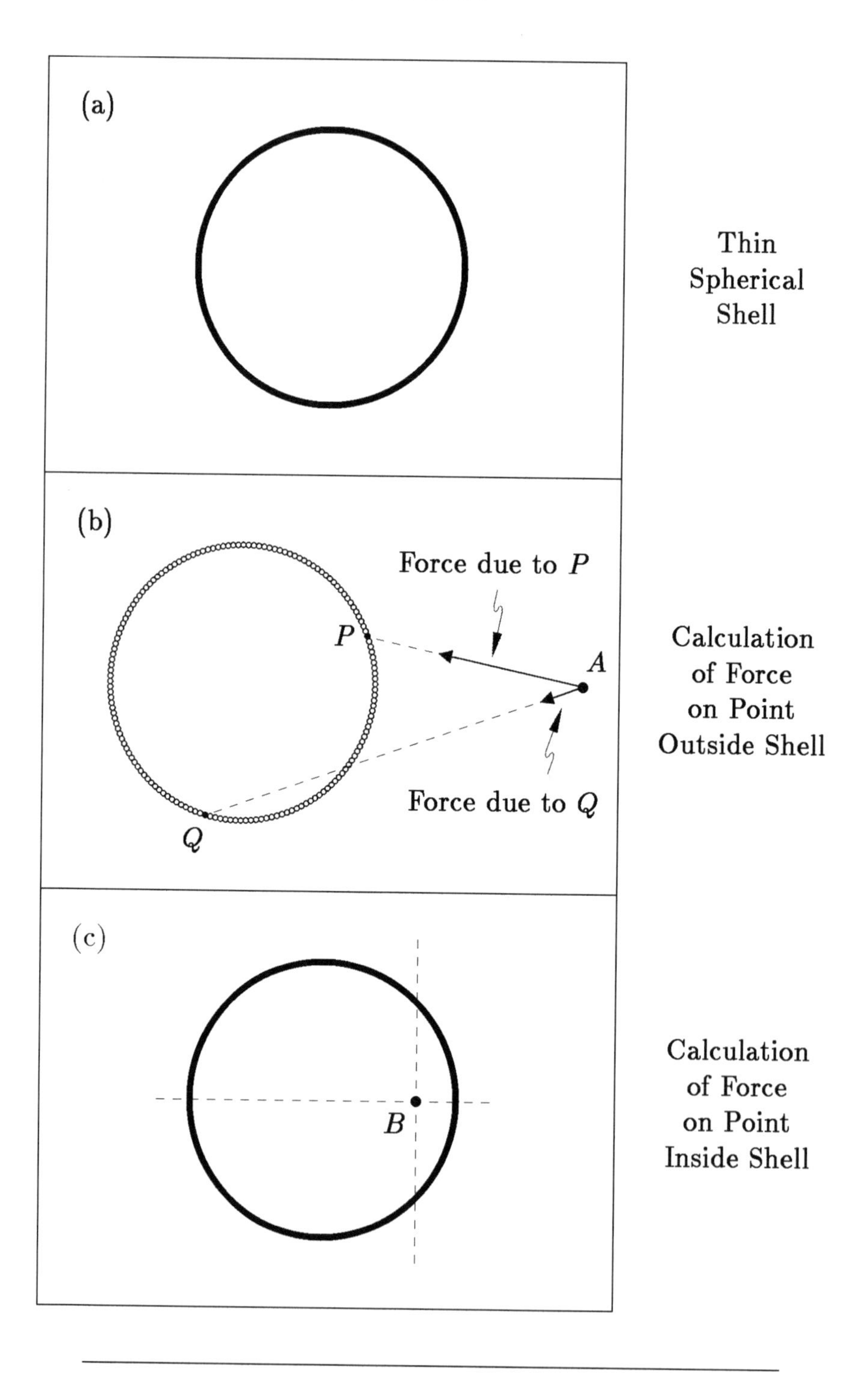
(a)
Thin Spherical Shell
(b)
Force due to P
P
A
Q
Force due to Q
Calculation of Force on Point Outside Shell
(c)
B
Calculation of Force on Point Inside Shell

an infinite number of quantities is the main subject of what is called integral calculus, which was largely developed by Newton himself.

The result for the gravitational field of a spherical shell of mass was first calculated by Newton, and it is the sort of calculation that is likely to show up in any college-level course in physics. Newton found that, outside the shell, the gravitational field at any point is directed radially inward toward the center of the shell. This answer could have been anticipated from the symmetry of the problem: there is no other reasonable answer to the question "What direction could it possibly point?" The strength of the gravitational field outside the shell can be described with surprising simplicity: the gravitational field has exactly the same strength as it would if all of the mass were concentrated at the center point of the shell.

What is the gravitational field inside the spherical shell? Consider, for example, the force on a particle at position *B*, as shown in part (c). By symmetry the force will be along the horizontal broken line, because the upward force caused by attraction toward the mass in the upper half of the diagram will be canceled by an opposing downward attraction toward the mass in the lower half of the diagram. We still must decide, however, if the force will point to the right or to the left. There is a persuasive argument which says that the force should be to the right: since the matter to the right is much closer than the matter to the left, the inverse square law should mean that the attraction to the matter on the right should dominate. There is also, however, a persuasive argument which says that the force should be to the left: There is more matter on the left, so the attraction toward it should dominate.

Which of the arguments in the previous paragraph is correct? Newton showed that these two arguments are equally valid, and in fact the forces cancel out exactly for a particle placed at any point inside the spherical cavity.

Figure A.1 **(facing page) The gravitational field of a thin spherical shell.** Part (a) shows the thin shell. The shell is really a three-dimensional sphere, but the diagram shows only a two-dimensional slice through the center of the sphere. In part (b) the shell has been divided into a large number of point masses, for the purpose of calculating the gravitational field at the point *A*. In principle there should be an infinite number of mass points, but only a finite number can be shown. The arrows indicate the gravitational attraction toward the mass points labeled *P* and *Q*. The attraction toward *P* is larger, because it is closer. Part (c) shows a point *B*, inside the spherical cavity, at which the gravitational field is to be calculated. The mass to the right of the vertical broken line is closer than the mass to the left, but there is more mass to the left.

To proceed with the discussion of gravitational energy, we need to answer one more question: How will gravity affect the spherical shell itself? Each point on the spherical shell will be attracted gravitationally toward each of the other points on the shell, and the net effect is a force pulling each point toward the center of the sphere. If the matter from which the shell is constructed is soft and compressible, then gravity will cause the shell to contract.

The situation is illustrated in Figure A.2, where part (a) shows the thin shell and the gravitational field that it generates. Outside the shell the gravitational field points inward, and inside the gravitational field is zero. Now imagine what would happen if the shell were allowed to uniformly contract, keeping its spherical shape. One can imagine, for example, extracting energy by tying ropes to each piece of the shell, as is illustrated in part (b). These ropes can be used to drive electric generators as each piece is lowered to its new position. Part (c) shows the sphere after the new radius is attained. The dashed circle indicates the original radius of the shell, and outside of the dashed circle the gravitational field is identical to that in part (a). (Recall that the field outside is the same as if all the mass were concentrated at the center, so it does not depend on the radius of the shell.) Inside the shell in its new position, the gravitational field remains zero. However, in the shaded region between the original and new positions of the shell, a gravitational field now exists where no field had existed before. The net effect of this operation is to *extract* energy, and to create a *new* region of gravitational field. Thus, energy is *released* when a gravitational field is created. The energy contained in the shaded region must therefore decrease, just as the water level in a tank decreases if water is released. Since the region began with no gravitational field and hence no energy, the final energy must be *negative*. In most physical processes the exchange of gravitational energy is much smaller than the rest energy (mc^2) of the particles involved, but cosmologically the total gravitational energy can be very significant.

***Figure** A.2* **(facing page) Thought experiment to understand the energy of gravity.** Part (a) shows a hollow spherical shell of mass, and the gravitational field lines that it produces. There is a force on each piece of the shell, pulling inward. Part (b) shows how energy can be extracted as the shell is allowed to uniformly contract. Each piece of the shell is tied by a rope to an electrical generator, producing power as the piece is "lowered" toward its final position. Part (c) shows the final configuration, which includes a gravitational field in the shaded region where no field existed before. Thus, the creation of the gravitational field is associated with the release of energy.

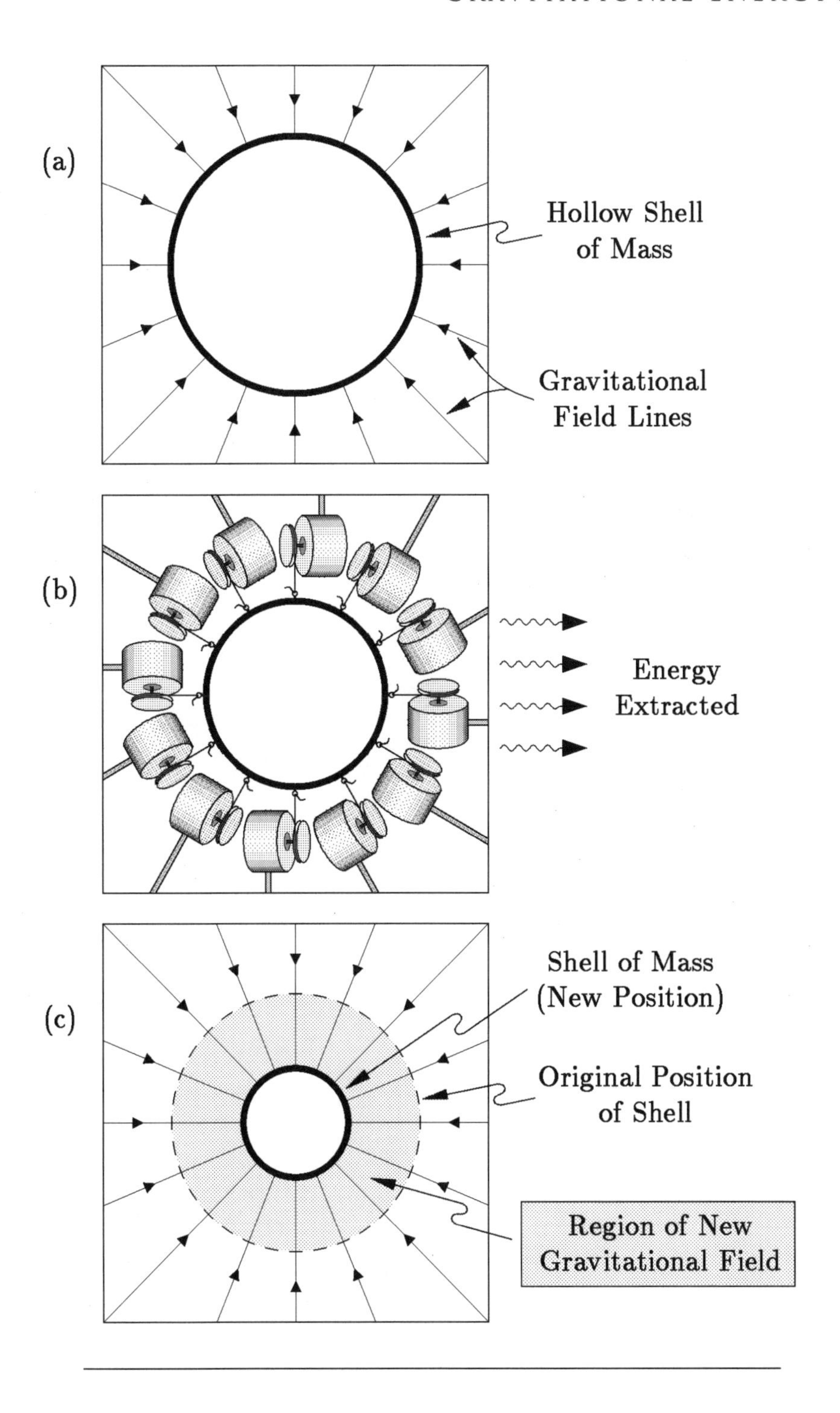
(a)
Hollow Shell of Mass
Gravitational Field Lines
(b)
Energy Extracted
(c)
Shell of Mass (New Position)
Original Position of Shell
Region of New Gravitational Field

APPENDIX B

NEWTON AND THE INFINITE STATIC UNIVERSE

The notion of a static universe is very appealing, and seems to have been the dominant belief until the work of Slipher and Hubble in the 1920s. There is a problem, however, because Newton's law of gravity implies that everything in the universe would attract everything else, so a static universe would not remain static. Newton had realized that any *finite* distribution of mass would collapse by gravitational attraction. He went on, however, to erroneously conclude that the collapse would be avoided if the universe were infinite and filled with matter throughout. In that case, Newton reasoned, there would be no center at which the mass might collect. Newton discussed this issue in a series of letters he wrote to the young theologian, Richard Bentley, during 1692–93 [1]:

> But if the matter was evenly disposed throughout an infinite space, it could never convene into one mass; but some of it would convene into one mass and some into another, so as to make an infinite number of great masses, scattered at great distances from one to another throughout all that infinite space. And thus might the sun and fixed stars be formed, supposing the matter were of a lucid nature. (Cambridge, December 10, 1692)

(The last phrase, by the way, refers to Newton's belief that the sun and stars are composed of "lucid" matter, distinct in kind from the matter found on Earth.)

Newton took this argument so seriously that he completely revised his view of the cosmos, abandoning his previous belief that the fixed stars occupy a finite region in an infinite void [2].

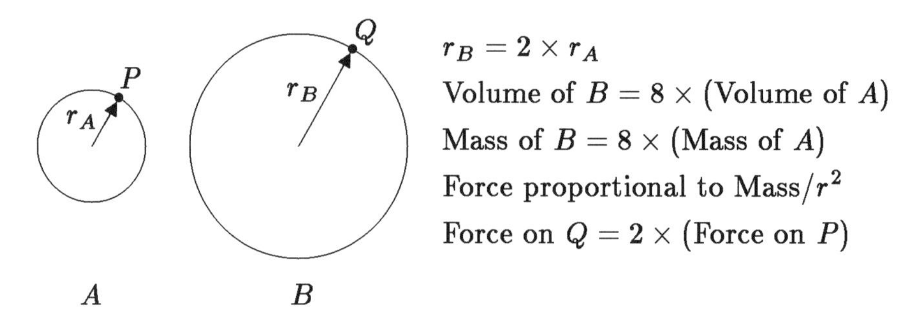

Figure B.1 **The gravitational collapse of two spheres.** Both spheres have the same density, but sphere *B* has twice the radius of sphere *A*. To compare the amount of time it will take each sphere to collapse, we can compare the gravitational accelerations of the points *P* and *Q* on the surfaces of the two spheres, as shown. Since the volume of a sphere is proportional to the cube of its radius, sphere *B* has 8 times the volume of sphere *A*, and therefore 8 times the mass. On the other hand, the point *Q* is twice as far away from the center of its sphere as is the point *P*. Since the force of gravitation falls off as the square of the distance, the extra distance of the point *Q* implies that the force is weakened by a factor of 4. Combining the effects of the extra mass and the extra distance, the force on *Q* is found to be twice as great as the force on *P*. It follows that the acceleration of *Q* will be twice as large as that of *P*, so at any given time the velocity of *Q* will be twice as large as that of *P*. Since the distance that *Q* must move while collapsing to the center is also twice as large, the conclusion is that both spheres will collapse in exactly the same amount of time!

The failure of Newton's reasoning is an illustration of how careful one has to be in thinking about infinity. From the modern viewpoint, an infinite distribution of matter under the influence of Newtonian gravity would unquestionably collapse. One way to correctly understand the situation is to imagine approaching the infinite distribution of matter by considering a succession of finite spheres with larger and larger radii. Suppose that two spheres of mass, *A* and *B*, have the same density of matter, but sphere *B* has twice the radius of sphere *A*. Suppose further that each sphere consists of a distribution of particles, such as stars, that are very small compared to the distances between them. Since the stars will not start to press against each other when the spheres begin to contract, there will be no pressure forces to resist the contraction. It can then be shown that gravity will cause both spheres to collapse in exactly the same amount of time! We can imagine doubling and redoubling the size of the sphere as many times as we like, but the time required for the collapse will not change. Since an infinite distribution of matter can be defined as the limit of a sphere when the radius is increased indefinitely, it follows that the infinite distribution of matter will collapse in the same length of time as any of the finite spheres.

For the benefit of mathematically inclined readers, the mathematical argument relating the two spheres *A* and *B* is shown in Figure B.1. The argument may seem a bit complex, but remember: If you get through it, you will have grasped a profound feature of gravity that eluded Isaac Newton himself!

The discussion here has not yet addressed the question raised directly in Newton's letter: If the matter is spread evenly throughout an infinite space, how would it choose a center about which to collapse? The answer is that, contrary to intuition, the matter can contract uniformly without choosing a center. As in the case of uniform expansion discussed in Chapter 3, each observer will see himself as the center. No matter where the observer might be in the infinite space, he would see all the rest of the matter in the universe collapsing towards him.

Such thoughtless 'passing to limit' is surprising in so sophisticated an author. An infinitely thick spherical shell defines a constant but infinite potential in its interior.

APPENDIX C

BLACKBODY RADIATION

The concept of blackbody radiation is very important to physicists, since it lies at the heart of an exciting chapter in the transition to quantum physics that took place at the beginning of the twentieth century. This transition was so important that today physicists use the word "classical" not to refer to ancient Greece or Rome, but instead to the physics that was known before this transition. Classical physics includes Newton's mechanics and even Einstein's theories of special and general relativity.

Prior to its application in physics, the word "quantum" was used to mean "share," or "portion," such as a "quantum of evidence." The name "quantum physics" originated from the discovery that many physical quantities, such as the energy of an atom, cannot take on any conceivable value, as had previously been thought. Instead the energy of an atom is *quantized*, meaning that the only possible values come from a list that is determined by the laws of physics for that particular type of atom. If a certain value is not on the list of energy levels for oxygen, then it is simply not possible for an oxygen atom to have that energy. As we will see below, this crucial notion of quantization began with the study of blackbody radiation.

To understand qualitatively why electromagnetic radiation in thermal equilibrium should have a spectrum like Figure 4.1 (on page 65), we need to understand a few facts about thermal equilibrium, and a few facts about electromagnetic radiation.

Starting with thermal equilibrium, it is well known today that temperature is really just a measure of the random motion of atoms and molecules. To make this notion quantitative, one would like to know how to relate the average energy of these random motions to the temperature.

For simplicity, I will begin with an approximate description. The average energy per atom of random motion can be estimated by multiplying the temperature in degrees Kelvin by a number called the Boltzmann constant,* which has the value 8.617×10^{-5} electron volts per degree Kelvin. For example, at 100°K a typical atom will be found jiggling through space with an energy of about 8.617×10^{-3} electron volts ($100 \times 8.617 \times 10^{-5}$). (An electron volt is a unit of energy equal to the energy that is released when a single electron passes through an ideal one-volt battery. It is equivalent to 1.602×10^{-12} ergs, and conversions to other units of energy are given in Appendix D.) The Boltzmann constant is usually abbreviated by the letter k,† and the temperature is denoted by the letter T. The average energy of random motion is then approximately the product of these two, denoted by kT.

While the energy of any one atom in random motion is unpredictable, the average energy of an atom in a large collection of atoms can be measured or calculated with precision. The precise value of the average energy of random motion will depend on the details of the system being described, but it will always be comparable to the product kT. Under a wide range of circumstances, the rules of thermal physics can be summarized by saying that each *degree of freedom*—each elementary type of motion—will have an average kinetic energy of $\frac{1}{2}kT$. For example, in a low-density gas of helium or argon (for which each molecule consists of a single atom), each atom will have an average kinetic energy of $\frac{3}{2}kT$, corresponding to one degree of freedom for motion in each of the three dimensions of space. Molecules with more than one atom, on the other hand, will have additional degrees of freedom associated with rotations.

Now consider again the electromagnetic waves inside a closed box at a uniform temperature. The electromagnetic waves will reach an equilibrium pattern of motion, just like the jostling atoms of a gas, and again the general rule will be an average energy of $\frac{1}{2}kT$ per degree of freedom. The full theory of statistical mechanics would give us a precise set of rules for calculating the number of degrees of freedom, but that would be the subject of a different book. Here I will give only a qualitative description of the calcula-

* The constant is named for Ludwig Boltzmann, the nineteenth century Austrian physicist who was one of the founders of statistical mechanics. Boltzmann's suicide in 1906 is said to have been due at least in part to his depression over the skepticism with which his work was viewed during his lifetime.

† If you ask me why, I am reduced to quoting Tevye from *Fiddler on the Roof*: "Tradition!" And how did this tradition get started, you might ask. Well, I'll tell you. . . . I don't know. (Dennis Overbye, however, informs me that the letter k and the name "Boltzmann's constant" were both introduced by Max Planck, in his paper on the blackbody radiation formula. Overbye cited J.L. Heilbron, *The Dilemmas of an Upright Man: Max Planck as Spokesman for German Science*, University of California Press, Berkeley, 1986.)

tion. For low frequencies, one can see from Figure 4.1 that the energy density is small. The counting of degrees of freedom falls off sharply at low frequencies, simply because there are not many ways that the long-wavelength waves associated with these frequencies* can fit into a box. As one considers higher frequencies and shorter wavelengths, the number of ways that the wave can fit inside the box increases. The counting of degrees of freedom therefore implies that the energy density should increase with higher frequencies. A detailed calculation indicates that the energy density should be proportional to the square of the frequency.

Figure C.1 shows the result obtained by counting degrees of freedom, plotted as a dashed line. The graph also shows the actual spectrum for blackbody radiation at 10°K, exactly as it was shown in Figure 4.1. The dashed line is seen to agree with the actual spectrum, but only for low frequencies. At higher frequencies, the two curves completely disagree. The actual energy density reaches a peak and then falls off at higher frequencies, while the dashed curve simply grows indefinitely with frequency.

The earliest accurate measurements of the blackbody spectrum were made by Lummer and Pringsheim in 1899, who found behavior in agreement with the solid line of Figure C.1. (Their measurements were not at 10°K, however, but instead were made at a variety of temperatures, all above 1000°K.) The following year Lord Rayleigh (born John William Strutt) worked out the classical theory that was described in the previous two paragraphs, leading to the dashed line of the figure. It was immediately obvious that something was wrong with Rayleigh's calculation. Not only did the predicted spectrum disagree with the experimental results of Lummer and Pringsheim, but the prediction was patently absurd. If the equilibrium energy density really grew without limit for high frequencies, it would mean that an infinite amount of energy would be necessary for thermal equilibrium to be attained. Since an infinite energy is never available, thermal equilibrium could never be reached—any energy in the box would shift endlessly to higher and higher frequencies, so that the energy remaining at any given frequency would get smaller and smaller. We would all freeze to death as the thermal energy in our houses was siphoned into radiation in the infrared,

* The frequency and wavelength of a light are connected by the fact that the wavelength is equal to the speed of light divided by the frequency. If this is not familiar, consider for example a frequency of 10 cycles per second. The time for each cycle is then one-tenth of a second, or in general one divided by the frequency. The source of light waves can be thought of as a pulse generator, generating one wave crest for each cycle. When the second wave crest is generated, the first wave crest has been moving for one cycle. The distance it has moved is then the speed of light multiplied by the time of one cycle, which is the speed of light divided by the frequency. The distance between these crests is precisely what is meant by the wavelength. (For further clarification, look again at Figure 3.5 on page 48.)

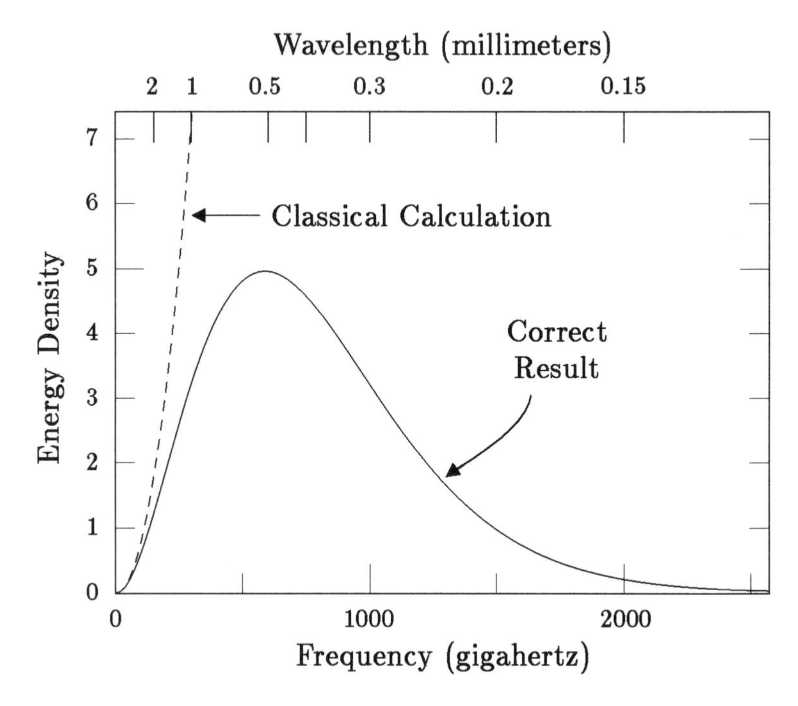

Figure C.1 **The spectrum of blackbody radiation: classical calculation versus the correct result.** Classical physics gives a nonsensical result for the equilibrium spectrum of blackbody radiation, as shown by the dashed line. The classically predicted energy density grows indefinitely with frequency, implying that an equilibrium could never be reached. The solid line shows the result of a quantum mechanical calculation, which is also known to agree with experiment. Both curves are shown for a temperature of 10°K, and the axes are labeled in the same way as Figure 4.1.

then visible, then ultraviolet, then X ray, and then gamma ray regions of the electromagnetic spectrum.

Yet it was clear that Rayleigh had not made a trivial mistake. He had correctly found the predictions of classical physics. Fortunately, however, we do not live in the frozen wasteland of a classical universe; we live instead in a warmer, happier, and much more hospitable quantum mechanical universe. In the quantum mechanical universe, we are somehow protected from this draining of energy into high frequency radiation.

The paradox discovered by Rayleigh, and later emphasized by Sir James Hopwood Jeans, became known as the *ultraviolet catastrophe*. It was the first clearly recognized failure of classical physics, and its resolution came with incredible swiftness. It is sometimes said that quantum theory was born on December 14, 1900, when Max Planck delivered a lecture to the *Physikalische Gesellschaft* (the German physical society) explaining his derivation of the correct formula for the blackbody spectrum.* Four years later Albert Einstein proposed the existence of *photons*, an understanding which leads to a much simpler derivation of Planck's formula.

Einstein's proposal, which has become part of modern quantum theory, holds that the energy of an electromagnetic wave is concentrated in bundles called photons. The energy of each photon is equal to the frequency of the wave multiplied by a universal number called Planck's constant. The frequency is typically measured in cycles per second (*hertz*), and the value of Planck's constant is found to be 4.136×10^{-15} electron volts per hertz. The energy that can be present at any particular frequency is then no longer an arbitrary number, but must instead be a multiple of the energy of a single photon. That is, a closed box might contain four photons or five photons, but it cannot contain four and a half photons.

For high frequencies the energy of a single photon can be very large compared to the typical thermal energy kT, and then the energy density of such photons is strongly suppressed. In terms of an analogy, it is as if you went to the bank to open a savings account, but discovered that the minimum deposit was much larger than the amount of money that you brought. You would be forced to go home, without having deposited anything. Thus, the average energy in high frequencies is cut off by the minimum-deposit effect, as is shown by the solid line in Figure C.1. For low frequencies, on the other hand, the energy of a single photon is very small compared to the typical thermal energy kT. Reasoning again by analogy, we can consider the task of measuring a pound of water. Since an individual molecule of water is so small, we can measure a pound of water to extraordinary precision without ever noticing that the mass of the water has to be a multiple of the mass of a molecule. Thus, the energy density behaves according to the classical calculation for small frequencies, growing as the square of the frequency. For high frequencies the energy density goes down, due to the quantum-mechanical minimum-deposit–effect. The maximum of the energy density occurs at the frequency for which the energy of a single photon is roughly equal to kT.†

* See, for example, B.L. van der Waerden, *Sources of Quantum Mechanics* (Dover Publications, Inc., New York, 1968).

† More precisely, it is calculated to be about $2.82kT$.

APPENDIX D

UNITS AND MEASURES

The SI (*Système International d'Unités*) system of units in principle provides a worldwide standard, but in practice even the scientific community has not yet fully adapted to the use of these standards. The purpose of this appendix is to provide the reader with the tools needed to translate from one system of units to another, so that he can make sense out of the quantities discussed in this book and in other sources.

Scientific Notation:

10^5 is read as "ten to the fifth power," and it means a 1 followed by 5 zeros, or 100,000. The superscript 5 is called the exponent. An exponent can also be negative, so 10^{-3} means a decimal point, 2 zeros, and then a 1, or .001. The SI system provides a lexicon of standard prefixes to denote various powers of ten, as shown in Table D.1. For example, the prefix "centi" means 10^{-2} and is abbreviated "c", so a centimeter is 0.01 meters and is abbreviated "cm."

Time:

The SI unit of time is the second, which is defined by an atomic clock. Specifically, a second is defined to be the duration of 9,192,631,770 cycles of the microwave radiation emitted when an atom of cesium undergoes a transition from its second lowest energy level to its lowest energy level. Many prefixed forms are in common use, including the picosecond, nanosecond, microsecond, and millisecond.

$$1 \text{ year} = 3.156 \times 10^7 \text{ second}$$

Length:

The SI unit of length is the meter, which is defined as the distance that light travels (in vacuum) in 1/299,792,458 second. Many prefixed forms are in common use: kilometer, centimeter, millimeter, micrometer (sometimes called a micron), nanometer, picometer, and femtometer (sometimes called a fermi, named for Enrico Fermi). Other units in common use for various purposes include:

1 inch = 2.540 cm

1 foot = 30.480 cm

1 mile = 5280 feet = 1.609 kilometer

1 nautical mile = 6080 feet

1 light-year = 9.461×10^{15} m

1 parsec = 1 pc = 3.262 light-year = 3.086×10^{16} m

1 megaparsec = 10^6 pc = 3.086×10^{22} m

1 Angstrom = 1 Å = 10^{-10} m

Volume:

1 liter = 1000 cm^3

1 U.S. gallon = 3.785 liter

1 Imperial gallon = 4.546 liter

Mass:

The SI unit of mass is the kilogram, which is defined as the mass of a particular cylinder of platinum-iridium alloy, kept at the International Bureau of Weights and Measures at Sèvres, near Paris. Other units include:

1 pound = 16 ounce = 453.6 gram

1 short ton = 2,000 pound

1 long ton = 2,240 pound

1 metric ton = 1,000 kilogram

1 solar mass = 1.99×10^{30} kilogram

1 atomic mass unit = 1 u = 1.661×10^{-27} kg

SI PREFIXES		
Multiple or Submultiple	Prefix	Symbol
10^{18}	exa	E
10^{15}	peta	P
10^{12}	tera	T
10^{9}	giga	G
10^{6}	mega	M
10^{3}	kilo	k
10^{2}	hecto	h
10	deca	da
10^{-1}	deci	d
10^{-2}	centi	c
10^{-3}	milli	m
10^{-6}	micro	mu
10^{-9}	nano	n
10^{-12}	pico	p
10^{-15}	femto	f
10^{-18}	atto	a

Table D.1 SI prefixes to denote powers of 10

Temperature:

Temperatures are commonly denoted by Fahrenheit (°F), Celsius (or centigrade) °C, or Kelvin (K) scales. According to official SI conventions, temperatures on the Kelvin scale are stated as "300 kelvins" and not "300 degrees Kelvin," and they are abbreviated as "300 K", and not "300°K". If T_C, T_K, and T_F denote the temperatures on the Celsius, Kelvin, and Fahrenheit scales, then the relationships are:

$$T_F = \frac{9}{5}T_C + 32° \qquad T_C = \frac{5}{9}(T_F - 32°)$$
$$T_K = T_C + 273.15$$

Force:

A dyne is the force necessary to accelerate a mass of 1 gram at an acceleration of 1 centimeter per second per second. A newton is the force necessary to accelerate 1 kilogram at an acceleration of 1 meter per second per second. Some useful relations are:

1 newton = 10^5 dyne

1 pound = 1 lb = 4.448 newton

Energy:

An erg is the energy necessary to exert a force of 1 dyne through a distance of 1 centimeter. It can also be described as twice the energy of a 1 gram mass moving at 1 centimeter per second. A joule is the energy required to exert a force of 1 newton through a distance of 1 meter, and it is twice the energy of a 1 kilogram mass moving at 1 meter per second. A foot-pound is the energy necessary to exert a force of 1 pound through a distance of 1 foot.

1 joule = 10^7 erg

1 foot-pound = 1.356 joule

1 calorie = 4.186 joule

1 British thermal unit = 1 BTU = 1055 joule

1 kilowatt-hour = 3.60×10^6 joule

1 electron volt = 1 eV = 1.602×10^{-19} joule

1 MeV = 10^6 eV

1 GeV = 10^9 eV

Power:

1 watt = 1 W = 1 joule/second

1 horsepower = 746 watt

1 BTU/hour = 0.293 watt

Frequency:

1 hertz = 1 cycle per second

Spin:

The spin of subatomic particles (and atoms as well) is often measured in units of $\hbar$, which is Planck's constant divided by 2π ($\pi = 3.14159265\ldots$). Numerically,

$$\hbar = 1.0546 \times 10^{-27} \text{ gram-cm}^2\text{/second}$$

That is, $\hbar$ represents the amount of spin of a 1 gram mass, moving in a circle of radius 1 cm, at a speed of 1.0546×10^{-27} cm/second. The spin of a subatomic particle is always either an integer multiple of $\hbar$ (usually 0, $\hbar$, or at most $2\hbar$), or a half-integral multiple (usually $\frac{1}{2}\hbar$ or $\frac{3}{2}\hbar$). The integer-spin particles are called *bosons*, while the half-integer spin particles are called *fermions*. Fermions obey the Pauli exclusion principle, which holds that no two identical particles can exist in the same state. There is no limit, however, to the number of identical bosons that can exist in the same state.

Hubble Constant:

Astronomers usually quote the value of the Hubble constant in the units of kilometers per second per megaparsec, while in this book I used kilometers per second per million light-years. The conversion is

$$H \text{ (in km per sec per megaparsec)}$$
$$= 3.262 \times H \text{ (in km per sec per million light-years)}$$

For purposes of thinking about the age of the universe, it is useful to measure $1/H$ in billions of years, for which the conversions are

$$\frac{1}{H} \text{ (in billions of years)}$$
$$= \frac{977.8}{H \text{ (in km per sec per magaparsec)}}$$
$$= \frac{299.8}{H \text{ (in km per sec per million light-years)}}$$

NOTES

Preface:

1. For the reader who might want to pursue these lines of research, I include a list of references:

   E.B. Gliner & I.G. Dymnikova, *Pis'ma v Astronomicheskii Zhurnal*, vol. 1, p. 7 (1975). English translation in *Soviet Astronomy Letters*, vol. 1, p. 93 (1975).

   A.A. Starobinsky, "A New Type of Isotropic Cosmological Models Without Singularity," *Physics Letters*, vol. 91B, pp. 99–102 (1980).

   Demosthenes Kazanas, "Dynamics of the Universe and Spontaneous Symmetry Breaking," *Astrophysical Journal*, vol. 241, pp. L59–L63 (1980).

   K. Sato, "First-order Phase Transition of a Vacuum and the Expansion of the Universe," *Monthly Notices of the Royal Astronomical Society*, vol. 195, p. 467 (1981).

   R. Brout, F. Englert, and E. Gunzig, "The Creation of the Universe as a Quantum Phenomenon," *Annals of Physics (NY)*, vol. 115, p. 78 (1978).

   V.F. Mukhanov and G.V. Chibisov, "Quantum Fluctuations and a Nonsingular Universe," *Pis'ma Zhurnal Eksperimentalnoi i Teoreticheskoi Fiziki*, vol. 33, pp. 549–553 (1981). English translation in *Journal of Experimental and Theoretical Physics (JETP) Letters*, vol. 33, pp. 532–535 (1981).

Chapter 1:

1. The gravitational force between two spherical bodies is equal to GM_1M_2/r^2, where G is Newton's constant, 6.6726×10^{-8} centimeter^3 per gram per second^2, M_1 and M_2 are the two masses, and r is the distance between them. The gravitational potential energy is equal to $-GM_1M_2/r$. If the masses are measured in

grams and the distance in centimeters, then these formulas give the force in dynes and the energy in ergs.

2. An excellent account of the life of Benjamin Thompson can be found in *Benjamin Thompson, Count Rumford,* by Sanborn C. Brown (The MIT Press, Cambridge, Massachusetts, 1979).
3. The most plausible method for the energy to escape was in the form of undetected gamma rays, which are actually light rays at a higher-than-visible frequency. C.D. Ellis and W.A. Wooster [*Proceedings of the Royal Society,* vol. A117, p. 109 (1927)] tested this hypothesis by measuring the total heat given off by a beta-radioactive material, in an experiment designed to trap any gamma rays that might be produced. They found, however, that the heat was in agreement with the known energy of the electrons, with no sign of hidden energy in any form.
4. Edward P. Tryon, "Is the Universe a Vacuum Fluctuation?" *Nature*, vol. 246, pp. 396–7 (1973).

Chapter 2:

1. P. Curie, "On the possible existence of magnetic conductivity and free magnetism," *Séances de la Société Française de Physique,* pp. 76–7 (1894). P.A.M. Dirac, "Quantised Singularities in the Electromagnetic Field," *Proceedings of the Royal Society of London,* Vol. A133, pp. 60–72 (1931). G. 't Hooft, "Magnetic Monopoles in Unified Gauge Theories," *Nuclear Physics,* Vol. B79, pp. 276–84 (1974). A.M. Polyakov, "Particle Spectrum in Quantum Field Theory," *JETP Letters,* Vol. 20, pp. 194–5 (1974).

Chapter 3:

1. English translation in H.A. Lorentz, A. Einstein, H. Minkowski, and H. Weyl, *The Principle of Relativity: A Collection of Original Memoirs on the Special and General Theory of Relativity.* Translated by W. Perrett and G.B. Jeffery (Dover Publications, New York, 1952, p. 184). Also in J. Bernstein and G. Feinberg, eds., *Cosmological Constants: Papers in Modern Cosmology* (Columbia University Press, New York, 1987, p. 22).
2. Arthur Eddington, "On the Instability of Einstein's Spheroidal World," *Monthly Notices of the Royal Astronomical Society,* vol. 90, pp. 668–78 (1930).
3. A. Friedmann, "On the Curvature of Space," *Zeitschrift für Physik*, vol. 10, pp. 377–86 (1922); "On the Possibility of a World with Constant Negative Curvature," *Zeitschrift für Physik*, vol. 21, pp. 326–32 (1924). Both articles are translated in *Cosmological Constants*, See Note 1.
4. A brief biography of Alexander Friedmann, by P. Ya. Polubarinova-Kochina, appears in the July 1963 issue of the Soviet physics journal *Uspekhi Fizicheskihk Nauk*, an issued devoted entirely to the commemoration of the seventy-fifth anniversary of Friedmann's birth. It is translated into English as *Soviet*

Physics Uspekhi, vol. 6, pp. 467–72 (1964). More recently a detailed biography has been written by Eduard A. Tropp, Viktor Ya. Frenkel, and Artur D. Chernin, *Alexander A. Friedmann: The Man Who Made the Universe Expand* (Cambridge University Press, Cambridge, 1993).

5. A. Einstein, "Remark on the Work of A. Friedmann (Friedmann 1922) 'On the Curvature of Space,'" *Zeitschrift für Physik*, vol. 11, p. 326 (1922). Translated in *Cosmological Constants*, See Note 1.

6. A. Einstein, "A Note on the Work of A. Friedmann 'On the Curvature of Space,'" *Zeitschrift für Physik*, vol. 16, p. 228 (1923). Translated in *Cosmological Constants*, See Note 1.

7. *The Collected Papers of Albert Einstein,* unpublished document I-026. Cited with the permission of the Hebrew University of Jerusalem by Alan Lightman and Roberta Brawer, *Origins: The Lives and Worlds of Modern Cosmologists* (Harvard University Press, Cambridge, Massachusetts, 1990, p. 499).

8. Further information about the cosmic distance ladder can be found in Chapter 3 of *The Big Bang,* by Joseph Silk (W.H. Freeman and Company, New York, 1989), or in *The Cosmological Distance Ladder: Distance and Time in the Universe,* by Michael Rowan-Robinson (W.H. Freeman And Company, New York, 1985).

9. Edwin Hubble, "A Relation Between Distance and Radial Velocity Among Extra-galactic Nebulae," *Proceedings of the National Academy of Science,* vol. 15, pp. 168–73 (1929).

10. Jerome Kristian, Allan Sandage, and James A. Westphal, "The Extension of the Hubble Diagram, II," *Astrophysical Journal*, vol. 221, pp. 383–94 (1978).

11. *The Measure of the Universe: A History of Modern Cosmology,* by J. D. North (Oxford University Press, Oxford, 1965; reprinted by Dover Publications, New York, 1990, pp. 228–9).

12. The model in Lemaître's 1927 paper is closely related to the original static model of Einstein, with a cosmological constant. Recall that the Einstein model was unstable—if the universe were given a small nudge outward, it would start to expand without limit. Lemaître's model was precisely the static Einstein model, modified by a small nudge. At very early times the model is almost indistinguishable from the static Einstein model, and so it has an infinitely long history and no big bang. The small nudge, however, causes the universe to start expanding, with a Hubble constant that rapidly increases and then levels off at a value determined by the cosmological constant. While this model avoided a true big bang, since it did not start from zero size, later Lemaître learned to use the cosmological constant to construct big bang models with large ages.

13. Georges Lemaître, "A Homogeneous Universe of Constant Mass and Increasing Radius Accounting for the Radial Velocity of Extra-galactic Nebulae," *Monthly Notices of the Royal Astronomical Society,* vol. 91, pp. 483–90 (1931).

14. *New Pathways in Science,* by Sir Arthur Eddington (The MacMillan Company, New York, 1935, p. 211).

15. Duncan Aikman, "Lemaître Follows Two Paths to Truth," *The New York Times Magazine,* February 19, 1933, p. 3.

Chapter 4:

1. An interview with Peebles appears in Alan Lightman and Roberta Brawer, *Origins: The Lives and Worlds of Modern Cosmologists* (Harvard University Press, Cambridge, Massachusetts, 1990, pp. 214–31). See also Dennis Overbye, *Lonely Hearts of the Cosmos: The Story of the Scientific Quest for the Secret of the Universe* (HarperCollins, New York, 1991, Chapter 7).
2. Peebles' submission was titled "Cosmology, Cosmic Black Body Radiation, and the Cosmic Helium Abundance." The original version referred to a review article on the earlier work (R.A. Alpher and R.C. Herman, *Annual Reviews of Nuclear Science,* vol. 2, p. 1 (1953)), but mentioned nothing in the text concerning its significance. On revision Peebles added three more footnotes: G. Gamow, *Reviews of Modern Physics,* vol. 21, p. 367 (1949); C. Hayashi, *Progress in Theoretical Physics,* vol. 5, p. 224 (1950); and R.A. Alpher, J.W. Follin, and R.C. Herman, *Physical Review,* vol. 92, p. 1347 (1953). He also added to the text the comment that the work had been "envisaged by Gamow and others." The footnotes seem to be adequate, but the referee apparently viewed as unacceptable the absence of any real description of the earlier contributions.
3. A.A. Penzias and R.W. Wilson, *Astrophysical Journal,* vol. 142, pp. 419–21 (1965).
4. R.H. Dicke, P.J.E. Peebles, P.G. Roll, and D.T. Wilkinson, *Astrophysical Journal,* vol. 142, pp. 414–19 (1965).
5. *The New York Times,* May 21, 1965.
6. P.G. Roll and David T. Wilkinson, "Cosmic Background Radiation at 3.2 cm—Support for Cosmic Black-Body Radiation," *Physical Review Letters,* vol. 16, pp. 405–7 (1966). This paper also mentions the recalibration of the original Penzias and Wilson measurement from 3.5°K to 3.1°K, which is otherwise unpublished.
7. A.A. Penzias and R.W. Wilson, "A Measurement of the Background Temperature at 1415 MHz," *Astrophysical Journal,* vol. 72, p. 315 (1967).
8. G.B. Field and J.L. Hitchcock, *Physical Review Letters,* vol. 16, p. 817 (1966). P. Thaddeus and J.F. Clauser, *Physical Review Letters,* vol. 16, p. 819 (1966). V.J. Bortolot, J.F. Clauser, and P. Thaddeus, *Physical Review Letters,* vol. 22, p. 307 (1969).
9. Dirk Muehlner and Rainer Weiss, "Balloon Measurements of the Far-Infrared Background Radiation," *Physical Review,* vol. D7, pp. 326–44 (1973).
10. D.P. Woody, J.C. Mather, N.S. Nishioka, and P.L. Richards, "Measurement of the Spectrum of the Submillimeter Cosmic Background," *Physical Review Letters,* vol. 34, pp. 1036–9 (1975).
11. H. Alfvén and O. Klein, "Matter and Antimatter Annihilation in Cosmology," *Arkiv för Fysik,* vol. 23, pp. 187–94 (1962).
12. T. Matsumoto, S. Hayakawa, H. Matsuo, H. Murakami, S. Sato, A.E. Lange, and P.L. Richards, "The Submillimeter Spectrum of the Cosmic Background Radiation," *Astrophysical Journal,* vol. 329, pp. 567–71 (1988).
13. At the time of the COBE proposal, John Mather was at the Goddard Institute for Space Studies in New York, but moved two years later to the NASA God-

dard Space Flight Center in Greenbelt, Maryland. The group led by Mather included Patrick Thaddeus (Mather's postdoctoral advisor in New York), Michael Hauser and Robert Silverberg (NASA Goddard), Rainer Weiss, Dirk Muehlner, and David Wilkinson. The second proposal, from the Lawrence Berkeley Laboratory of the University of California, was led by Luis Alvarez, and the third proposal was from the Jet Propulsion Laboratory of the California Institute of Technology, led by Samuel Gulkis.

14. The members of the original COBE study group, which came to be known as the COBE Science Working group, were Mather, Hauser, Weiss, Wilkinson, Gulkis, and George Smoot (of the Lawrence Berkeley Laboratory). A "Principal Investigator" was chosen to guide the development of each of the three instruments: Mather for the FIRAS, George Smoot for the DMR, and Michael Hauser for the DIRBE. The group decided to drop plans for a fourth instrument, which would have measured nonuniformities of the cosmic background radiation at frequencies above the spectral peak to complement the lower frequency measurements of the DMR. Oversight of the daily affairs of the entire project remained the responsibility of John Mather. Unlike his more senior colleagues with university positions, Mather had the full-time availability and the NASA affiliation that the job required. Rainer Weiss was chosen to be chairperson of the Science Working Group, which would eventually grow to nineteen members spread out across the nation.

15. The COBE Team: J.C. Mather, E.S. Cheng, R.E. Eplee, Jr., R.B. Isaacman, S.S. Meyer, R.A. Shafer, R. Weiss, E.L. Wright, C.L. Bennett, N.W. Bogess, E. Dwek, S. Gulkis, M.G. Hauser, M. Janssen, T. Kelsall, P.M. Lubin, S.H. Moseley, Jr., T.L. Murdock, R.F. Silverberg, G.F. Smoot, and D.T. Wilkinson, "A Preliminary Measurement of the Cosmic Microwave Background Spectrum by the Cosmic Background Explorer (COBE) Satellite," *Astrophysical Journal Letters,* vol. 354, pp. L37–40 (1990).

16. The COBE Team: J.C. Mather, E. S. Cheng, D.A. Cottingham, R.E. Eplee, Jr., D.J. Fixsen, T. Hewagama, R.B. Isaacman, K.A. Jensen, S.S. Meyer, P.D. Noerdlinger, S.M. Read, L.P. Rosen, R.A. Shafer, E.L. Wright, C.L. Bennett, N.W. Boggess, M.G. Hauser, T. Kelsall, S.H. Moseley, Jr., R.F. Silverberg, G.F. Smoot, R. Weiss, and D.T. Wilkinson, "Measurement of the Cosmic Microwave Background Spectrum by the COBE FIRAS," *Astrophysical Journal*, vol. 420, pp. 439–44 (1994).

Chapter 5:

1. The thermal equilibrium ratio of protons to neutrons is equal to $e^{(m_n - m_p)c^2/kT}$, where e = 2.71828 is the base of the natural logarithms, m_n and m_p are the masses of the neutron and proton, respectively, c is the speed of light, k is the Boltzmann constant, and T is the absolute temperature. Numerically $(m_n - m_p)$ c^2 = 1.293 million electron volts, and $k = 8.617 \times 10^{-5}$ electron volts per degree Kelvin. This formula assumes that 1) the other relevant particles—electrons, positrons, neutrinos, and antineutrinos—are in thermal equilibrium; 2) the density of neutrinos is equal to that of antineutrinos; and 3) the density of electrons is equal to that of positrons. (The third assumption was not quite true for the early universe, since electrical neutrality implies that the density of electrons

equaled the density of positrons plus that of protons. At this time, however, there was only one proton for every billion positrons, so the imbalance between electrons and positrons was negligibly small.) Although the equilibrium ratio of protons to neutrons is determined by the rates of the reactions shown in Figure 5.2, a detailed knowledge of these reaction rates is not needed to calculate the answer. General principles of quantum theory guarantee that the rates of the reactions shown in Figure 5.2 are related to each other in such a way that the equilibrium ratio of protons to neutrons is given by the formula quoted above.

2. The story of Gamow's life is told in his unfinished autobiography, *My World Line* (Viking Press, New York, 1970). The work of Gamow and his group on nucleosynthesis is summarized very well in *Cosmology, Fusion & Other Matters: George Gamow Memorial Volume,* edited by Frederick Reines (Colorado Associated University Press, 1972). See especially the two articles by Alpher and Herman.

3. *Physics Today,* February, 1968.

4. G. Gamow, "Expanding Universe and the Origin of Elements," *Physical Review,* vol. 70, pp. 572–3 (1946).

5. G. Gamow, "The Origin of Elements and the Separation of Galaxies," *Physical Review,* vol. 74, pp. 505–7 (1948).

6. Ralph A. Alpher and Robert Herman, "Evolution of the Universe," *Nature,* vol. 162, pp. 774–5 (1948). Ralph A. Alpher and Robert C. Herman, "Remarks on the Evolution of the Expanding Universe," *Physical Review,* vol. 75, pp. 1089–95 (1949).

7. See, for example, Steven Weinberg, *The First Three Minutes: A Modern View of the Origin of the Universe* (Basic Books, New York, 1977, p. 130).

8. Ralph A. Alpher, James W. Follin, Jr., and Robert C. Herman, "Physical Conditions in the Initial Stages of the Expanding Universe," *Physical Review,* vol. 92, pp. 1347–61 (1953).

9. Chushiro Hayashi, "Proton-Neutron Concentration Ratio in the Expanding Universe at the Stages Preceding the Formation of the Elements," *Progress in Theoretical Physics,* vol. 5, pp. 224–35 (1950).

10. H.A. Bethe, "Energy Production In Stars," *Physical Review,* vol. 55, p. 434 (1939).

11. Martin Schwarzschild and Barbara Schwarzschild, "A Spectroscopic Comparison Between High- and Low-Velocity F Dwarfs," *Astrophysical Journal,* vol. 112, pp. 248–65, (1950).

12. Edwin E. Salpeter, "Nuclear Reactions in Stars Without Hydrogen," *Astrophysical Journal,* vol. 115, pp. 326–8 (1952).

13. E. Margaret Burbidge, G.R. Burbidge, William A. Fowler, and F. Hoyle, "Synthesis of the Elements in Stars," *Reviews of Modern Physics,* vol. 29, pp. 547–650 (1957).

14. Fred Hoyle and Roger J. Tayler, *Nature,* vol. 203, pp. 1108–10 (1964).

15. Private communication.

16. The original paper by Dicke, Peebles, Roll, and Wilkinson on the interpretation of the cosmic background radiation included citations to R.A. Alpher, H. Bethe,

and G. Gamow, "The Origin of Chemical Elements," *Physical Review*, vol. 73, pp. 803–4 (1948), and also to the papers by Alpher, Follin, and Herman and by Hoyle and Tayler cited above.

17. P.J.E. Peebles and David T. Wilkinson, "The Primeval Fireball," *Scientific American*, vol. 216, pp. 28–38 (June 1967).
18. Robert V. Wagoner, William A. Fowler, and F. Hoyle, "On the Synthesis of Elements at Very High Temperatures," *Astrophysical Journal*, vol. 148, pp. 3–49 (1967).
19. Most of the data for this graph was adapted from *The Early Universe*, by Edward W. Kolb and Michael S. Turner (Addison-Wesley, Redwood City, California, 1990, pp. 87–113). The helium data was adapted from Terry P. Walker, "BBN Predictions for ^{4}He," in *Proceedings of the 16th Texas Symposium on Relativistic Astrophysics and 3rd Symposium on Particles, Strings, and Cosmology*, December 1992, Berkeley, CA, C. Akerlof and M. Srednicki, eds. (Annals of the New York Academy of Sciences, 1993).

Chapter 6:

1. Ralph A. Alpher and Robert Herman, "On Nucleon-Antinucleon Symmetry in Cosmology," *Science*, vol. 128, p. 904 (1958).
2. J.H. Christenson, J.W. Cronin, V.L. Fitch, and R. Turlay, "Evidence for the 2π Decay of the K_2^0 Meson," *Physical Review Letters*, vol. 13, pp. 138–40 (1964).
3. Howard Georgi and S.L. Glashow, "Unity of All Elementary-Particle Forces," *Physical Review Letters*, vol. 32, pp. 438–41 (1974).
4. Steven Weinberg, "The Quantum Theory of Massless Particles," in *Lectures on Particles and Field Theory: 1964 Brandeis Summer Institute in Theoretical Physics, Volume* 2, S. Deser and K. Ford, eds. (Prentice-Hall, Englewood Cliffs, N.J., 1965, p. 482).
5. A.D. Sakharov, "Violation of CP Invariance, C Asymmetry, and Baryon Asymmetry of the Universe," translated in *Soviet Physics, JETP Letters*, vol. 5, pp. 32–5 (1967).
6. Motohiko Yoshimura, "Unified Gauge Theories and the Baryon Number of the Universe," *Physical Review Letters*, vol. 41, pp. 281–4 (1978), and erratum at vol. 42, p. 746 (1979).

Chapter 7:

1. Steven Weinberg, "A Model of Leptons," *Physical Review Letters*, vol. 19, pp. 1264–6 (1967).
2. Abdus Salam, "Weak and Electromagnetic Interactions," in *Elementary Particle Theory*, Proceedings of the 8th Nobel Symposium, N. Svartholm, ed. (Almqvist and Wiksell, Stockholm, 1968, pp. 367–77).
3. G. 't Hooft, "Renormalization of Massless Yang-Mills Fields," *Nuclear Physics*, vol. B33, pp. 173–99 (1971).

4. Sidney Coleman, "The 1979 Nobel Prize in Physics," *Science*, vol. 206, pp. 1290–2 (1979).
5. In 1972 't Hooft wrote a more detailed paper, in collaboration with his thesis advisor, Martinus (Tini) Veltman, introducing a powerful new method which they used to put the argument on a rigorous basis (G. 't Hooft and M. Veltman, "Regularization and Renormalization of Gauge Fields," *Nuclear Physics*, vol. B44, pp. 189–213 (1972)). In the same year, Benjamin W. Lee (of the Fermi National Accelerator Laboratory and the State University of New York at Stony Brook) and Jean Zinn-Justin (of the Center for Nuclear Studies at Saclay, France) wrote a series of four articles, totaling forty-eight journal pages, which used somewhat different methods to substantiate the same conclusion (B.W. Lee and J. Zinn-Justin, "Spontaneously Broken Gauge Symmetries," I: *Physical Review*, vol. D5, pp. 3121–37 (1972); II: vol. D5, pp. 3137–55; III: vol. D5, pp. 3155–60; IV: vol. D7, pp. 1049–56 (1973)).
6. Data taken from The Particle Data Group, "Review of Particle Properties," *Physical Review D*, vol. 50, no. 3, part I, August 1, 1994.
7. M. Gell-Mann, "A Schematic Model of Baryons and Mesons," *Physics Letters*, vol. 8, pp. 214–5 (1964).
8. G. Zweig, CERN Preprints TH 401 and 412 (1964) (unpublished).
9. C.N. Yang and R.L. Mills, "Conservation of Isotopic Spin and Isotopic Gauge Invariance," *Physical Review*, vol. 96, pp. 191–5 (1954). The original paper proposed a specific theory, but the generalization to a large class of "Yang-Mills" theories is straightforward. In the text I called these "gauge theories," but more precisely they should be called "non-abelian gauge theories." The full set of gauge theories is a bit larger, including also the electromagnetic theory of Maxwell. The adjective "non-abelian" indicates that the theory describes more than one kind of charge, and that the different kinds of charge mutually interact.
10. David J. Gross and Frank Wilczek, "Ultraviolet Behavior of Non-Abelian Gauge Theories," *Physical Review Letters*, vol. 30, pp. 1343–6 (1973). This paper was quickly followed by "Asymptotically Free Gauge Theories, I," *Physical Review*, vol. D8, pp. 3633–52 (1973).
11. H. David Politzer, "Reliable Perturbative Results for Strong Interactions?" *Physical Review Letters*, vol. 30, pp. 1346–9 (1973).
12. The suggestion that quarks come in three colors had been made in 1964 by O.W. Greenberg (of the University of Maryland), "Spin and Unitary-Spin Independence in a Paraquark Model of Baryons," *Physical Review Letters*, vol. 13, pp. 598–602 (1964). The original motivation was to reconcile the quark theory with the Pauli exclusion principle of quantum theory, which states that no two quarks of the same type can occupy the same state. There was evidence, however, that all three up quarks in a short-lived particle called the delta plus plus (Δ^{++}, see Figure 7.1) *are* in the same state, but the contradiction disappeared if the quarks were all of different types, distinguished by their color. In addition, by 1973 it was known that the lifetime of a particle called the neutral pion, and the rate at which high energy collisions of electrons and positrons produced strongly interacting particles, were predicted correctly by the quark theory *only* if quarks came in three colors. The idea that each flavor of quark comes in three

variations had been championed by Murray Gell-Mann, who invented the name "color" and later the name "quantum chromodynamics."

13. The reader who would like to know more should look at any of several *Scientific American* articles from recent years: "Quarks with Color and Flavor," by Sheldon Lee Glashow, October 1975; "Elementary Particles and Forces," by Chris Quigg, April 1985; "Gauge Theories of the Forces Between Elementary Particles," by Gerard 't Hooft, June 1980; "A Unified Theory of Elementary Particles and Forces," by Howard Georgi, April 1981. All of these articles are contained in two excellent reprint volumes, both edited by Richard A. Carrigan, Jr. and W. Peter Trower. The first two articles are in *Particles and Forces: At the Heart of the Matter* (Freeman, 1990), and the second two articles are in *Particle Physics in the Cosmos* (Freeman, 1989). For readers familiar with the mathematical definition of a matrix, I mention that $SU(3)$ refers to the group of 3×3 complex matrices that are *special*, meaning that they have determinant equal to one, and *unitary*, meaning that $M^\dagger \cdot M = I$, where I is the identity matrix, the dot indicates matrix multiplication, and $M^\dagger$ is the matrix obtained from M by interchanging rows and columns and taking the complex conjugate of each element. The interactions of the Yang-Mills theory are determined by this *symmetry group*.
14. In the language of quantum theory, the red/anti-red combination is *superimposed* with the blue/anti-blue and green/anti-green combinations. An observer would find either a red quark and its antiquark, or a blue quark and its antiquark, or a green quark and its antiquark.
15. Steven Weinberg, "Non-abelian Gauge Theories of the Strong Interactions," *Physical Review Letters*, vol. 31, pp. 494–7 (1973).
16. B.J. Bjorken and S.L. Glashow, *Physical Review Letters*, vol. 10, p. 531 (1963).
17. S.L. Glashow, J. Iliopoulos, and L. Maiani, "Weak Interactions with Lepton-Hadron Symmetry," *Physical Review*, vol. D2, pp. 1285–92 (1970).
18. Murray Gell-Mann, Marvin L. Goldberger, Norman M. Kroll, and Francis E. Low, "Amelioration of Divergence Difficulties in the Theory of Weak Interactions," *Physical Review*, vol. 179, pp. 1518–27 (1969).
19. Steven Weinberg, *Dreams of a Final Theory: The Search for the Fundamental Laws of Nature* (Pantheon Books, New York, 1992, p. 90).
20. Sheldon Lee Glashow, "Towards a Unified Theory: Threads in a Tapestry," *Reviews of Modern Physics*, vol. 52, pp. 539–43 (1980).

Chapter 8:

1. Howard Georgi and S.L. Glashow, "Unity of all Elementary-Particle Forces," *Physical Review Letters*, vol. 32, pp. 438–41 (1974).
2. H. Georgi, H.R. Quinn, and S. Weinberg, "Hierarchy of Interactions in Unified Gauge Theories," *Physical Review Letters*, vol. 33, pp. 451–4 (1974).
3. Paul Langacker and Nir Polonsky, "Uncertainties in Coupling Constant Unification," *Physical Review*, vol. D47, pp. 4028–45 (1993). I thank the authors for sending me the data in numerical format.

Chapter 9:

1. Ya. B. Zeldovich and M. Yu. Khlopov, "On the Concentration of Relic Magnetic Monopoles in the Universe," *Physics Letters*, vol. 79B, pp. 239–41 (1978).
2. John P. Preskill, "Cosmological Production of Superheavy Magnetic Monopoles," *Physical Review Letters*, vol. 43, pp. 1365–8 (1979).
3. T.W.B. Kibble, "Topology of Cosmic Domains and Strings," *Journal of Physics*, vol. A9, pp. 1387–98 (1976).
4. Paul Langacker, "Grand Unified Theories and Proton Decay," *Physics Reports*, vol. 72C, p. 185 (1981).

Chapter 10:

1. Georges Lemaître, "Note on de Sitter's Universe," *Journal of Mathematics and Physics*, vol. 4, pp. 188–92 (1925).
2. Sidney Coleman, "The Fate of the False Vacuum, 1: Semiclassical Theory," *Physical Review D*, vol. 15, pp. 2929–36 (1977).
3. Alan H. Guth and S.-H.H. Tye, "Phase Transitions and Magnetic Monopole Production in the Very Early Universe," *Physical Review Letters*, vol. 44, pp. 631–4, errata at p. 963 (1980).
4. Martin B. Einhorn, D.L. Stein, and Doug Toussant, "Are Grand Unified Theories Compatible with Standard Cosmology?" *Physical Review D*, vol. 21, p. 3295 (1980).

Chapter 11:

1. Alan H. Guth, "The Inflationary Universe: A Possible Solution to the Horizon and Flatness Problems," *Physical Review D*, vol. 23, pp. 347–56 (1981).
2. Alan H. Guth and Erick J. Weinberg, "Could the Universe Have Recovered from a Slow First Order Phase Transition?" *Nuclear Physics*, vol. B212, pp. 321–64 (1983).
3. Stephen Hawking, Ian Moss, and John Stewart, "Bubble Collisions in the Very Early Universe," *Physical Review D*, vol. 26, pp. 2681–93 (1982).

Chapter 12:

1. This and subsequent Linde quotations in this section are taken from *Origins: The Lives and Worlds of Modern Cosmologists*, by Alan Lightman and Roberta Brawer (Harvard University Press, Cambridge, Massachusetts, 1990).
2. The version of the anecdote told here is based on an E-mail message that I received from Linde on November 29, 1995. It differs in several minor ways from the version that Linde told Alan Lightman in 1987, as published in *Origins*. In 1987 Linde described the phone conversation with Rubakov as taking place "in the summer, maybe even in the late spring of 1981," while in 1995 he

said "I made it in the late spring." In 1987 Linde said "After that [the phone conversation] the whole picture had crystallized," while in the E-mail message Linde said, "Before calling I already knew the scenario, but I wanted to check whether he had any objections since he studied related issues."

3. Andrei D. Linde, "A New Inflationary Universe Scenario: A Possible Solution of the Horizon, Flatness, Homogeneity, Isotropy, and Primordial Monopole Problems," *Physics Letters*, vol. 108B, pp. 389–92 (1982).
4. Andreas Albrecht and Paul J. Steinhardt, "Cosmology for Grand Unified Theories with Radiatively Induced Symmetry Breaking," *Physical Review Letters*, vol. 48, pp. 1220–3 (1982).

Chapter 13:

1. Gary W. Gibbons and Stephen W. Hawking, "Cosmological Event Horizons, Thermodynamics, and Particle Creation," *Physical Review D*, vol. 15, pp. 2738–51 (1977).
2. This consensus led to the publication of four separate articles: Stephen W. Hawking, "The Development of Irregularities in a Single Bubble Inflationary Universe," *Physics Letters*, vol. 115B, pp. 295–7 (1982); Alexei A. Starobinsky, "Dynamics of Phase Transition in the New Inflationary Universe Scenario and Generation of Perturbations," *Physics Letters*, vol. 117B, pp. 175–8 (1982); Alan H. Guth & So-Young Pi, "Fluctuations in the New Inflationary Universe," *Physical Review Letters*, vol. 49, pp. 1110–3 (1982); and James M. Bardeen, Paul J. Steinhardt, & Michael S. Turner, "Spontaneous Creation of Almost Scale-Free Density Perturbations in an Inflationary Universe," *Physical Review D*, vol. 28, pp. 679–93 (1983). The last of these is probably the most informative.
3. John Barrow and Michael Turner, "The Inflationary Universe—Birth, Death and Transfiguration," *Nature*, vol. 298, pp. 801–5 (August 26, 1982).

Chapter 14:

1. M.R. Krishnaswamy et al., "Fully Confined Events Indicative of Proton Decay in the Kolar Gold Fields Detector," *Physics Letters*, vol. 115B, pp. 349–58 (1982).
2. G. Blewitt et al., "Experimental Limits on the Free-Proton Lifetime for Two- and Three-Body Decay Modes," *Physical Review Letters*, vol. 55, pp. 2114–7 (1985).
3. This graph was later published as George F. Smoot et al., "Structure in the COBE Differential Microwave Radiometer First-Year Maps," *Astrophysical Journal Letters*, vol. 396, pp. L1–5 (1992).
4. E.L. Wright et al., "Interpretation of the Cosmic Microwave Background Radiation Anisotropy Detected by the COBE Differential Microwave Radiometer," *Astrophysical Journal Letters*, vol. 396, pp. L13–18 (1992).

Chapter 15:

1. Andrei Linde, "The Eternally Existing, Self-Reproducing Inflationary Universe," in the Proceedings of the Nobel Symposium on *Unification of Fundamental Interactions* (June 2–7, 1986, Marstrand, Sweden), edited by L. Brink, R. Marnelius, J.S. Nilsson, P. Salomonson, and B.-S. Skagerstam (World Scientific, Singapore, 1987); also published as *Physica Scripta*, vol. T15, p. 169 (1987).
2. The eternal nature of new inflation was first pointed out by Alexander Vilenkin, in "Birth of Inflationary Universes," *Physical Review*, vol. D27, pp. 2848–55 (1983). Andrei Linde, in "Eternal Chaotic Inflation," *Modern Physics Letters*, vol. A1, p. 81 (1986), extended the idea of eternal inflation to a wider class of inflationary theories, and in particular to the "chaotic inflationary theory" that is discussed in the Epilogue of this book.
3. Two important articles about this subject are the following: A.S. Goncharov, A.D. Linde, and V.F. Mukhanov, "The Global Structure of the Inflationary Universe," *International Journal of Modern Physics*, vol. A2, pp. 561–91 (1987); Mukunda Aryal and Alexander Vilenkin, "The Fractal Dimension of Inflationary Universe," *Physics Letters*, vol. 199B, pp. 351–7 (1987).
4. Arvind Borde and Alexander Vilenkin, "Eternal Inflation and the Initial Singularity," *Physical Review Letters*, vol. 72, pp. 3305–9 (1994).

Chapter 16:

1. Steven K. Blau, Eduardo I. Guendelman, and Alan H. Guth, "The Dynamics of False Vacuum Bubbles," *Physical Review D*, vol. 35, pp. 1747–66 (1987).
2. Edward H. Farhi and Alan H. Guth, "An Obstacle to Creating a Universe in the Laboratory," *Physics Letters*, vol. 183B, p. 149 (1987).
3. Roger Penrose, *Physical Review Letters*, vol. 14, p. 57 (1965).
4. Edward H. Farhi, Alan H. Guth, and Jemal Guven, "Is It Possible To Create a Universe in the Laboratory by Quantum Tunneling?" *Nuclear Physics*, vol. B339, pp. 417–90 (1990).
5. Willy Fischler, Daniel Morgan, and Joseph Polchinski, "Quantization of False Vacuum Bubbles: A Hamiltonian Treatment of Gravitational Tunneling," *Physical Review D*, vol. 42, pp. 4042–55 (1990).

Chapter 17:

1. Alexander Vilenkin, "Creation of Universes from Nothing," *Physics Letters*, vol. 117B, pp. 25–8 (1982).
2. James B. Hartle and Stephen W. Hawking, "Wave Function of the Universe," *Physical Review D*, vol. 28, pp. 2960–75 (1983).

Epilogue:

1. Andrei D. Linde, "Chaotic Inflation," *Physics Letters*, vol. 129B, pp. 177–181 (1983).
2. Daile La and Paul J. Steinhardt, "Extended Inflationary Cosmology," *Physical Review Letters*, vol. 62, pp. 376–378 (1989), and erratum at vol. 62, p. 1066 (1989).
3. Avishai Dekel, David Burstein, and Simon D.M. White, "Measuring Omega," to be published in *Critical Dialogues in Cosmology*, edited by Neil Turok (World Scientific). This paper is also available on the Worldwide Web from the Los Alamos e-print server, http://xxx.lanl.gov/, as astro-ph/9611108.
4. Brian Chaboyer, Peter J. Kernan, Lawrence M. Krauss, and Pierre Demarque, "A Lower Limit on the Age of the Universe," *Science*, vol. 271, p. 957 (1996). Available from the Los Alamos e-print server as astro-ph/9509115.
5. Wendy L. Freedman, "Determination of the Hubble Constant," to be published in *Critical Dialogues in Cosmology*, edited by Neil Turok (World Scientific). Available from the Los Alamos e-print server as astro-ph/9612024.
6. Allan Sandage and G.A. Tammann, "Evidence for the Long Distance Scale with $H_0 < 65$," to be published in *Critical Dialogues in Cosmology*, edited by Neil Turok (World Scientific). Available from the Los Alamos e-print server as astro-ph/9611170. Note that "$H_0 < 65$" refers to the astronomers' favorite units of kilometers per second per megaparsec. This translates into $H_0 < 20$ kilometers per second per million light-years.
7. Paul L. Schechter *et al.*, "The Quadruple Gravitational Lens PG1115+080: Time Delays and Models," *Astrophysical Journal Letters*, vol. 475, p. L85 (1997). Available from the Los Alamos e-print server as astro-ph/9611051, and also available in published format on the Worldwide Web through the NASA Astrophysical Data System, http://adswww.harvard.edu/.
8. James Glanz, "Could Stellar Ash Revise Cosmic Ages?" *Science*, vol. 274, p. 1837 (December 13, 1996). Sweigart's original article is published in *Astrophysical Journal Letters*, vol. 474, pp. L23–L26 (1997), and is available through the NASA Astrophysical Data System.
9. Bharat Ratra and P.J.E. Peebles, "Inflation in an Open Universe," *Physical Review*, vol. D52, pp. 1837–1894 (1995).
10. A plausible solution to the cosmological constant problem was proposed in 1988 by Sidney Coleman of Harvard, but it rests on the shaky foundation of our imperfect understanding of how to combine general relativity with quantum theory. Sidney Coleman, "Why there is Nothing rather than Something: A Theory of the Cosmological Constant," *Nuclear Physics*, vol. B310, pp. 643–668 (1988).

Appendix B:

1. The original letters are still kept at Trinity College, Cambridge, and are published in H. W. Turnbull, ed., *The Correspondence of Isaac Newton, Volume III, 1688–1694* (Cambridge University Press, Cambridge, England,

1961, p. 233). They are also reprinted in Milton K. Munitz, ed., *Theories of the Universe: From Babylonian Myth to Modern Science* (The Free Press, New York, 1957, p. 211).

2. Newton's involvement in this problem is discussed in a fascinating article by Edward Harrison, "Newton and the Infinite Universe," *Physics Today*, February 1986, p. 24.

GLOSSARY

Note: If a word within a definition is *italicized*, then it also appears in the glossary.

accelerator: See *particle accelerator.*

action-at-a-distance: A description of a force, such as Newton's law of gravity, in which two separated bodies are said to directly exert forces on each other. In the modern description, the bodies produce a *gravitational field,* which in turn exerts forces on the two bodies. See *gravitational field.*

aether: See *luminiferous ether.*

alpha decay: The disintegration of an atomic nucleus, in which the final products are an *alpha particle* and a nucleus with two fewer protons and two fewer neutrons than the original.

alpha particle: A particle consisting of two protons and two neutrons bound together. The nucleus of a helium atom is an alpha particle.

antibaryon: The *antiparticle* of a *baryon.*

antimatter: Matter that is not composed of protons, neutrons, and electrons, like the matter we see around us, but is instead composed of *antiprotons*, *antineutrons*, and *positrons* (the *antiparticles* of electrons).

antineutrino: The *antiparticle* of a *neutrino.*

antineutron: The *antiparticle* of a neutron. A neutron and antineutron both have the same mass and zero electric charge, but can be differentiated by their interactions: a neutron and an antineutron can annihilate into *gamma rays*, while two neutrons cannot.

antiparticle: For every known type of particle, there exists an antiparticle with exactly the same mass, but with the opposite electric charge. When a particle and its antiparticle come together, they can always annihilate to form *gamma rays.* The antiparticle of an electrically neutral particle is sometimes the same as the original particle (e.g., photons) and sometimes it is distinct (e.g., neutrons).

antiproton: The *antiparticle* of the proton.

antiquark: The *antiparticle* of the *quark.*

asymptotic freedom: A force of interaction between particles is said to be asymptotically free if it becomes weaker as the energy of the interacting particles increases. Empirically the force between *quarks* in a proton or neutron is found to be asymptotically free, a feature that can be explained by assuming that the force is described by a *Yang-Mills* theory.

baryogenesis: The hypothetical process by which the early universe acquired a large positive *baryon number.*

baryon number: The total number of *baryons* in a system, minus the number of *antibaryons*, is called the baryon number of the system. In particle physics experiments the total baryon number of a system is always found to be conserved, but our theories predict that baryon number would not be conserved at the extraordinarily high temperatures that prevailed in the early universe.

baryon: Protons and neutrons, which comprise most of the mass of ordinary matter, as well as a number of short-lived particles such as the lambda, sigma and delta, are all called baryons. These particles have in common the fact that they are each composed of three *quarks.*

beta decay: The disintegration of an atomic nucleus, in which an electron (which was historically called a beta particle) and an *antineutrino* are emitted. Since the electron carries away one unit of negative charge, the final nucleus has a charge one greater than the initial nucleus.

big crunch: If the universe has a mass density exceeding the *critical mass density,* then gravity will eventually reverse the expansion, causing the universe to recollapse into what is often called the big crunch. See also *closed universe.*

big-bang nucleosynthesis: The process, which took place between one second and 3-4 minutes after the beginning, in which the protons and neutrons of the primordial soup condensed to form the lightest atomic nuclei: *deuterium*, helium-3, helium-4, and lithium-7. See *isotope* and *lithium.*

blackbody radiation: If a closed box made of any material is heated to a uniform temperature, the interior will become filled with electromagnetic radiation (*photons*) with an intensity and spectrum determined by the temperature alone, independent of the composition of the box. Radiation with this intensity and spectrum is called blackbody, or thermal radiation. The intensity and spectrum are determined by the criterion of *thermal equilibrium*; i.e., only for this intensity and spectrum will the absorption and emission of photons by the walls be in balance for each wavelength, so the intensity for each wavelength can be independent of time. The universe today appears to be permeated with blackbody radiation at a temperature of 2.73°K—the *cosmic background radiation*—which we interpret as a remnant afterglow of the heat of the early universe.

black hole: An object with such a strong *gravitational field* that even light cannot escape. Matter can fall into a black hole, but according to classical physics no matter or energy can leave it. (Hawking has used quantum theory to show that black holes emit blackbody radiation, but the effect is significant only for black holes much smaller than those that are expected to form by the collapse of stars, which have masses of several solar masses or more.)

blueshift: If a star or galaxy is moving towards us, the radiation from the star or galaxy appears shifted towards shorter *wavelengths*, or towards the blue end of the spectrum. See *Doppler shift*.

bottom: A flavor of quark. See *flavor.*

bubble: The *false vacuum* decays in a manner similar to the way water boils, forming bubbles of normal matter in the midst of the false vacuum, just as bubbles of steam form in the midst of water heated past its boiling point. (The bubbles that form when the false vacuum decays should not be confused with *false vacuum bubbles*, which have false vacuum rather than normal matter on the inside.)

chaotic inflationary universe theory: A version of the *inflationary universe theory*, proposed by Andrei Linde in 1983, for which the energy density diagram for the fields driving inflation can be as simple as a bowl, with a unique minimum at the center. If the initial randomly chosen value of the fields corresponds to a point high up the hill on the side of the bowl, then sufficient inflation can occur as the fields roll towards the state of minimum energy density.

charmed: A flavor of quark. See *flavor.*

charmonium: A bound state consisting of a *charmed quark* and a charmed *antiquark*. A major impetus for the quark theory was the discovery in 1974 of the J/ψ particle, a particle whose properties closely matched the predictions for charmonium.

closed universe: A *homogeneous*, *isotropic* universe is said to be temporally closed if gravity is strong enough to eventually reverse the expansion, causing the universe to recollapse. It is said to be spatially closed if gravity is strong enough to curve the space back on itself, forming a finite volume with no boundary. Triangles would contain more than 180°, the circumference of a circle would be less than π times the diameter, and a traveler intending to travel in a straight line would eventually find herself back at her starting point. If Einstein's *cosmological constant* is zero, as is frequently assumed, then a universe which is temporally closed is also spatially closed, and vice versa.

color: Each *flavor* of *quark* can exist in three variations, called colors, usually labeled as red, green, and blue. The color of a quark has no relation to its visual appearance, but the word color is used because there are *three* variations, in analogy with the three primary colors. Measurable properties of the quarks, such as electric charge and mass, depend on the flavor but not the color, but the color is responsible for the interactions that bind the quarks together (see *Yang-Mills theories*). Individual quarks cannot exist independently, but are forever *confined* within *baryons* or *mesons*, each of which is colorless. Baryons achieve colorlessness by being composed of three quarks, one of each color, while mesons achieve colorlessness by pairing each colored quark with its corresponding antiquark.

confinement: The property of *quarks* which implies that they cannot exist as free particles, but are forever bound into protons, neutrons, etc. See *color.*

cosmic background radiation: See *blackbody radiation.*

cosmic distance ladder: A method for estimating distances to the remote galaxies by using a sequence of techniques. Each technique in the sequence is calibrated by the previous techniques, and extends the range of measurement to greater distances.

cosmic strings: Microscopically thin, spaghetti-like objects which, according to some theories of elementary particles, could form randomly during a phase transition in the early universe. Cosmic strings could provide the seeds for structure formation in the universe, as an alternative to the possibility that the seeds originated as quantum fluctuations during *inflation.*

cosmological constant: A parameter that determines the strength of the *cosmological term* in the equations of general relativity. This term was added by Einstein because he thought the universe was static, and the term provided a repulsive gravitational force that was needed to prevent the universe from collapsing under the force of ordinary gravity. The *false vacuum* of inflationary models creates a similar repulsive gravitational force, except that it prevails for only a brief period in the early universe. The cosmological constant is often assumed to be zero, but it might make a significant contribution to the evolution equations of our universe.

cosmological constant problem: The puzzle of why the cosmological constant has a value which is either zero, or in any case roughly 120 *orders of magnitude* or more smaller than the value that particle theorists would expect. Particle theorists interpret the cosmological constant as a measure of the energy density of the *vacuum*, which they expect to be large because of the complexity of the vacuum. See *vacuum*.

Cosmological Principle: A term introduced by E.A. Milne in 1933 to describe the assumption that the universe is both *homogenous* and *isotropic*.

cosmological term: See *cosmological constant*.

critical mass density: If the *cosmological constant* is assumed to vanish, then the critical mass density is that density which puts the universe just on the border between eternal expansion (*open universe*) and eventual collapse (*closed universe*).

dark matter: Matter that is detected only by its gravitational pull on visible matter. At least 90%, and possibly 99% of the matter in the universe is dark. The composition is unknown; it might consist of very low mass stars or supermassive black holes, but *big-bang nucleosynthesis* calculations limit the amount of such *baryonic* matter to a small fraction of the *critical mass density*. If the mass density is critical, as predicted by the simplest versions of *inflation*, then the bulk of the dark matter must be a gas of weakly interacting non-baryonic particles, sometimes called WIMPS (Weakly Interacting Massive Particles). Various extensions of the *standard model of particle physics* suggest specific candidates for the WIMPs.

decay of the false vacuum: Since the universe is not *inflating* today, the inflationary theory depends on the fact that the *false vacuum* which drives inflation is *metastable*, so it will eventually decay. The theory requires that this decay happen after more than 100 doubling times of the *exponential expansion*, but it still happens when the universe is much less than one second old. If the *inflaton fields* are in a local valley of the energy density diagram (see Figure 10.1 on page 168), then the decay happens via the formation of *bubbles*, like the boiling of water, and the randomness of the bubble formation destroys the *homogeneity* of the universe and causes the theory to fail. This failure is called the *graceful exit problem* of the original inflationary theory. If, however, the inflaton field is at the top of a local plateau (see Figure 12.1 on page 209), as assumed in the *new inflationary universe theory*, then inflation continues as the fields roll gently towards a minimum energy value, and a single bubble becomes large enough to encompass the entire observed universe.

density perturbations: Nonuniformities in the density of matter in the universe. Such nonuniformities are *gravitationally unstable*, which means that they are amplified by gravity. Any region with a mass density slightly higher than average produces a stronger-than-average gravitational field, which pulls in more matter and further increases the mass density. Thus, very mild nonuniformities in the early universe can serve as seeds for the formation of galaxies.

deuterium: An *isotope* of hydrogen in which each nucleus contains one proton and one neutron, instead of only one proton as in normal hydrogen. Water containing deuterium instead of ordinary hydrogen, called heavy water, is sometimes used as a moderator in nuclear reactors. See also *tritium.*

down: A flavor of quark. See *flavor.*

Doppler shift: The shift in the received *frequency* and *wavelength* of a sound wave or *electromagnetic wave* that occurs when either the source or the observer are in motion. Approach causes a shift toward shorter wavelengths and higher frequencies, called a *blueshift.* Recession has the opposite effect, called a *redshift.*

dyne: The force necessary to cause a mass of one gram to accelerate at one centimeter per second per second.

electromagnetic wave: A pattern of electric and magnetic fields that moves through space. Depending on the *wavelength*, an electromagnetic wave can be a radio wave, a microwave, an infrared wave, a wave of visible light, an ultraviolet wave, a beam of X rays, or a beam of gamma rays. See *photon.*

electromagnetism: The phenomena associated with electrical and magnetic forces. Electrical and magnetic forces are intimately related, since a changing electric field produces a magnetic field, and vice versa. *Electromagnetic waves* are an example of electromagnetism.

electron volt: The energy released when a single electron passes through an ideal one-volt battery.

electroweak interactions: The unified description of the *weak interactions* and *electromagnetism*, developed between 1967 and 1970 by Sheldon Glashow, Steven Weinberg, and Abdus Salam.

erg: Twice the amount of energy necessary to accelerate a one gram mass to a speed of one centimeter per second.

exponential expansion: An expansion described by a fixed doubling time. The size doubles after one doubling time, quadruples after two doubling times, octuples after three doubling times, etc.

extended inflationary universe theory: A version of the *inflationary universe theory* proposed in 1989 by Paul Steinhardt and Daile La. Its key new feature was the suggestion that a new field interacts directly with the gravitational field, causing the strength of gravity to change with time. This causes the expansion of the universe to slow down, allowing the *bubbles* forming at the end of inflation to catch up with the expansion and smoothly fill the universe.

false vacuum: The false vacuum is a peculiar form of matter which is predicted to exist by many (probably most) current theories of elementary particles. Particle physicists describe the false vacuum, like all forms of matter, in terms of *fields*. Originally the phrase "false vacuum" was used to describe a region in which one or more fields have a set of values that do not minimize the energy density, but which are in a local valley of the energy density diagram (as in Figure 10.1 on page 168). Classically such a state would be absolutely stable, since there would be no energy available for the fields to jump over the hills that surround the valley. By the rules of quantum theory, however, the fields can *tunnel* through the hill of the energy density diagram and settle eventually in the *true vacuum*, the state of lowest energy density. The false vacuum is nonetheless *metastable*, and in fact it can endure for a period which is long by early universe standards. The false vacuum has a negative pressure which creates a repulsive gravitational field, capable of driving the universe into a period of *exponential expansion* called *inflation*. Similar effects occur if the fields are at the top of a broad plateau (as in Figure 12.1 on page 209), instead of in a valley, so in this book I have stretched the phrase "false vacuum" to include this case as well.

false vacuum bubble: A bubble which has *false vacuum* on the inside, and *true vacuum* on the outside. In principle, the creation of such a bubble offers the possibility of creating a new universe in a hypothetical laboratory.

field: A quantity that can be defined at each point in space, such as an electric, magnetic, or gravitational field. When the rules of quantum theory are applied to the electromagnetic field, one concludes that the energy is concentrated into indivisible lumps, called *photons*, which behave essentially as particles. In present-day particle theory the photon is used as the prototype for all particles, so a field is introduced for the electron, for each type of *quark*, etc. All particles are viewed as the quantized lumps of energy in a field.

first order: A *phase transition* is called first order if it occurs in a manner similar to the way water boils. Bubbles of the new phase (steam) form in the midst of the old phase (water), so that temporarily the two distinct phases (steam and water) coexist. In a second order phase transition, by contrast, one phase evolves into the other as the temperature changes, so the two phases never coexist.

flat universe: A *homogeneous*, *isotropic* universe is called flat if it is just on the borderline between being spatially *closed* and spatially *open*, so the geometry is precisely Euclidean. If Einstein's *cosmological constant* is zero, then a flat universe will go on expanding forever, but the velocity of recession between any two objects would approach zero at large times.

flatness problem: A problem of the traditional big bang theory (without inflation) related to the precision required for the initial value of *omega*, the ratio of the actual mass density to the *critical mass density.* If the description is started at one second after the big bang, for example, omega must have been equal to one to an accuracy of fifteen decimal places, or else the resulting universe would not resemble our own. Yet the traditional big bang theory offers no explanation for this special value, which must be incorporated as an arbitrary postulate about the initial conditions. See also *horizon problem.*

flavor: The known *quarks* exist in six different types, or flavors: up, down, charmed, strange, top, and bottom. The up and down quarks belong to the first *generation*, the charmed and strange quarks belong to the second, and the top and bottom quarks belong to the third. The up, charmed, and top quarks each have an electrical charge 2/3 that of a proton, while the down, strange, and bottom quarks have a charge –1/3 that of a proton. See Table 7.1 on page 120.

force carriers: Particles that act as the transmitters of forces. The best known example is the *photon*, which transmits electromagnetic forces. The *gluons* are the transmitters of the *strong interactions*, and the W^+, W^-, and Z^0 particles are the transmitters of the *weak interactions*. See Table 7.1 on page 120.

fractal: A geometric figure in which a pattern is repeated ad infinitum on smaller and smaller scales. A classic example is Von Koch's snowflake, for which the construction begins with an equilateral triangle. Trisect each side, and replace the middle section by two sides of a smaller equilateral triangle, bulging outward. The snowflake is obtained by repeating this process for each side of the resulting figure, then for each side of the subsequent figure, and continuing forever.

free parameter: A number which is needed to define a theory well enough so that predictions can be made, but which must be determined by experiment or observation.

frequency: The number of peaks (often called crests) of a propagating *wave* that cross a given point in a unit of time. For example, if 1000 peaks cross a given point in one second, one says that the frequency is 1000 cycles per second, or 1000 hertz.

gamma ray: An *electromagnetic wave* with a *wavelength* in the range of 10^{-13} to 10^{-10} meter, corresponding to *photons* with energy in the range of 10^4 to 10^7 *electron volts*.

gauge theories: See *Yang-Mills theories*. (Technically speaking, electromagnetism is also an example of a gauge theory.)

generation: The fundamental particles of the *standard model of particle physics* can be grouped into *quarks*, *leptons*, *force carriers*, and the *Higgs particle*. The quarks and leptons have been found to exist in three generations, each of which is a copy of the first, except that the masses increase with each generation. See Table 7.1 on page 120. (The first generation includes all the building blocks of ordinary matter—electrons and the up and down quarks that make up protons and neutrons. The particles of the other generations were discovered by studying high energy collisions of first generation particles.)

gluons: The force-carrying particles associated with the *strong interactions,* the forces which bind *quarks* inside of protons and neutrons. For more details, see *Yang-Mills theories*.

graceful exit problem: A problem of the original formulation of the inflationary theory, in which the formation of *bubbles* at the end of inflation destroys the *homogeneity* of the universe. See *decay of the false vacuum*, and *percolation*.

grand unified theories: A speculative class of theories of particle *interactions*, first developed in 1974, which attempt to describe *electromagnetism*, the *weak interactions*, and the *strong interactions* in a fully unified theory. Of the known forces, only *gravitation* is omitted.

gravitation: The mutual attraction between any two masses, as was first described accurately by Newton. Gravity appears strong because it has infinite range and it is always attractive (except for a *false vacuum*), but on a subatomic level gravity is the weakest of the known interactions; the gravitational force between a proton and an electron is 2×10^{39} times weaker than the electrical attraction.

gravitational energy: Same as *gravitational potential energy.*

gravitational field: Instead of describing gravity as an *action-at-a-distance* force, modern physicists describe it in terms of a gravitational field. At each point of space, the field is defined as the force that would be experienced by a standard mass, if the mass were positioned at that point. While Newton's law of gravity can be expressed equally well in terms of an action-at-a-distance or a field, Einstein's theory of general relativity, which is now the accepted description of gravity, can be formulated only in terms of fields.

gravitational field lines: A method of depicting a *gravitational field* by drawing lines. The direction of the field is indicated by the direction of the lines, and the strength of the field is indicated by how closely the lines are spaced.

gravitational potential energy: When we lift a weight from the floor to a tabletop, we clearly put energy into it. The energy is not lost, however, because we can retrieve it by allowing the weight to fall back to the floor. While the weight is on the table, we say that the energy is stored as gravitational potential energy. The energy is stored in the *gravitational field*.

gravitationally unstable: See *density perturbations*.

Great Wall: A sheet of galaxies which stretches more than 500 million light-years across the sky.

hierarchy problem: In the context of grand unified theories, the hierarchy problem is our inability to understand theoretically why the energy scale at which the unification becomes apparent, about 10^{16} GeV (billion electron volts), is so much higher than other energy scales of relevance to particle physics, such as the mass/energy of a proton, which is only 1 GeV.

Higgs fields: Higgs *fields* are part of the *standard model of particle physics*, and analogous fields, also called Higgs fields, are part of *grand unified theories*. In both cases, the Higgs fields have nonzero values in the vacuum, and they serve to create distinctions between particles that would otherwise be identical. The standard model contains four Higgs fields. There are many combinations of nonzero values of the four fields that equally well minimize the energy density, so the combination of nonzero values in the vacuum is chosen randomly as the universe cools. Since the other fields of the theory interact with the Higgs fields, their behavior is affected by this random choice. In the standard model, it is the interactions with the Higgs fields that are responsible for the difference between electrons and neutrinos. In grand unified theories, Higgs fields are responsible for all the differences between electrons, neutrinos, and quarks.

Higgs particle: The particle or particles associated with the bundles of energy in the *Higgs field*. Such particles are analogous to the *photons* that are associated with the electromagnetic field. The *standard model of particle physics* predicts one electrically neutral Higgs particle which has not yet been found, but which will be sought in upcoming *particle accelerator* experiments. The grand unified theories predict many Higgs particles, but they are too massive to be accessible at existing or foreseeable accelerators.

homogeneous: A universe is called homogeneous if it would look the same to all observers, no matter where they were located. The real universe is not precisely homogeneous, but it appears to be homogeneous on large scales.

That is, if we averaged the mass density or other property of the matter in the universe over cubes of a few hundred million light-years on each side, all such cubes would be very similar to each other. See also *isotropic.*

homogeneous expansion: To a good approximation, our universe appears to be undergoing homogeneous expansion, which means that successive snapshots of a given region would each look like a photographic blowup of the first snapshot. Homogeneous expansion is also called Hubble expansion, since it implies *Hubble's law.* See Figure 2.1 on page 21.

horizon distance: Since the big bang theory assigns a finite age to the universe, at any given time there is a maximum distance that light could have traveled since the beginning, called the horizon distance. Since no signal propagates faster than light, the horizon distance is the maximum distance over which any information can be obtained.

horizon problem: A problem of the traditional big bang theory (without inflation) related to the large scale uniformity of the observed universe. The problem is seen most clearly in the *cosmic background radiation*, which is believed to have been released at about 300,000 years after the big bang, and has been observed to have the same temperature in all directions to an accuracy of one part in 100,000. Calculations in the traditional big bang theory show that the sources of the background radiation arriving today from two opposite directions in the sky were separated from each other, at 300,000 years after the big bang, by about 100 *horizon distances*. Since no energy or information can be transported further than one horizon distance, the observed uniformity can be reconciled only by postulating that the universe began in a state of near-perfect uniformity. See also *flatness problem.*

Hubble constant: According to Hubble's law, discovered by Edwin Hubble in 1929, distant galaxies are receding from us, on average, with a speed equal to the product of the Hubble constant and the distance to the galaxy. Hubble's "constant" is independent of distance, but actually decreases slowly in time as the expansion is slowed by the gravitational pull of each galaxy on all the others. The present value is somewhere between 15 and 30 kilometers per second per million light-years.

Hubble time: The Hubble time is one divided by the Hubble constant, which gives a number from 10 to 20 billion years. For a *flat universe* with no *cosmological constant*, the age of the universe is two-thirds of the Hubble time.

Hubble's law: See *Hubble constant.*

inflation: The phenomenon by which the universe is driven into *exponential expansion* by the repulsive gravitational field created by a *false vacuum*. The inflation would end with the *decay of the false vacuum.*

Although the inflation would occur in far less than a second, it could account for the "bang" of the big bang theory, it could explain the origin of essentially all the matter in the observed universe, and it can solve the *horizon problem* and the *flatness problem*. It could also generate the *density perturbations* that would later become the seeds for galaxy formation.

inflationary universe theory: See *inflation.*

inflaton: The name given to whatever *fields* are responsible for driving *inflation.*

infrared radiation: See *electromagnetic wave.*

interaction: Particle physicists use the word "interaction" to refer to anything that particles can do, which includes decaying, scattering, and annihilating. The known interactions of nature are divided into four classes, which from the weakest to the strongest are *gravitation*, the *weak interactions*, *electromagnetism*, and the *strong interactions.*

ionized: If the atoms of a gas have lost one or more electrons, they are said to be ionized. The gas in the early universe is believed to have been ionized until about 300,000 years after the big bang. Since ionized gases interact very strongly with *photons*, the photons of the *cosmic background radiation* could not move freely until after the first 300,000 years.

isotope: Two nuclei with the same number of protons but different numbers of neutrons are said to represent the same element, but different isotopes. For example, helium-3, with two protons and one neutron in each nucleus, and helium-4, with two protons and two neutrons, are two different isotopes of helium. For another example, see *deuterium.*

isotropic: A universe is said to be isotropic, from the point of view of a given observer, if it looks the same in all directions. The isotropy of the real universe is seen most strikingly in the *cosmic background radiation*, which has the same temperature in all directions to an accuracy of about one part in 100,000. See also *homogeneous.*

kinetic energy: The energy that a body has due to its motion.

leptons: A class of non-strongly-interacting particles that includes the electron, muon, tau, and their associated neutrinos. See Table 7.1 on page 120.

lithium: The chemical element with atomic number 3, meaning that there are three protons in the nucleus. The most common isotope is lithium-7 (92.5% of naturally occurring lithium), which has four neutrons in its nucleus.

Local Group: A conglomeration of about twenty galaxies including our own Milky Way.

Local Supercluster: A loosely knit assemblage of some 100 clusters of galaxies, including the Local Group.

luminiferous ether: A material that, prior to special relativity, was believed to permeate all of space, allowing the propagation of light. With the advent of relativity, we now believe that light can propagate without a medium.

magnetic monopole: A magnet with an isolated north (or south) pole, rather than a pair of equal-strength north and south poles, as in conventional magnets. Magnetic monopoles have never been observed, but they are predicted to exist by *grand unified theories*.

magnetic monopole problem: A problem, discovered by John Preskill in 1979, concerning the compatibility of *grand unified theories* with standard cosmology. Preskill showed that if standard cosmology were combined with grand unified theories, far too many *magnetic monopoles* would have been produced in the early universe.

mass: The measure of the inertia of an object, determined by observing the acceleration when a known force is applied. The gravitational force acting on an object is found to be proportional to its mass, as is the gravitational force that it exerts on other objects.

meson: A *strongly interacting* particle consisting of a *quark* and an *antiquark*.

metastable: A state which is not stable, but which lives long enough to have significance, is called metastable.

microwave: An *electromagnetic wave* with a *wavelength* between one millimeter and 30 centimeters, or sometimes one meter.

neutrino: An electrically neutral, very weakly interacting particle, with a rest energy which is either zero or very small. The particle was predicted in 1931 as a means of reconciling the measurements of *beta decays* with the conservation of energy, but it was not directly detected until 1956.

new inflationary universe theory: The first fully successful version of the inflationary theory, discovered independently by Andrei Linde in the Soviet Union, and by Andy Albrecht and Paul Steinhardt in the U.S. See *inflation*, *false vacuum*, and *decay of the false vacuum*.

omega: Cosmologists use the Greek letter omega (Ω) to denote the ratio of the actual mass density of the universe to the *critical mass density*.

one-standard-deviation uncertainty: An estimate of the uncertainty of a measurement which is specified so that the probability of the true value of the measured quantity lying within the uncertainty interval is two out of three. There is one chance in three that the true value lies outside the interval.

open inflationary universe theory: A version of the *inflationary universe theory*, proposed by Bharat Ratra and Jim Peebles in 1995, which produces an *open universe*. The usual inflationary prediction of a *flat universe* is avoided by proposing an energy diagram for the inflation-driving field with a hill of just the right length, so inflation ends before the universe is driven to flatness.

open universe: A *homogeneous*, *isotropic* universe is said to be temporally open if gravity is not strong enough to eventually reverse the expansion, so the universe goes on expanding forever. It is said to be spatially open if it curves the opposite way from a *closed universe*, so that triangles would contain less than 180°, the circumference of a circle would be more than π times the diameter, and the volume would be infinite. If Einstein's *cosmological constant* is zero, as is frequently assumed, then a universe which is temporally open is also spatially open, and vice versa.

order of magnitude: A factor of ten. Two orders of magnitude indicate a factor of 100, etc.

orthorhombic crystal: A crystal in which the atoms are arranged in a rectangular solid, for which each of the three principal lengths are different. Such a crystal provides a simple example of *spontaneous symmetry breaking*, since the *rotational invariance* of the underlying physical laws is broken by the randomly chosen orientation of the crystal. Inside such a crystal there are three distinct speeds of light, depending on which of the three axes the light is following.

particle accelerator: A device using electric and magnetic fields to accelerate beams of particles—usually electrons, *positrons*, protons, or *antiprotons*—to high energies for experimental purposes. Modern accelerators are often very large: the main ring at Fermilab, for example, is 4 miles in circumference.

percolation: A random system defined mathematically on an infinite space, such as the random formation of bubbles in the early universe, is said the percolate if the objects merge to form an infinite cluster. In the original inflationary theory the bubbles that ended the inflation were found not to percolate, implying that they could never be homogeneous enough to describe our universe. The failure of the original inflationary theory became known as the *graceful exit problem*. See *decay of the false vacuum*.

Perfect Cosmological Principle: A term introduced to describe the *steady-state universe theory*, in which the appearance of the universe is assumed to be independent not only of the position of the observer (*homogeneity*) and the direction she is looking (*isotropy*), but also the time at which she is making the observations. See also *Cosmological Principle*.

period: The period of a *wave* is the time interval between the receipt of two successive peaks (often called crests) of the propagating disturbance.

perturbation expansion: A method of successive approximations that is used to obtain predictions from a theory such as *quantum electrodynamics*, which cannot be solved exactly.

phase transition: A sudden change in the behavior of a material as the temperature is varied. Examples include the boiling and freezing of water. *Grand unified theories* predict that the hot matter of the early universe would have undergone a phase transition at about 10^{-37} seconds after the instant of the big bang.

photon: According to quantum theory, the energy of an electromagnetic wave (such as a light wave) is concentrated into indivisible lumps, called photons, which are essentially particles of light.

Planck energy: An energy of 1.22×10^{19} GeV (billion electron volts), at which the strength of the gravitational interactions of fundamental particles becomes comparable to that of the other interactions. It is believed that the quantum effects of gravity become important at approximately this energy.

plasma: A gas in which the atoms are *ionized.*

positron: The *antiparticle* of an electron.

QCD: Abbreviation for *quantum chromodynamics.*

QED: Abbreviation for *quantum electrodynamics.*

quantum chromodynamics: Abbreviated QCD, this is the accepted theory of the forces that bind *quarks* together to form protons, neutrons, and other strongly interacting particles. See *Yang-Mills theories.*

quantum electrodynamics: Abbreviated QED, this is the accepted theory of electromagnetic interactions, including all the effects of relativity and quantum theory. The *photon* acts as the carrier of the electromagnetic force.

quantum tunneling: A process by which a quantum system can suddenly and discontinuously make a transition from an initial configuration to a final one, even if the system does not have enough energy to classically attain the configurations between the two.

quarks: The particles which are the constituents of protons, neutrons, and other *strongly interacting* particles (*baryons* and *mesons*). For more details, see *flavor* and *color.*

recombination: At about 300,000 years after the big bang, the *plasma* of free electrons and nuclei condensed to form a neutral gas, in a process called recombination. The prefix "re-" is not meaningful here, however, since according to the big bang theory the electrons and protons were combining for the first time ever.

red giant: A phase in the life cycle of a not-too-heavy star, in which the star expands tremendously and then blows off its outer shells into space. The star then contracts to become a stable object called a *white dwarf.*

redshift: If a star or galaxy is moving away from us, the radiation from the star or galaxy appears shifted towards longer *wavelengths*, or towards the red end of the spectrum. See *Doppler shift.*

relativity: The theory of special relativity, proposed by Albert Einstein in 1905, is based on the premise that the speed of light would be measured at the standard value c by all observers who move at a constant speed in a fixed direction, even if those observers are in motion relative to each other. Einstein showed that this postulate is consistent if we accept that measurements of distance and time are relative, meaning that they depend on the motion of the observer and the object observed. In particular, moving objects appear to contract in the direction of motion, and moving clocks run slower than normal. The laws of physics, however, appear the same to all observers. General relativity, discovered by Einstein in 1915, is a theory of gravity that is consistent with special relativity. Gravity is not described as an ordinary force, but is instead described as the bending, twisting, or stretching of space, which is assigned a geometry which is no longer assumed to be Euclidean.

renormalization: When a theory such as *quantum electrodynamics* is approximated by a *perturbation expansion*, the first approximation is found to give answers that agree well with experiment. The second approximation, however, is found to produce mathematical expressions that are infinite, and hence meaningless. Renormalization is a technique for reformulating the theory so that the infinities are avoided, developed independently by Feynman, Tomonaga, and Schwinger in the 1940s.

rest energy: The energy which a particle has even when it is at rest. According to the famous relation $E = mc^2$ of special relativity, this rest energy is equal to the rest mass of the particle—the mass it has when a rest—times the square of the speed of light. If the mass is in grams and the speed of light in centimeters per second ($c = 2.998 \times 10^{10}$ centimeters per second), then the energy is given in ergs.

rotational invariance: The property of being unchanged by a rotation. A sphere is rotationally invariant, but a rectangle is not. As far as we know the fundamental laws of physics are rotationally invariant.

scale-invariant: Most inflationary models predict that the *spectrum* of *density perturbations* is nearly scale-invariant, meaning essentially that each wavelength has the same strength. This spectrum is also called the Harrison-Zeldovich spectrum, named for two astrophysicists who proposed the spectrum a decade before inflation was invented.

singularity: If the standard big bang theory is extrapolated all the way back to time zero, one reaches an instant of infinite density, infinite pressure, and infinite temperature—an instant that is frequently called the initial singularity. This singularity is sometimes said to mark the beginning of time, but it is more realistic to recognize that an extrapolation to infinite density cannot be trusted.

spectrum: A spectrum of a sound *wave* or a light wave is obtained by decomposing the wave into contributions of different *wavelengths*, and then drawing a graph of the intensity versus the wavelength. A similar process works for *density perturbations*, although one cannot use a prism to decompose the wave into different wavelengths, as one often does for light. Instead one relies on mathematical techniques to express the graph of how density varies with position as a sum of curves of a standard form, each of which has a definite wavelength. Finally, one constructs a graph which shows how strongly each wavelength contributes to the sum.

spontaneous symmetry breaking: The breaking of an exact symmetry of the underlying laws of physics by the random formation of some object. For example, the *rotational invariance* of the laws of physics can be broken by the randomly chosen orientation of an *orthorhombic crystal* that condenses as the material is cooled. In the *standard model of particle physics*, the symmetry between electrons and neutrinos is spontaneously broken by the values that are randomly chosen by the *Higgs fields*. In *grand unified theories*, the symmetry between electrons, neutrinos, and quarks is spontaneously broken by the values chosen randomly by the Higgs fields.

standard model of particle physics: A theory of particle *interactions*, developed in the early 1970's, which successfully describes *electromagnetism*, the *weak interactions*, and the *strong interactions*. The theory consists of two parts, *quantum chromodynamics* to describe the strong interactions, and the unified *electroweak* theory to describe the electromagnetic and weak interactions.

steady-state theory: A theory of cosmology, advanced in 1948 by Bondi, Gold, and Hoyle, which held that the universe appears the same at all times, with new matter being created to fill in the expanding gaps between the separating galaxies. See *Perfect Cosmological Principle*.

strange: A flavor of quark. See *flavor*.

strong interactions: The interactions which bind *quarks* together to form protons, neutrons, and other particles. The residual effects of these forces are responsible for the forces between protons and neutrons. See *Yang-Mills theories*.

supercooling: The phenomenon in which a system can be cooled below the normal temperature of a *phase transition* without the phase transition taking place. The original form of the inflationary universe theory was based on the assumption that the universe supercooled below the temperature of the *grand unified theory* phase transition. See *phase transition*.

superstring theory: A proposal for the ultimate laws of nature, a "theory of everything," stemming primarily from discoveries in the mid 1980's. The fundamental entity in this theory is an ultramicroscopic string-like object, with a length of typically 10^{-33} centimeters and effectively zero thickness. At present our understanding of string theory is very limited. The simplest predictions of superstring theory concern processes at the *Planck energy*, and so far very little is known about the consequences of string theory at lower energies.

superunified: While *grand unified theories* attempt to describe three of the four known interactions of nature—the weak, strong, and electromagnetic interactions—in a unified way, the fourth interaction, gravity, is omitted. Theories which attempt to include gravity as well, such as *superstrings*, are called superunified.

thermal equilibrium: A simple theoretical description of the early universe is possible because the primordial soup had time to reach thermal equilibrium, the state for which the relative abundances of different constituents is determined not by the initial conditions, but instead by the criterion that the production and annihilation rates for any particular constituent are in balance. As long as there are reactions that can convert one constituent to another, the ratios of those two constituents will be determined by thermal equilibrium. For example, protons and neutrons can interconvert by the reactions

$$\text{proton} + \text{electron} \leftrightarrow \text{neutron} + \text{neutrino},$$

where the double arrow indicates that either the left-hand side or the right-hand side can represent the initial particles. If we imagine that the universe began with only protons, then this reaction would cause some to convert to neutrons. As the density of neutrons increased, the rate at which neutrons are converted back to protons would increase. The reactions would continue until there were just enough neutrons so that the number of neutrons converting to protons each second would equal the number of protons converting to neutrons. As another example, thermal equilibrium implies that the *photons* in the early universe did not have some arbitrary distribution, but were instead described by the intensity and spectrum of *blackbody radiation*.

top: A flavor of quark. See *flavor*.

toy theory: A theory which is known to be too simple to describe reality, but which is nonetheless useful for theorists to study because it incorporates some important features of reality. For example, most of what is known about magnetic monopoles in grand unified theories was discovered first in a toy theory that includes only three Higgs fields, while the simplest realistic grand unified theory includes twenty-four of them.

tritium: A short-lived *isotope* of hydrogen in which each nucleus contains one proton and two neutrons, instead of only one proton as in normal hydrogen. See also *deuterium.*

true vacuum: This phrase has the same meaning as *vacuum*, with the word "true" being used only to emphasize the distinction with the *false vacuum.*

tunneling: See *quantum tunneling.*

up: A flavor of quark. See *flavor.*

vacuum: Roughly speaking the vacuum is a space devoid of matter, but this definition suffers from the ambiguity of the word "matter." Particle physicists therefore define the vacuum as the state of lowest possible energy density. The vacuum is not simple, since the inherently probabilistic nature of quantum theory implies that unpredictable events, such as the chance materialization of an electron and its antiparticle, the *positron*, can occur at any time. Such pairs have a fleeting existence of perhaps 10^{-21} seconds, and then annihilate into nothingness. In addition, particle theories suggest that there are *Higgs fields* which have nonzero values in the vacuum.

vacuum fluctuation: An unpredictable event occurring in the vacuum as a result of the inherently probabilistic nature of quantum theory. Particles can materialize in conjunction with their antiparticles, and fields can undergo fluctuations in their values. See *vacuum.*

vacuum Higgs value: The value that the Higgs field(s) have in the *vacuum.* See *Higgs fields.*

Virgo Cluster: An aggregation including about 200 bright galaxies which is centered roughly 60 million light-years away.

wave: A propagating pattern of disturbance. One example is a sound wave, in which a pattern of alternating high and low pressures propagates through air. Another example is an *electromagnetic wave*, in which a pattern of electric and magnetic fields propagates through empty space.

wavelength: The distance between two successive peaks (often called crests) of the propagating disturbance of a *wave.*

weak interactions: Interactions with a typical range of about 1/100 the size of an atomic nucleus, which are responsible for the *beta decay* of some kinds of atomic nuclei. Neutrinos are subject to only the weak interactions and the much weaker force of gravity.

white dwarf: See *red giant.*

white hole: The time-reversal of a *black hole*. A white hole is a singularity from which matter emerges unpredictably, but into which matter cannot enter. The initial singularity of the standard big bang theory is an example of a white hole. It can be shown that the creation of a new universe from a *false vacuum bubble* in the context of classical general relativity would require a white hole singularity, which means essentially that it cannot be done, even in principle. However, a false vacuum bubble could conceivably grow to become a new universe through a process of *quantum tunneling*.

wormhole: An intriguing solution to the equations of general relativity which describes a neck that can connect two completely separate universes. Wormholes arise in the discussion of the creation of a universe in the laboratory, because the new universe disappears through a wormhole and completely detaches itself from the parent universe.

Yang-Mills theories: Also known as gauge theories, these theories were invented in 1954 by Chen Ning Yang and Robert Mills. In 1973 David Gross, Frank Wilczek, and David Politzer showed that these theories possess a property called *asymptotic freedom*, just what was needed for a theory of how *quarks* bind to form protons and neutrons. The new theory, dubbed *quantum chromodynamics* or *QCD*, proposed that the *color* of the quarks acts as the charge of the Yang-Mills interactions.

ylem: The word used by Gamow and his collaborators for the primordial material of the big bang. In most of his work Gamow assumed that the ylem consisted entirely of neutrons. In inflationary cosmology, the role of the ylem is played by the *false vacuum*.

CREDITS

Figures:

Figure 1.1: Courtesy of the Rumford Historical Association, Woburn, Massachusetts.

Figure 3.5: The illustrations of the rabbit and the jet airplane are from the Micrografx ClipArt "Personal Collection," Micrografx, Inc., Richardson, Texas.

Figures 5.2 and 5.3: The image of the small explosion is from the Micrografx ClipArt "Personal Collection," Micrografx, Inc., Richardson, Texas.

Figure 8.3: The illustration of the sombrero is from the Micrografx ClipArt "Personal Collection," cited above.

Figure 10.2: The image of the hand is from the *Presentation Task Force* clip art collection, copyright, New Vision Technologies, Inc.

Figure 10.3: The image of the man is from the *Print House* collection by the Corel Corporation.

Figure 10.5: The image of the globe is from the *Presentation Task Force* clip art collection, copyright, New Vision Technologies, Inc.

Photographs:

Edward Tryon, p. 13, courtesy of Edward Tryon.

Robert Dicke, p. 19, by Robert Matthews, Princeton University.

Henry Tye, p. 27, by Bik Tye.

Arno Penzias and Robert Wilson, p. 59, courtesy of Bell Telephone Laboratories.

Jim Peebles, p. 63, by Robert Matthews, Princeton University.

David Wilkinson, p. 67, by Robert Matthews, Princeton University.

John Mather, p. 79, courtesy of Goddard Space Flight Center photo shop.

Steve Weinberg, p. 106, by Louise Weinberg.

John Preskill, p. 151, by Bob Paz, courtesy of California Institute of Technology.

Henry Tye, p. 181, by Bik Tye.

Andrei Linde, p. 204, by Neil Calder, courtesy of European Laboratory for Particle Physics (CERN).

Paul Steinhardt, p. 205, courtesy of Paul Steinhardt.

So-Young Pi, p. 222, courtesy of So-Young Pi.

Michael Turner, p. 225, courtesy of Michael Turner.

David Wilkinson, p. 240, by Robert Matthews, Princeton University.

Alex Vilenkin, p. 274, by Mark Morelli.

INDEX

Note: Numbers followed by the letter "f" indicate figures; numbers followed by the letter "n" indicate footnotes; numbers followed by the letter "t" indicate tables; numbers followed by the letter "g" indicate glossary entries.